Oxford Graduate Texts in Mathematics

Series Editors

OXFORD GRADUATE TEXTS IN MATHEMATICS

---

1. Keith Hannabuss: *An introduction to quantum theory*
2. Reinhold Meise and Dietmar Vogt: *Introduction to functional analysis*
3. James G. Oxley: *Matroid theory*
4. N. J. Hitchin, G. B. Segal, and R. S. Ward: *Integrable systems: twistors, loop groups, and Riemann surfaces*
5. W. Rossmann: *Lie Groups*
6. Qing Liu: *Introduction to algebraic curves over Dedekind domains*
7. Martin R. Bridson and Simon M. Salamon: *Invitations to geometry and topology*

Invitations to Geometry and Topology

Edited by

Martin R. Bridson
Professor of Topology, University of Oxford

and

Simon M. Salamon
Professor of Geometry, Politecnico di Torino

Great Clarendon Street, Oxford OX2 6DP

Oxford University Press is a department of the University of Oxford.
It furthers the University's objective of excellence in research, scholarship,
and education by publishing worldwide in

Oxford New York

Auckland Bangkok Buenos Aires Cape Town Chennai
Dar es Salaam Delhi Hong Kong Istanbul Karachi Kolkata
Kuala Lumpur Madrid Melbourne Mexico City Mumbai Nairobi
São Paulo Shanghai Taipei Tokyo Toronto

Published in the United States
by Oxford University Press Inc., New York

First published 2002

A catalogue record for this title is available from the British Library

Library of Congress Cataloging in Publication Data
(Data available)

ISBN 0 19 850772 0

10 9 8 7 6 5 4 3 2 1

Typeset by Cepha Imaging Pvt. Ltd.
Printed in Great Britain
on acid-free paper by Biddles Ltd, Guildford & King's Lynn

Preface

The purpose of this volume is to present an array of topics that introduce the reader to key ideas in active areas of geometry and topology. The material is presented in a manner that graduate students should find both accessible and enticing. Each chapter is written by a former student of Brian Steer. Brian has inspired generations of Oxford students with his intellectual generosity, refreshingly idiosyncratic teaching and perceptive guidance. This volume is dedicated to him by his students with deep affection; it was conceived at a gathering to mark his sixtieth birthday and is being published as he begins his retirement.

We hope that this book is true to the ethos that Brian instilled in us: it is meant to be useful to students and to inspire them; it is not a celebratory ornament. The topics covered here range from Morse theory and complex geometry to geometric group theory. The text is accompanied by exercises that are designed to deepen the readers' understanding and to guide them in exciting directions for future investigation.

Oxford M. R. B.
Turin S. M. S.
August 2001

Contents

Contributors

A. J. (Jon) Berrick	Department of Mathematics, National University of Singapore, Singapore 119260, Republic of Singapore.
Martin R. Bridson	Department of Mathematics, Imperial College of Science Technology & Medicine, 180 Queen's Gate, London, SW7 2BZ, U.K.
Michael Crabb	Department of Mathematical Sciences, University of Aberdeen, Aberdeen, AB24 3UE, U.K.
Michael Eastwood	Department of Pure Mathematics, University of Adelaide, South Australia 5005, Australia.
Martin A. Guest	Department of Mathematics, Tokyo Metropolitan University, Minami-Ohsawa 1-1, Hachioji, Tokyo 192-0397, Japan.
Nigel Hitchin	Mathematical Institute, University of Oxford, 24-29 St. Giles', Oxford OX1 3LB, U.K.
Anthony Potter	Department of Mathematical Sciences, University of Aberdeen, Aberdeen, AB24 3UE, U.K.
Simon M. Salamon	Dipartimento di Matematica, Politecnico di Torino, Corso Duca degli Abruzzi 24, 10129 Torino, Italy.
Justin Sawon	Mathematical Institute, University of Oxford, 24-29 St. Giles', Oxford, OX1 3LB, U.K.
José Seade	Instituto de Matemáticas, Universidad Nacional Autónoma de México, Unidad Cuernavaca, Apartado Postal 273-3, C.P. 62251 Cuernavaca, Morelos, México.

1
A topologist's view of perfect and acyclic groups

A. J. BERRICK

Introduction

Perfect groups, and in particular non-abelian simple groups, are important objects of study in finite group theory in their own right. In contrast, for a topologist or geometer, perfect groups tend to be those that arise in certain interesting situations, for example via the study of fundamental groups. In the first section below, we try to explain why this might be so. The second section outlines what general results about perfect groups are known, or conjectured. While the author's biases are naturally reflected in the choice of material in these parts, it is in the third that they become rampant. The reader will find there an exhibition of specimens of a key class of perfect groups, namely the acyclic groups.

1 Motivation

1.1 Algebraic motivation

This is very intuitive. The basic point is that abelian groups are much simpler objects to consider than arbitrary groups. For example, finitely generated abelian groups are easily described, since each is the direct product of a finite number of cyclic groups, whose orders therefore determine the finitely generated abelian group (up to isomorphism). On the other hand, as soon as one allows more than one element in a generating set, finitely generated non-abelian groups get wretchedly difficult to describe. For instance, it is known that every countable group embeds in a two-generator simple group, and there are at least $2^{\aleph_0}$ such groups [68, IV.3].

It follows that a good invariant of any group G is its abelianization G_{ab}. This is also known as its first homology group $H_1(G)$ (for homology we use trivial integer coefficients unless otherwise stated), or commutator quotient group G/G'. Here $G' = [G, G]$ is the first derived or commutator subgroup, the subgroup of G generated by all commutators $[g_1, g_2] = g_1 g_2 g_1^{-1} g_2^{-1}$. The invariant G_{ab} is characterized by the universal property that every group homomorphism from G to an abelian group factors uniquely through the epimorphism $G \to G_{\mathrm{ab}}$.

The beauty of this invariant is that it enables one to obtain an abelian group from a group that may be truly horrible in its complexity. Now although there is much that can be said (e.g. [40]) about abelian groups, they are indeed easier objects to deal with than arbitrary groups. There is however always a price to be paid for simple invariants. That price is the information lost in passing to the invariant. There are two ways of looking at this situation.

First, the building blocks of the theory will be those groups whose invariant is trivial. Here that means that the group G has trivial abelianization and so is equal to its commutator subgroup. In other words, G is generated by its commutators, and every element of G can be written as a product of commutators. This is the definition of a *perfect* group. In fact, any attempt to describe an arbitrary group G by means of abelian groups inevitably leads to a perfect group (possibly trivial), as we shall now see.

In the general case, the information lost in abelianizing is measured by the kernel G' of the epimorphism $G \to G_{\text{ab}}$. This poses the problem of describing the first derived group G'. Since abelianization was considered a good method for attacking the original group, it should also be good enough for the group G'. How much information is lost in passing to its abelianization? This is given by its commutator subgroup, the second derived subgroup

$$G^{(2)} = (G')' = [G', G'].$$

And so on ... In this way one obtains the *derived series*

$$G = G^{(0)} \supseteq G' = G^{(1)} \supseteq G^{(2)} \supseteq \cdots$$

with $G^{(n+1)} = [G^{(n)}, G^{(n)}]$, spinning off an abelian group $G^{(n)}_{\text{ab}}$ at each stage. This process either reaches a perfect group and terminates:

$$G^{(n)} \text{ perfect} \iff G^{(n+1)} = G^{(n)};$$

or it doesn't. In the latter case there is a new subgroup $G^{(\omega)} = \bigcap_{n\geq 0} G^{(n)}$, and one may apply the whole procedure to it. This leads to the *transfinite derived series* for G, with $G^{(\beta)}$ defined for each ordinal β as follows. If β is the successor of an ordinal α, then define $G^{(\beta)} = [G^{(\alpha)}, G^{(\alpha)}]$. If, on the other hand, β is not a successor ordinal, then put $G^{(\beta)} = \bigcap_{\gamma<\beta} G^{(\gamma)}$. Now, since G is after all a set, its cardinality gives an upper bound for how far this series can descend. Eventually it must reach an ordinal ν for which $G^{(\nu+1)} = G^{(\nu)}$. In other words, it terminates at the (possibly trivial) perfect group $G^{(\nu)}$, the intersection of the transfinite derived series of G.

Hence we have obtained a whole chain of invariants of an arbitrary group G. All except the last are abelian subquotient groups of G, while the last is a perfect subgroup of G. Now the knowledge of all these groups need not allow total reconstruction of the original group, since the groups G' and G_{ab} may fit into a group extension

$$G' \rightarrowtail H \twoheadrightarrow G_{\text{ab}}$$

with G' isomorphic to the kernel of a group epimorphism $H \twoheadrightarrow G_{\text{ab}}$ for some group H that is not isomorphic to G. But that is not the point. The point is that the simple process of extracting all the abelian invariants from a given group G inevitably leads to a perfect subgroup of G that is just as fundamental to a description of G as say the most obvious abelian group, G_{ab}.

1.2 Geometric motivation

The above fact that for any discrete group G, $H_1(G) = G_{\text{ab}}$, has the important topological generalization that for any topological space X

$$H_1(X) = \pi_1(X)_{\text{ab}}.$$

The case where X is the classifying space of G, the Eilenberg–Mac Lane space $K(G,1)$, reduces to

$$H_1(G) = H_1(K(G,1)) = \pi_1(K(G,1))_{\text{ab}} = G_{\text{ab}}.$$

Then the above considerations suggest that for classes of spaces X for which homology groups are easily calculated, perfect subgroups of the fundamental group have a role to play in calculation of $\pi_1(X)$. As an extreme case, if it is known that $H_1(X)$ is the zero group, then $\pi_1(X)$ is perfect. So perfect groups arise as, for example, the fundamental groups of homology spheres (see (2.4.2) below).

For relatively easy visualization, let's do some low-dimensional topology. To see which surfaces have perfect fundamental groups, we start by looking for commutators. For a simple loop on a surface to represent a commutator, it must bound a punctured torus (also known as a bridged annulus [46, p. 20]), as follows.

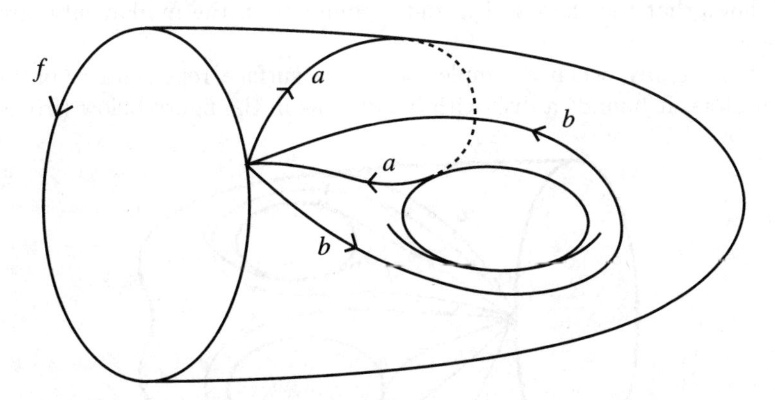

Fig. 1 Punctured torus

The fact that the boundary curve of a punctured torus is indeed a commutator is perhaps most easily checked on the model for the quotient space.

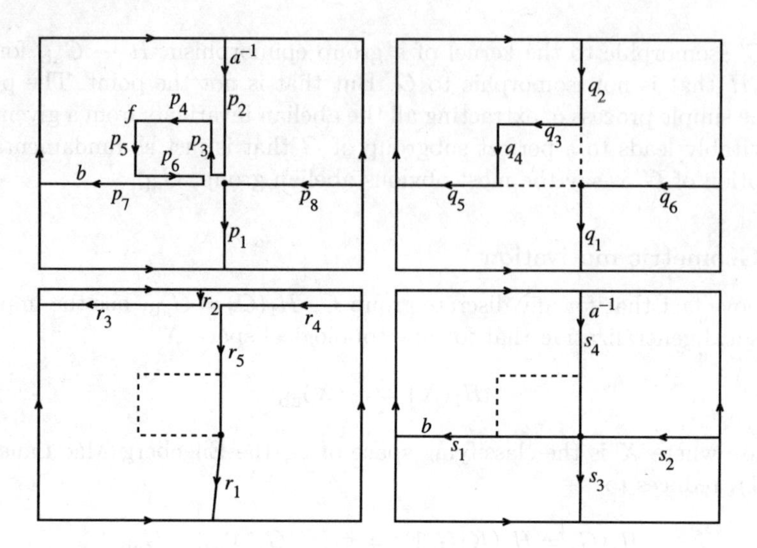

Fig. 2 Homotopy

Here

$$\begin{aligned} a^{-1}fb &= (p_1p_2)(p_3p_4p_5p_6)(p_7p_8) \\ &\simeq q_1q_2q_3q_4q_5q_6 \\ &\simeq r_1r_2r_3r_4r_5 \\ &\simeq (s_1s_2)(s_3s_4) = ba^{-1}. \end{aligned}$$

Hence

$$[f] = [a][b][a]^{-1}[b]^{-1}.$$

This argument is probably more familiar to readers in the form where the puncture that the loop f bounds is filled in. Then f becomes trivial and the proof shows that the classes of a and b commute in the fundamental group of the torus.

More generally, when a simple loop on a surface represents a product of commutators, it bounds a disk-with-handles, as in the figure below (from [26]).

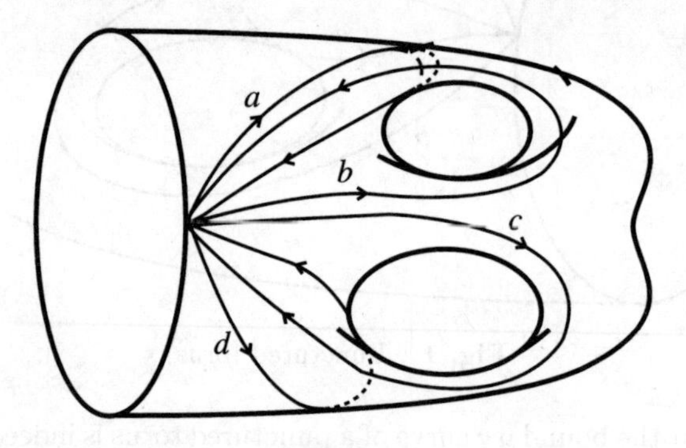

Fig. 3

Now, in the situation of a perfect fundamental group, every element that features in a product of commutators is itself equal to a product of commutators. This can be achieved by making each non-trivial loop such as a, b, c, d in turn bound a disk-with-handles. This procedure has to be iterated infinitely often, resulting in the concept of a *grope* [90], as illustrated below.

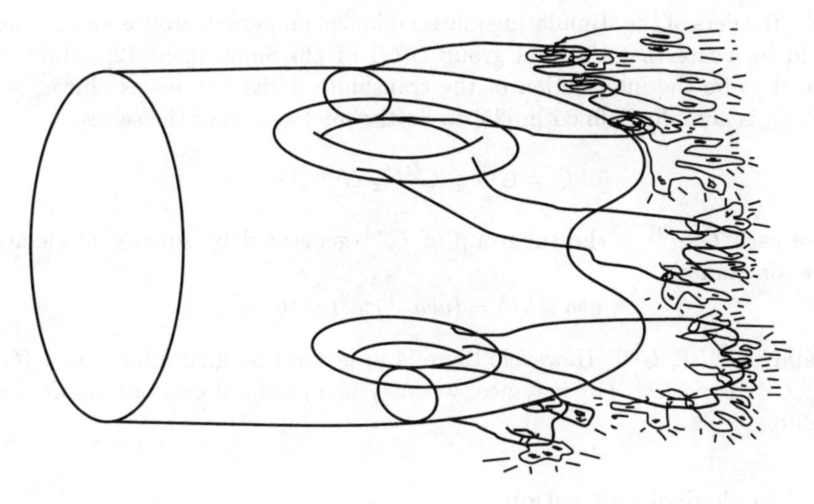

Fig. 4

The formal description of a grope [31] is the direct limit L of a nested sequence of compact 2-dimensional polyhedra

$$L_0 \hookrightarrow L_1 \hookrightarrow L_2 \hookrightarrow \cdots$$

obtained as follows. Let S_g denote an oriented compact surface of positive genus g from which an open disk has been deleted. (So $g = 1$ gives a punctured torus.) Take L_0 as some S_g. To form L_{n+1} from L_n, for each loop a in L_n that generates the group $H_1(L_n)$, attach to L_n some S_{g_a} by identifying the boundary of S_{g_a} with the loop a. Since the fundamental group of S_g punctured is a free group on $2g$ generators, this procedure embeds each $\pi_1(L_n)$, and thus each finitely generated subgroup of $\pi_1(L)$, as a subgroup of a free group, with each generator a of $\pi_1(L_n)$ becoming a product of g_a commutators in $\pi_1(L_{n+1})$. Hence $\pi_1(L)$ is a countable, perfect, locally free group.

The most economical example is the *minimal grope* M^*, for which one takes only one genus one surface at each step. Homotopically, each L_n is in this case just a bouquet of finitely many circles. So M^* is the classifying space of its fundamental group, which is easily seen to have the following description. Let Σ denote the set of all (non-empty) words of finite length on the two symbols $0, 1$. Then $\pi_1(M^*)$ is the group generated by symbols x_w for each $w \in \Sigma$, subject only to all relations $x_w = [x_{w0}, x_{w1}]$. Thus the x_w with w of length n generate the free group $\pi_1(L_n)$, but each is a commutator when embedded in the larger free group $\pi_1(L_{n+1})$.

The 3-dimensional version of this construction occurs in [39, Theorem 2], where a nested sequence of handle-bodies leads to a locally free fundamental group, and conversely.

Aside. Readers of the stimulating material in [26] on perfect groups and geometry should be aware that the wild group $\omega(G)$ of [26, Supplement 12], while there claimed to be the intersection of the transfinite derived series as above (hence perfect), is actually defined in [25] to be the intersection of the series

$$G = G^{[0]} \supseteq G^{[1]} \supseteq G^{[2]} \supseteq \cdots$$

where each $G^{[n+1]}$ is the subgroup of $G^{[n]}$ generated by squares of elements. Since for example

$$aba^{-1}b^{-1} = (aba^{-1})^2 a^2 (a^{-1}b^{-1})^2,$$

certainly $G^{(n)} \subseteq G^{[n]}$. However, there is in general no guarantee that $\omega(G) = \bigcap_{n=0}^{\infty} G^{[n]}$ is perfect. For instance, when G is a group of odd order, $\omega(G) = G$ is soluble.

1.3 Topological motivation

There is in algebraic topology an important theorem of Kan and Thurston that represents homotopy types of spaces as pairs consisting of a discrete group and a distinguished perfect normal subgroup [61]. The key to this representation is Quillen's *plus-construction*, invented to build higher K-theory of rings [79]. We work in **Ho**, the pointed homotopy category of connected CW-complexes. (One can think of this as the topological category of based, connected CW-complexes (called *spaces* below) in which a formal inverse is adjoined to each based homotopy equivalence.) For each space X and perfect normal subgroup N of $\pi_1(X)$, the (relative) plus-construction is a map $q : X \to X_N^+$ that is *acyclic* (meaning that its homotopy fibre F_q is an acyclic space: $\tilde{H}_*(F_q) = 0$) and has the following universal property [9, 5.2].

Lemma 1.3.1 *Let X be any connected space. If N is a perfect normal subgroup of $\pi_1(X)$, then any map $f : X \to Y$ such that $f_*(N)$ is trivial factors through the plus-construction $q : X \to X_N^+$, uniquely up to homotopy under X.*

The Kan–Thurston theorem states that any space Y has the homotopy type of some X_N^+, where X may be chosen to be the classifying space of a discrete group T. Thus the group T and perfect normal subgroup N of T determine Y as BT_N^+. In fact, one can go further and show that the plus-construction induces an equivalence of categories. The codomain category is **Ho**. The domain is obtained from the category of pairs (G, N) of group G and perfect normal subgroup N as a category of fractions, by formally inverting those group homomorphisms that induce isomorphisms of the quotient G/N and also induce isomorphisms of G in homology with coefficients induced from G/N [6]. This equivalence of categories means that many notions and constructions in homotopy theory have parallels in group theory, in a way that crucially involves perfect groups [53].

1.4 Examples

Where are good places to go hunting for perfect groups?

A group theorist would instantly point out that any non-abelian simple group G must be perfect, since the normal subgroup G' cannot be trivial. Thus the great corpus of knowledge of finite simple groups provides information about finite perfect groups [56], although perfectness has not been a key concept in the classification of finite simple groups.

Many finite simple groups occur as groups of matrices over finite fields, for example, matrices of determinant 1. (Actually, this gives a perfect group; one has to divide out by the centre, comprising the scalar matrices, to obtain a simple group.) An $n \times n$ matrix may be regarded as an $(n+1) \times (n+1)$ matrix by adjoining an entry 1 in the $(n+1, n+1)$ position, and zeroes elsewhere in the $(n+1)$-st row and column. In this way we can take the union of all the groups of determinant 1 matrices to obtain the infinite special linear group. Note that this group is centreless, for the only infinite scalar matrix that has some diagonal entries equal to 1 is the identity matrix. For about a hundred years it has been known that this group is simple, whenever the entries lie in a field; moreover the field need not be finite [80, 3.2]. These results are essentially algebraic, although much of their motivation came from projective geometry, in particular groups of collineations of projective spaces (e.g. [67, VII.5], [51, 7.1.1]).

Combinatorial topology provided an essential input that led to the generalization of these examples from fields to arbitrary rings, and thus gave birth to the subject of algebraic K-theory. This developed from J.H.C. Whitehead's attempt [97], [98] to describe a homotopy equivalence of compact spaces as an iteration of elementary collapses and expansions (a map homotopic to such an iteration being called a *simple homotopy equivalence*). Each elementary collapse/expansion is determined by the attaching map of an attaching r-cell. (It was for this purpose that CW-complexes were invented.) The complete endeavour is worthy of a book ([28] is unhesitatingly recommended) and too technical even to outline in the space available here. However, some informal remarks hint at how it inexorably leads to an important class of perfect groups.

First, by means of mapping cylinders one manœuvres into the situation of a finite r-dimensional complex K, with subcomplex L which is a strong deformation retract of K. Moreover, by a geometric normalization argument ('trading cells') one may assume that K is determined from a subcomplex K' by n attaching maps $f_i : S^{r-1} \to K'$, and K' in turn is formed from L by the attachment of n $(r-1)$-cells, via maps $g_i : S^{r-2} \to L$. Now the strong deformation retraction kills the first and last groups in the homotopy exact sequence of the triple (K, K', L):

$$\pi_r(K, L) \to \pi_r(K, K') \xrightarrow{\partial} \pi_{r-1}(K', L) \to \pi_{r-1}(K, L).$$

Thus the connecting homomorphism ∂ is an isomorphism. This is an isomorphism of $\mathbb{Z}[\pi_1(L)]$-modules because the sequence respects the natural action of $\pi = \pi_1(L)$ on the relative homotopy groups [88, 7.3.10]. (Recall that when a group G acts on an abelian group A compatibly with addition in A, then A becomes a module over the *group ring* $\mathbb{Z}[G]$. $\mathbb{Z}[G]$ consists of finite formal sums $\sum m_i x_i$ ($m_i \in \mathbb{Z}$, $x_i \in G$) with

$$\left(\sum m_i x_i\right) \cdot \left(\sum n_j y_j\right) = \sum_{i,j} m_i n_j x_i y_j,$$

as well as the obvious formal addition.) In the present setup, both $\pi_r(K, K')$ and $\pi_{r-1}(K', L)$ are free $\mathbb{Z}[\pi]$-modules (isomorphic to the direct sum of n copies of $\mathbb{Z}[\pi]$), with generating classes represented by the f_i and g_i respectively. Hence ∂ defines an $n \times n$ change-of-basis matrix, an element of the group $\mathrm{GL}_n(\mathbb{Z}[\pi])$ of invertible $n \times n$ matrices over the group ring.

What drives the whole theory is the effect of elementary collapses and expansions on this matrix. They correspond variously to products of the following operations:

(i) embedding $\mathrm{GL}_n(\mathbb{Z}[\pi])$ in $\mathrm{GL}_{n+1}(\mathbb{Z}[\pi])$, by adjunction of 1 in the $(n+1, n+1)$-slot, as previously described;

(ii) multiplication by an elementary matrix $e_{ij}(r)$ (an identity matrix, save for a unique off-diagonal entry r in the (i, j)-slot), corresponding to the operation of adding a multiple of one row or column to another;

(iii) multiplication by an invertible diagonal matrix (diagonal entries comprising elements of π and their negatives, these necessarily being units of $\mathbb{Z}[\pi]$).

So, to decide whether a given homotopy equivalence is a simple homotopy equivalence one has to determine whether a matrix in $\mathrm{GL}_n(\mathbb{Z}[\pi])$ is reducible to the identity matrix by a sequence of the above operations. Now define, for any ring R (associative, with identity 1), the subgroup $E_n R$ of $\mathrm{GL}_n R$ to be that generated by all $n \times n$ elementary matrices. Combining (i) and (ii) above gives the subgroup ER of $\mathrm{GL}R$.

An important lemma (the Whitehead lemma) asserts that ER is the commutator subgroup of $\mathrm{GL}R$. Therefore the group $K_1 R := \mathrm{GL}R/ER$ is abelian. Hence the obstruction to a homotopy equivalence being a simple homotopy equivalence (the Whitehead torsion) lies in the abelian group (the *Whitehead group*) $\mathrm{Wh}(\pi)$ obtained from $K_1(\mathbb{Z}[\pi])$ by factoring out the subgroup of diagonal matrices specified in (iii) above. Conversely, every element of $\mathrm{Wh}(\pi)$ is geometrically realizable as such an obstruction. The calculation of $\mathrm{Wh}(\pi)$, even for finite π, is an active area of research [78].

In general, K_1R is but one of a whole sequence of abelian group invariants of a ring R, whose study comprises the topic of algebraic K-theory [5], [9], [81]. The group ER plays a crucial role in this subject. Since for distinct i, j, k

$$e_{ij}(r) = [e_{ik}(r), e_{kj}(1)] \tag{1.4.1}$$

each E_nR, $n \geq 3$, is generated by commutators and so perfect. Hence ER too is perfect, and indeed is the intersection of the transfinite derived series of $\mathrm{GL}R$. We referred above to Quillen's use of the plus-construction to define higher K-groups of a ring R. Specifically, for $i \geq 1$, he defined K_iR to be the ith homotopy group of the space $B\mathrm{GL}R^+_{ER}$. It follows from Lemma 1.3.1 that for $i = 1$ this agrees with the previous definition of K_1R.

The observation that ER is perfect can be pushed further. For any ideal I of R, EI is defined to be the subgroup of ER generated by all $e_{ij}(a)$ with $a \in I$, and $E(R, I)$ has generators of the form $\alpha e_{ij}(a)\alpha^{-1}$ where $\alpha \in ER$ and $e_{ij}(a) \in EI$. It is readily shown to be normal in $\mathrm{GL}R$. Now suppose that I is an idempotent ideal. This means that we can write

$$a = \sum_p a'_p a''_p,$$

a finite sum with each $a'_p, a''_p \in I$. So for any k distinct from both i and j, as in Equation 1.4.1

$$\begin{aligned} \alpha e_{ij}(a)\alpha^{-1} &= \textstyle\prod_p [\alpha e_{ik}(a'_p)\alpha^{-1}, \alpha e_{kj}(a''_p)\alpha^{-1}] \\ &\in [E(R, I), E(R, I)]. \end{aligned}$$

This makes $E(R, I)$ a perfect normal subgroup of ER. With more work one can show the converse, and thereby characterize the perfect normal subgroups of $\mathrm{GL}R$ as precisely those $E(R, I)$ with I an idempotent ideal of R [15]. In favourable cases, such as R being a commutative domain, it is known that there are no proper non-zero idempotent ideals, making ER the sole non-trivial perfect normal subgroup of $\mathrm{GL}R$.

2 Basic facts

2.1 Preliminaries

If all the general theorems on perfect groups were laid end-to-end, they would scarcely reach a conclusion. What we compile here amounts to little more than a collection of observations, which at least makes for easy reading.

Since perfect groups are those generated by their commutators, and the homomorphic image of a commutator is again a commutator, any homomorphic image of a perfect group must also be perfect. In the other direction, let

$$N \hookrightarrow G \twoheadrightarrow Q$$

be a group extension with both N and Q perfect. Then the associated homology exact sequence

$$N/[N,G] \to H_1(G) \to H_1(Q)$$

has first and last terms zero, leaving G perfect. (The alternative, barehanded calculation is a simple exercise.) So the class of perfect groups is extension-closed. Evidently it is also closed under the formation of finite direct and free products. Further, the direct limit of a direct system of perfect groups is perfect too, since every element in the limit can be traced back to a product of commutators somewhere.

If one restricts to subgroups of a given group G, then again all conjugates and products of perfect subgroups are perfect. So the product of all perfect subgroups is both normal and perfect. It must contain every perfect subgroup of G, and so is the *maximum perfect subgroup*, or *perfect radical*, $\mathcal{P}G$ of G. Because whenever $\mathcal{P}G \leq H$, a subgroup of G,

$$\mathcal{P}G = [\mathcal{P}G, \mathcal{P}G] \leq [H, H],$$

$\mathcal{P}G$ must lie inside each term of the derived series, and then within each term of the transfinite derived series of G. Thus $\mathcal{P}G$ is contained in the intersection of the transfinite derived series. However, we have already noted that this intersection is itself perfect, hence necessarily contained in $\mathcal{P}G$. So we have an alternative description of $\mathcal{P}G$ as the intersection of the transfinite derived series of G. Its elements are thereby those elements of G that can be expressed as products of commutators of elements that are themselves products of commutators of elements that are in turn ..., and so (transfinitely) on. Since the image of a perfect group is perfect, any homomorphism $G \to H$ maps $\mathcal{P}G$ inside $\mathcal{P}H$, so the construction $G \mapsto \mathcal{P}G$ is functorial.

2.2 Actions of perfect groups

Just as groups with trivial first derived group are called abelian, those with trivially intersecting transfinite derived series are sometimes called *hypoabelian*. So certainly soluble groups are hypoabelian. Abelianization is a universal construction for obtaining an abelian quotient of an arbitrary group. Likewise there is a process that might be termed *hypoabelianization* (by those who can tolerate a mixture of Greek, Norwegian and Latin in a single word). For, let P be a subgroup of G that has perfect image $\overline{P}$ in the quotient $G/\mathcal{P}G$. After multiplying by $\mathcal{P}G$, we may assume that P contains $\mathcal{P}G$. Then the group extension

$$\mathcal{P}G \hookrightarrow P \twoheadrightarrow \overline{P}$$

forces P to be a perfect subgroup of G. Maximality of $\mathcal{P}G$ makes P equal to $\mathcal{P}G$, and so $\overline{P}$ is trivial. On the other hand, any map from G to a group H with $\mathcal{P}H = 1$ must kill $\mathcal{P}G$ and so factor through $G/\mathcal{P}G$. Therefore the epimorphism $G \twoheadrightarrow G/\mathcal{P}G$ has the universal property of being initial in the category of all maps from G to hypoabelian groups. An example is given by the plus-construction $X \to X^+ = X^+_{\mathcal{P}\pi_1(X)}$ with respect to the maximum perfect subgroup of $\pi_1(X)$. It follows from (1.3.1) that on fundamental groups it induces the hypoabelianization $\pi_1(X) \twoheadrightarrow \pi_1(X)/\mathcal{P}\pi_1(X)$.

A much-used corollary of the above is that any homomorphism from a perfect group to a hypoabelian (for example, soluble) group is trivial. Another 'industrial lemma' is as follows. (Recall the notations $[H, K]$ for the subgroup generated by all commutators $[h, k]$ with $h \in H$, $k \in K$, and ${}^hk = hkh^{-1}$.)

Lemma 2.2.1 *Let N, P be subgroups of a group G, and suppose that P is perfect. If $[[N,P],P] = 1$, then $[N,P] = 1$.*

Proof From the Hall–Witt identity

$$^a[[a^{-1},b],c] \cdot {}^c[[c^{-1},a],b] \cdot {}^b[[b^{-1},c],a] = 1$$

(a brute-force calculation), one has the Three Subgroup lemma: For any three subgroups N, P, Q of a group G, $[[P,Q],N]$ lies in every normal subgroup of G that contains both $[[N,P],Q]$ and $[[N,Q],P]$. Here we put $Q = P = [P,P]$.

This forms the basis for an induction argument. When P is a group that acts on a group N, then N and P may both be taken to be subgroups of their semidirect product $G = N \rtimes P$. One then calls the original action of P on N *trivial* if $[N,P] = 1$ in G and *nilpotent* if some

$$[[\cdots[[N,P],P]\cdots],P] = 1.$$

The extended argument thus shows that if a perfect group acts nilpotently on another group, then it acts trivially.

This result has implications for the study of fibrations in algebraic topology. A fibration $F \to E \to B$ of path-connected spaces is called *orientable* if $\pi_1 B$ acts trivially on the homology groups of F, and *quasi-nilpotent* if the induced action is nilpotent. Normally one spends some effort in showing that theorems about orientable fibrations extend to quasi-nilpotent fibrations too. However, when the fundamental group of B is perfect, we have just seen that the extension is vacuous.

Staying with the group theory, it is interesting to focus on the case where N is a subgroup of P. Our result may be recast as the assertion that if N lies in the hypercentre (the union of the upper central series) of P, then N is contained in the centre $\mathcal{Z}(P)$ of P. Thus the group $P/\mathcal{Z}(P)$ is centreless. In fact, the theory of central subgroups of perfect groups is both simple and elegant.

2.3 Central extensions

Readers may like to treat the following assertions as exercises. Much of the material may be found in [76, ch. 5], proved by elementary means. For an alternative approach by easy obstruction theory, see [9, ch. 8], using the fact that a group extension

$$N \hookrightarrow G \twoheadrightarrow Q$$

is central ($N \subseteq \mathcal{Z}(G)$) if and only if the associated fibration

$$BN \to BG \to BQ$$

is principal. For many related topics, see [22], [47]. We also refer to the epimorphism $G \twoheadrightarrow Q$ (and occasionally G itself) as a central extension (over Q).

Lemma 2.3.1 *A composite of central extensions over perfect groups is a central extension.*

Lemma 2.3.2 *A group G is perfect if and only if for each homomorphism $\phi : G \to R$ and each central extension $\pi : H \twoheadrightarrow R$ there is at most one lift $\mu : G \to H$ with $\phi = \pi\mu$.*

Lemma 2.3.3 *If $G \twoheadrightarrow Q$ is a central extension with Q perfect, then $G' = [G, G]$ is perfect.*

Theorem 2.3.4 *The category of central extensions over a group Q, with morphisms as commuting triangles over Q, has an initial object (the* universal central extension*) if and only if Q is perfect.*

Lemma 2.3.5 *For Q perfect with universal central extension $S \twoheadrightarrow Q$, and any central extension $G \twoheadrightarrow Q$, the group G is perfect if and only if the unique map $S \to G$ over Q is surjective.*

Theorem 2.3.6 *Let $M \hookrightarrow S \twoheadrightarrow Q$ be a central extension. Then the following are equivalent.*

(i) $S \twoheadrightarrow Q$ is a universal central extension.

(ii) S is superperfect: *$H_1(S) = H_2(S) = 0$.*

(iii) S is perfect and every central extension over S splits.

(iv) S is perfect and M is isomorphic to $H_2(Q)$, the Schur multiplicator *of Q.*

(v) Any group extension $R \hookrightarrow F \twoheadrightarrow Q$ with free group F induces an isomorphism of exact sequences

$$\begin{array}{ccccc} (R \cap [F,F])/[R,F] & \hookrightarrow & [F,F]/[R,F] & \twoheadrightarrow & Q \\ \downarrow\cong & & \downarrow\cong & & \downarrow \mathrm{id} \\ M & \hookrightarrow & S & \twoheadrightarrow & Q \end{array}$$

For a pleasing analogous theory of extensions of S-modules when S is superperfect, see [23].

It turns out that every abelian group M is the Schur multiplicator of some perfect group Q as in this theorem; moreover, for this purpose S can be chosen (functorially) so that all its homology groups vanish [10] (see (3.1.13) below).

The smallest non-trivial perfect group is the simple group of order 60. Formulating it as the projective special linear group $\mathrm{PSL}_2(\mathbb{F}_5)$, one sees that the universal central extension is its double cover $\mathrm{SL}_2(\mathbb{F}_5)$, the binary icosahedral group. Thus the Schur multiplicator has order 2. Analogous statements hold for other finite simple groups of Lie type. Analysis of this situation led Steinberg [91] to construct, for a ring R, the *Steinberg group* $\mathrm{St}R$ (by mimicking the generators and relations, such as Equation 1.4.1, in ER). Then $\mathrm{St}R \twoheadrightarrow ER$ is the universal central extension of ER [65], whose kernel $H_2(ER)$ became the definition of $K_2(R)$ [76, ch. 5].

A more challenging exercise, based on the 5-term homology exact sequence of a group extension, is to show that if $N \hookrightarrow G \twoheadrightarrow Q$ is a (not necessarily central) extension with G perfect-by-soluble (in other words, $\mathcal{P}G = G^{(n)}$ for some finite n), then there is an exact sequence

$$H_2(\mathcal{P}G) \to H_2(\mathcal{P}Q) \to N/[\mathcal{P}G, N \cap \mathcal{P}G] \to G/\mathcal{P}G \twoheadrightarrow Q/\mathcal{P}Q.$$

2.4 Some open problems

2.4.1 Kervaire conjecture. In characterizing those groups that are the fundamental group of the complement of a smoothly embedded n-sphere in $\mathbb{R}^{n+2}$ (called a *higher knot group*), Kervaire [63] found a necessary condition to be that the group is generated by the conjugates of a single element and its inverse. He was led to conjecture that this condition is always violated by the free product $G * C_\infty$ of a non-trivial group G with the infinite cyclic group C_∞. In this event, let us call the group G *Kervaire*. Evidently, for G to fail to be Kervaire, the abelianization of $G * C_\infty$ must be cyclic, so that G is perfect. (Indeed, it follows from [63] that a superperfect group G is non-Kervaire precisely when $G * C_\infty$ is a higher knot group.)

The few direct assaults on this problem have yielded as Kervaire all locally residually finite [82] and locally indicable groups [58] ('locally indicable' means that every finitely generated subgroup maps onto C_∞), all groups having a faithful finite-dimensional unitary representation [68, p. 50], all torsion-free groups [66], and all groups having a quotient group that is Kervaire [20]. Because confirmation of the conjecture is equivalent to its verification for algebraically closed groups [20], it may be considered solely as a problem in combinatorial group theory, where one asks whether a system of equations in a group admits a solution in some larger group [60]. However, it gains interest from its relation to a number of problems in low-dimensional topology [24], [59], where it is related to other conjectures [41], [89]. One such, discussed in [57], is a problem posed by J.H.C. Whitehead: Is every subcomplex of an aspherical 2-complex also aspherical? The fundamental group of any counterexample must contain both a superperfect and a finitely generated perfect (non-trivial) subgroup.

2.4.2 Homology spheres. We have already noted that the fundamental group π of a homology n-sphere (say, a smooth n-dimensional manifold M with $H_*(M) \cong H_*(S^n)$) must have $H_1(\pi) = 0$. By standard classification of surfaces, we may assume that $n \geq 3$. Since $H_2(\pi)$ is the cokernel of the Hurewicz homomorphism from $\pi_2(M)$ to $H_2(M)$, it too is zero, making π a superperfect, finitely presented group. In [64], Kervaire uses surgery to show that these necessary conditions also suffice whenever $n \geq 5$. This is not the case for $n = 3$, since there the only non-trivial finite group π is the binary icosahedral group $SL_2(\mathbb{F}_5)$, the fundamental group of the Poincaré 3-sphere. In low dimensions there are the following implications. The first may be deduced from [33], while the others are in [64].

Theorem 2.4.3 *Among the following statements,*

$$\text{(i)} \quad \Longrightarrow \quad \text{(ii)} \quad \Longrightarrow \quad \text{(iii)} \quad \Longrightarrow \quad \text{(iv)}$$

(i) π is the fundamental group of a homology 3-sphere.

(ii) π is a perfect group and has a finite presentation with an equal number of generators and relations.

(iii) π is the fundamental group of a homology 4-sphere.

(iv) For each $n \geqslant 5$, π is the fundamental group of a homology n-sphere.

In [52], it is shown that the implication (iii) $\Rightarrow$ (iv) above is strict.*

2.4.4 μ-problem. For an arbitrary group G, we can define $\mu(G)$ to be the smallest cardinality of a non-empty subset X whose normal closure (the subgroup generated by all conjugates of elements of X and their inverses) is G. This notation is compatible with that of ring theorists. However, there is no convention. For example, in [63], μ is called the weight w. It is shown there that, for $n \geq 1$, when N is a connected n-dimensional submanifold of a simply-connected $(n+2)$-dimensional smooth manifold M, then $\mu(\pi_1(M - N)) = 1$. (Another notation, attributed to P. Hall, is $d_G(G)$, referring to the number of generators of G as a G-group, the G-operation being conjugation.)

Evidently, for any quotient Q of G, $\mu(G) \geq \mu(Q)$. The most interesting comparison is when $Q = G_{\text{ab}}$, since $\mu(G_{\text{ab}})$ is just the rank of the abelian group G_{ab}. The μ-problem asks when equality holds. Let us call

$$\mu_{\text{def}}(G) = \mu(G) - \mu(G_{\text{ab}})$$

the *μ-defect* of G. Since a perfect group P has $\mu(P_{\text{ab}}) = 1$, perfect groups are a promising place to look for positive μ-defects. For example, for any non-trivial perfect Kervaire group P, the free product $C_\infty * P$ has positive μ-defect. In fact, when P has finite composition length, then by Theorem 2.4.5 below, $\mu(P) = 1$, so that $\mu_{\text{def}}(C_\infty * P) = 2 - 1 = 1$. For more extreme examples, the

* *Added in proof:* The strictness of (i) $\Rightarrow$ (ii) follows from C.M. Campbell and E.F. Robertson: A deficiency 0 presentation of $SL(2,p)$, *Bull. London Math. Soc.* **12** (1980), 17–20. For the irreversibility of (ii) $\Rightarrow$ (iii), see: J.A. Hillman: An homology 4-sphere group with negative deficiency, to appear in *L'Enseignement Math.*

perfect, locally nilpotent groups of (3.1.13) below have infinite μ-defect, as does the countable, locally finite, superperfect group of [19] that has all subnormal subgroups normal, of infinite index.

On the other hand, the same logic suggests that, to find groups with zero μ-defect one should consider groups that are far from perfect on passage to quotients. This leads to the class of *imperfect* groups, namely those groups with no non-trivial perfect quotient. Examples are soluble groups and finite symmetric groups. This class relates to the μ-problem as follows.

Theorem 2.4.5 [19] *If G is an extension of an imperfect group by a group with finite composition length, then $\mu_{\text{def}}(G) = 0$.*

Many of the groups of homeomorphisms referred to in (3.1.6) below also have zero μ-defect. Indeed, much of the literature on these groups is devoted to establishing uniform bounds on the number of conjugates in the expression of an arbitrary element as a product of conjugates of a given element.

2.4.6 Number of commutators. Since in a perfect group G every element is a product of commutators, a very obvious problem, related to the above, is to ask how few commutators are required. The number may be unbounded, as for example, in the case of $E_n(\mathbb{C}[x])$ with $n \geq 3$ [30]. When there is a bound, one writes $c(G)$ for the minimum number of commutators in which any element of G' is expressible.

Linear groups have been the most studied from this point of view. For example, for any field F, $c(E_nF) \leq 2$ and $c(EF) = 1$ [93], $c(E_nR) \leq 5$ for a commutative ring R of stable rank 1 (such as $R = \mathbb{Z}/k$) [30], while for any R $c(ER) \leq 2$ [49].

There is an old conjecture of O. Ore that every finite simple group G has $c(G) = 1$. The binate groups encountered in 3 below all have $c = 1$. Much of the literature on non-binate homeomorphism groups aims to establish bounds on c. For example, the group of piecewise linear homeomorphisms of $\mathbb{R}$ with compact support is perfect with $c \leq 2$ [92].

2.4.7 Finite groups. There is of course a large number of conjectures and open problems concerning finite perfect groups. I mention here two that relate to the material in this chapter. The first asserts that perfect finite groups are rare, and thus stands in contrast to a result cited in (3.1.8) below. Let p_n and g_n denote the number of isomorphism classes of perfect, respectively all, groups of order n.

Conjecture 2.4.8 *For the multiplicative directed set structure on $\mathbb{N}$ where $d \preceq n \Leftrightarrow d \mid n$,*

$$\operatorname{dirlim}(p_n/g_n) = 0.$$

Strong supporting evidence has appeared in [56].

For the second, we call a group G *k-connected* (*k-acyclic* in [23]) if, for $1 \leq i \leq k$, $H_i(G) = 0$ (equivalently the space BG^+ is k-connected). The search for 3-connected finite groups culminated in Milgram's computation [75] that the Mathieu group M_{23} is 4-connected. This still leaves open the following. (Related material is surveyed in [1].)

Problem 2.4.9 Does there exist k such that no non-trivial finite group G is k-connected?

3 Acyclic groups

Our earlier discussion of the plus-construction showed the importance of the class of acyclic spaces. When these spaces are classifying spaces of discrete groups, as occurs in the context of the Kan–Thurston theorem, the group is also called *acyclic*. Acyclic groups therefore form a useful subclass of the class of all perfect groups. (An acyclic group G is one with $H_i(G) = 0$ for all $i \geq 1$; it must be perfect because $H_1(G) = 0$.) Below is a selection of my favourite examples. (This has been described as the 'zoo', a fair term since a little taxonomy is possible. It was partly exhibited in [11].) The list is more or less in chronological order. It is notable that it is drawn from a very wide range of mathematics. We shall see that there are some interesting, and surprising, patterns. One of these is worth highlighting before we begin.

A group G is called *binate* [11] if it is the direct limit of subgroups G_λ where for each λ there exist $\mu \geq \lambda$, $u_\lambda \in G_\mu - G_\lambda$ and $\phi_\lambda : G_\lambda \to G_\mu$ such that for all $g \in G_\lambda$

$$g = [u_\lambda, \phi_\lambda g]. \tag{3.0.1}$$

Evidently a binate group is perfect; in fact, every element is a commutator. Since any finitely generated subgroup of a binate group G is contained in some G_λ, it excludes u_λ. Thus

- binate groups cannot be finitely generated.

Deeper facts are:

- binate groups are acyclic [11];
- binate groups have no finite-dimensional representations over any field [2];
- every binate group contains infinitely many images of a universal binate group [20].

(Let us agree not to refer to the existence of trivial representations and images.)

3.1 Examples of acyclic groups

3.1.1 Higman's 4-generator, 4-relator group. This is a candidate for being the 'oldest' acyclic group in the literature, although its acyclicity was not proved until much later [32]. It has the presentation

$$\langle x_i \mid x_{i+1} = [x_i, x_{i+1}] \rangle_{i \in \mathbb{Z}/k}$$

where k equals 4 (or any larger integer; $k = 0$ works too, but $k = 1, 2$ or 3 makes the group trivial.) By combinatorial means, Higman [54] shows that it has no finite image and thus, since it is finitely generated, no finite-dimensional representation over any field (for finitely generated subgroups of $\mathrm{GL}_n(F)$ are residually finite). See [21] for generalizations. More geometrically, [32] describes the classifying space of this group as a 2-dimensional complex consisting of one 0-cell, four 1-cells and four 2-cells, and also observed that the group is the fundamental group of a homology 4-sphere. On the other hand, because the group is finitely presented, it must also be the fundamental group of a minimal, symplectic 4-manifold [42]. So it follows from claims of [29] that the free product of two copies of this group is a finitely presented acyclic group that is the fundamental group of a 4-manifold (the connected sum of those above) that has non-trivial Seiberg–Witten invariant yet admits no symplectic structure.

3.1.2 Algebraically closed groups. By analogy with algebraic closure for fields, a (non-trivial) group G is called *algebraically closed* [86] if every finite system of equations consistent with G has a solution in G itself. In particular Equation (3.0.1) always has a solution, and so G is binate. G is also simple [55]. Algebraically closed groups are especially relevant to the study of the Kervaire conjecture cited above [20].

3.1.3 Philip Hall's countable universal locally finite group. This group is built as the union of a nested sequence of finite groups, starting with G_0 of order at least 3. Given G_i, let G_{i+1} be the symmetric group of all permutations of the finite set $G_i \times G_i$, with $G_i \times G_i$ embedded in G_{i+1} by means of the (right) regular representation ρ on the first factor, and G_i in $G_i \times G_i$ by inclusion as the first factor. Then, by the proof of Lemma 1 of [48], any isomorphism between subgroups of $G_i \times G_i$ is expressible as conjugation in G_{i+1}. Let ϕ_i be the embedding of G_i in $G_i \times G_i$ (and thence in G_{i+1}) by inclusion as the second factor. Since this produces a subgroup of $G_i \times G_i$ isomorphic to the one obtained from the diagonal inclusion (the isomorphism is $(\mathrm{id} \times \bar{g}) \mapsto (\bar{g} \times \bar{g})$), we take u_i to be an element of G_{i+1}, conjugation by which restricts to this isomorphism. Then for any g in G_i we have

$$g = [u_i, \phi_i g].$$

It is clearly a countable, locally finite group satisfying P. Hall's condition for universality: that every finite group is embedded, and any two isomorphic finite subgroups are conjugate. Because it is both acyclic and locally finite, this group has the remarkable topological property [16] that there is no homotopically non-trivial map from the classifying space to its 2-skeleton, although of course both these spaces have the same fundamental group. For a generalization to universal locally finite groups made to have the same cardinality as, and to contain, an arbitrary, locally finite group, see [62].

3.1.4 Commutator subgroups of 2-generator, 1-relator groups. This appears to be the first proof of acyclicity of a group in the literature [8]. Let G be a group generated by two elements subject to a single relation. If G_{ab} is infinite cyclic and G' is perfect, then it is shown that G' is acyclic. Remarkably, an example is the geometrically-defined group whose acyclicity was demonstrated just weeks later. In [34], Epstein performs a variant of the grope construction, whereby a doubly-punctured torus is added at each stage. The two punctures bound different loops in the previous stage, one a toral meridian and one a longitude, as prescribed by taking the complement in S^3 of a regular neighbourhood of the 1-complex shown in Fig. 5.

Current work with my colleague Yan-Loi Wong [21], and M. Cencelj in Ljubljana, reveals that the fundamental group of the open aspherical 3-manifold so constructed is after all the commutator subgroup of the group

$$G = \langle x, y \mid x = [x, yx^{-1}y^{-1}]\,[x, y^{-1}xy]\rangle .$$

Moreover, Higman's group [54] is also of this type, at least when $k = 0$, with the other values of k corresponding to quotient groups.

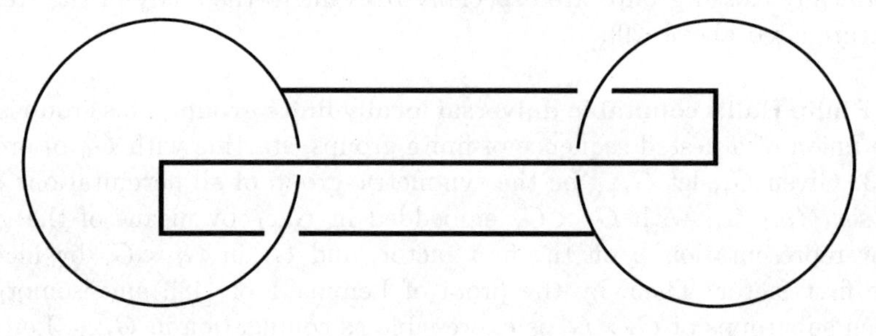

Fig. 5

3.1.5 Stitch groups. In [38], Fox and Artin introduced a wild arc corresponding to iteration (infinitely, towards a cluster point, in both directions) of an underlying pattern called a *stitch* in [21].

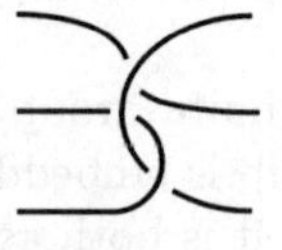

Fig. 6

Stitches also arise in [27], [37], [39], [74], [77]. In general, $\pi_1(S^3 - \kappa)$ is acyclic for any arc, and the commutator subgroup of a finitely presented group for any wild arc κ obtained from a stitch.

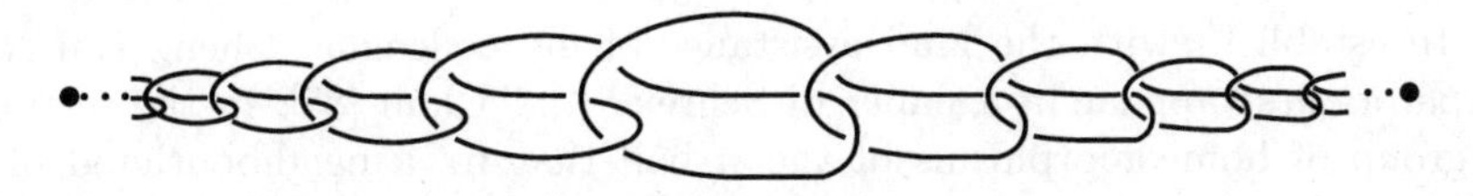

Fig. 7

3.1.6 Groups of self-homeomorphisms with support symmetries. Groups of this kind have been investigated for over six decades, primarily for their conjugacy properties. Many of them are algebraically simple. The first homological treatment was in [69], where the group of self-homeomorphisms of $\mathbb{R}^n$ with compact support was shown to be acyclic (and binate, alias pseudo-mitotic, in [95]). Recall that the support supp(h) of a self-map h of a topological space is the closure of the set of points x with $h(x) \neq x$. The requirements for a subgroup of a group H of homeomorphisms of a space X to be binate are typically as follows.

Suppose that X has a directed set of subsets B_λ, $\lambda \in \Lambda$. Put $G = \text{dirlim} G_\lambda$, where $G_\lambda = \{g \in H \mid \text{supp}(g) \subseteq B_\lambda\}$. Suppose that each λ has $\rho_\lambda \in G$, which we refer to as a *dissipator*, that satisfies:

(i) for all $i \geqslant 1$, $\rho^i_\lambda(B_\lambda) \cap B_\lambda = \emptyset$; and

(ii) for all $g \in G_\lambda$ there is a function in H defined by

$$\phi_\lambda(g) = \begin{cases} \rho^i_\lambda g \rho^{-i}_\lambda & \text{on } \rho^i_\lambda(B_\lambda),\ i \geq 1, \\ \text{id} & \text{elsewhere.} \end{cases}$$

A typical picture of a dissipator looks like this:

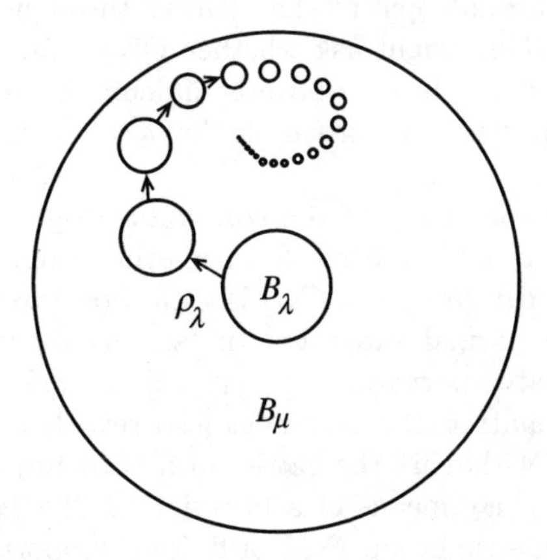

Fig. 8

Given these properties, which may be technically difficult to substantiate (!), then G is a binate group. For by (i), supp$(\phi_\lambda(g)) \subseteq$ supp$(\rho_\lambda) \subseteq$ some B_μ. Then by (i) again $\phi_\lambda(G_\lambda)$ commutes with G_λ in G_μ, so that ϕ_λ is a homomorphism. One can check that Equation 3.0.1 holds, with $u_\lambda = \rho^{-1}_\lambda$. So far as I have been able to establish (with the kind assistance of my colleague, Cheng Kai Nah), dissipators first appear in a paper of Schreier and Ulam [85], in the setting of the group of homeomorphisms of the n-ball that fix a neighbourhood of the boundary.

With the addition of a further set-theoretic property, G becomes the minimum normal subgroup of H. Thus when G is H itself, the group G is simple, the case of chief historical interest.

Other previously studied (discrete) groups of homeomorphisms that may retrospectively be shown to be binate by means of dissipators include:

- the (G, K)-rotational examples of [3], namely the Homeo(X) subgroup generated by all homeomorphisms of X that restrict to the identity on some non-empty open subset of X, where X is the rationals, the irrationals, the universal curve, the Cantor ternary set, or, in the case of orientation-preserving homeomorphisms, the universal plane curve or S^n, $n > 1$ (the case $n = 1$ is also claimed in [4], as are some subgroups of the above that leave invariant a given countable dense subset) – see [83] for some proofs, and [84] for some remarkable self-embedding properties;
- the group generated by those homeomorphisms of a topological n-manifold that have support contained in some internal closed n-cell [36];
- for a space X endowed with Euclidean neighbourhoods, the subgroup of Homeo(X) generated by those homeomorphisms that fix a member of a certain family of subsets of X (corresponding to neighbourhoods of the boundary in the n-ball B^n) [99];
- the group of those C^1-diffeomorphisms of a paracompact connected C^1-manifold M that are isotopic to the identity through compactly supported C^1 isotopies [35]. (One might ask whether C^r results also hold, but higher differentiability of $\phi_\lambda(g)$ in (ii) above is difficult to check. Nevertheless, the resulting group is perfect, in fact simple, for $r \neq 1 + \dim M$ [70], [71], [94].)

3.1.7 The general linear group of the cone on a ring. To study the algebraic K-theory of a ring R, [96] embeds R in another ring, its cone CR, whose direct limit general linear group $\mathrm{GL}(CR)$ is then shown to be acyclic, containing a copy of $\mathrm{GL}R$ as a normal subgroup. In fact, as shown in [11], this group is binate, with dissipator portrayed in [9, p. 85]. Because this dissipator is an infinite permutation matrix, the argument also reveals as binate the group of permutations of $\mathbb{N} \times \mathbb{N}$ that fix the entries of all but finitely many copies of $\mathbb{N}$ (regarded as columns). By means of a bijection of $\mathbb{N} \times \mathbb{N}$ with $\mathbb{N}$, this shows that the infinite symmetric group $\mathfrak{S}_\infty$, of finitely supported permutations of a countably infinite set, is a normal subgroup of an acyclic group.

3.1.8 The cone on a group. The cone CG on a group G is defined in [61] as the semi-direct product of $G^{\mathbb{Q}}$ by $\mathrm{Aut}(\mathbb{Q})$. Here $G^{\mathbb{Q}}$ is the set of functions from the rationals $\mathbb{Q}$ to G which map all numbers outside some finite interval to the identity; the group structure on G determines that on $G^{\mathbb{Q}}$. Likewise $\mathrm{Aut}(\mathbb{Q})$

denotes the restricted symmetric group on $\mathbb{Q}$ comprising those permutations with compact support. As the reader probably suspects, this group allows dissipators and so is binate. Again, see [11] for details. Observe that G normally embeds in $G^{\mathbb{Q}}$ by means of $g \mapsto b_g$, where $b_g : \mathbb{Q} \to G$ sends 0 to g and non-zero numbers to $1 \in G$. Thus G is embedded as a two-step subnormal subgroup of a binate group.

This raises the interesting issue of which groups can be normal subgroups of acyclic groups. (Some examples occur in (3.1.7) above.) An immediate necessary condition is based on the fact that the image of a perfect group must be perfect. So if N is normal in a perfect group G but the automorphism group of N is hypoabelian, then the map from G to $\mathrm{Aut}(N)$ induced by conjugation must be trivial. In particular, the action of N on itself by conjugation must be trivial, making N abelian.

3.1.9 Diffeomorphism groups for foliation theory. These acyclic groups have significant links to classifying spaces for foliations [94]. Let Y be the interior of a compact manifold $\overline{Y}$ with boundary, and X the complement in $\overline{Y}$ of a closed collar neighbourhood of the boundary of $\overline{Y}$, so that X is an open, relatively compact submanifold of Y diffeomorphic to Y. Then the group of those diffeomorphisms of Y that restrict to the identity on some neighbourhood of the closure of X is acyclic [87]. As noted above, in the attempt to construct dissipators, smoothness of $\phi_\lambda(g)$ is not readily verified. So the argument presented in [87] does not use dissipators explicitly. Likewise, acyclicity of the relative group of volume-preserving diffeomorphisms is treated in [72], [73]. A further variant of this technique is applied to the group of C^1 piecewise $\mathrm{SL}_2(\mathbb{R})$ homeomorphisms of $\mathbb{R}$ that restrict to the identity near $-\infty$ in [45]. Acyclic groups of orientation-preserving homeomorphisms of $\mathbb{R}$ that are germ-connected to the identity are discussed in [43].

3.1.10 Mitotic groups. These binate groups were first used in [6] to provide combinatorial, finitely generated and, later [7], finitely presented results analogous to those of [61]. Because of their combinatorial construction, they do not have dissipators. They do have the remarkable property that all quotients are also binate.

3.1.11 Perfect, locally free groups. We met such groups in (1.2) above as the fundamental groups of gropes. To deduce acyclicity, we combine the following three facts. Every group is the direct limit of its finitely generated subgroups. Homology preserves direct limits. Free groups have zero homology in all dimensions above the first (for their classifying spaces are bouquets of circles). The conclusion is that a locally free group will be acyclic precisely when it is perfect. The significance of this assertion was emphasized by a construction of Heller [53]: given any element x of a perfect group P, there is a countable, perfect, locally free group D and a homomorphism from D to P whose image contains x. Now, since free products of groups give rise to wedges of classifying spaces, a free product of acyclic groups must be acyclic. It follows that any perfect group is the image of a free product of a suitably large set of representatives of each isomorphism class of countable, perfect, locally free groups (see [16] for related topological subtleties). From one viewpoint, this says that there are enough normal-in-acyclic groups to guarantee the presentation of every perfect group as the image of an acyclic group.

3.1.12 Automorphism groups of large structures. Modelled on (3.1.7) above, a number of automorphism groups are shown, in effect, to have dissipators and so be binate [50]. Examples include the group of all continuous linear automorphisms, or of invertible isometries, of an infinite-dimensional Hilbert space, the group of invertible or of unitary elements in a properly infinite von Neumann algebra, the group of measure-preserving automorphisms of a Lebesgue measure space, and the group of permutations on an arbitrary infinite set. (Note however that the group of finitely supported permutations is not perfect, since the sign representation maps onto the group of order 2.) The class of examples of acyclic groups is greatly enlarged by showing that in many cases the group of unrestricted automorphisms is sufficiently rich in binate subgroups to be acyclic too. It turns out that none of these groups has a finite-dimensional linear representation [14].

3.1.13 Generalized McLain groups. For an arbitrary ring R, the group $M(\Lambda, R)$ of all upper unitriangular matrices over R is acyclic, provided that the ordered set Λ that indexes the entries is dense (in that, between any two elements there lies a third) [10]. Since finitely generated groups of upper unitriangular matrices are nilpotent, $M(\Lambda, R)$ is a locally nilpotent group. If R has a positive characteristic, then $M(\Lambda, R)$ is also locally finite. Moreover, when Λ has both a first and last element, the matrices have a central top-right-hand corner entry, which, together with all entries of the first row may be taken to lie in any right R-module A. This gives a more general group $M(\Lambda, (R, A))$, still acyclic, but having A as its centre. Then the group extension

$$A \hookrightarrow M(\Lambda, (R, A)) \twoheadrightarrow M(\Lambda, (R, A))/A$$

is by Theorem 2.3.6 a universal central extension. In particular, with $R = \mathbb{Z}$, we obtain any abelian group as normal-in-acyclic and as a Schur multiplicator. From our discussion in (3.1.8) above, we see that this construction cannot be generalized to embrace nilpotent groups. For example, the quaternion group of order 8 has hypoabelian automorphism group $\mathfrak{S}_4$, and so is not a normal subgroup of any perfect group.

3.1.14 Torsion-generated acyclic groups. Specific instances of acyclic groups generated by their elements of finite order have been encountered in (3.1.3), (3.1.13) above. It turns out that such torsion-generated acyclic groups share with binate groups the property of lacking finite-dimensional complex representations [12]. This places severe limitations on those groups that can be normal in such an acyclic group. In [18] is constructed a universal finitely presented acyclic group that is strongly torsion generated (normally generated by a single element of arbitrary finite order, hence perfect). (However, the construction does not reveal the number of generators or relations required.) Related results from [18] of interest include the following:

- every sequence of abelian groups is realizable as the higher homology groups of a strongly torsion generated group;
- if a torsion generated group has only finitely many non-zero homology groups, then it is perfect;
- if a locally finite group has only finitely many non-zero homology groups, then it is acyclic (H.-W. Henn).

See also [17] for p-primary refinements of the last two results.

3.1.15 An extension of the braid group. The braid group B_n on n strands can be described geometrically as the group of isotopy classes of orientation-preserving, boundary-fixing homeomorphisms of the disk B^2 that permute a distinguished set X_n of n interior points. Restriction to the permutation of the distinguished points gives an epimorphism from B_n to the symmetric group $\mathfrak{S}_n$. Algebraically, B_n has $n-1$ generators $\sigma_1, ..., \sigma_{n-1}$ and relations

$$\begin{aligned} \sigma_i\sigma_{i+1}\sigma_i &= \sigma_{i+1}\sigma_i\sigma_{i+1}, \\ \sigma_i\sigma_j &= \sigma_j\sigma_i \qquad\qquad |i-j|>1. \end{aligned}$$

Then σ_i maps to the transposition $(i \quad i+1)$ to give the epimorphism to $\mathfrak{S}_n$.

Compatibly with the inclusion of X_n in X_{n+1} there is an embedding of $\mathfrak{S}_n$ in $\mathfrak{S}_{n+1}$ and B_n in B_{n+1}. On passing to the direct limit one obtains the group B_∞, which maps onto $\mathfrak{S}_\infty$ and turns out to have the same homology as the double loop-space of S^3. Compatibly with the representation of $\mathfrak{S}_\infty$ as a normal-in-acyclic group in [96] (referred to in (3.1.7) above), there is an acyclic group in which B_∞ is normally embedded. It is constructed in [44] as a group of automorphisms of a generalized braid group for which the above set X_n is replaced by a tree whose vertices are indexed by the dyadic numbers between 0 and 1.

3.1.16 The universal binate tower. Although we have seen that it commonly occurs in the geometric setting of dissipators, the characteristic equation (3.0.1) for binate groups may be viewed as a purely algebraic phenomenon. This suggests a universal example [11]. Starting with any group H as base (the trivial group gives the extreme case), one can construct a binate tower by means of HNN-extensions. Let $H_0 = H$ and for each $i \geq 0$

$$H_{i+1} = \mathrm{gp}\left\langle H_i \times H_i, u_i \mid (g,g) = u_i(1,g)u_i^{-1} \text{ for each } g \in H_i \right\rangle .$$

With H_i embedded in H_{i+1} as $H_i \times 1$ and $\phi_i(g) = (1,g)$, the binate equation is readily checked, making the direct limit a binate group (even though the intermediate groups have hyperexponentially-growing homology). This tower is the initial object in a category of binate towers with base group H [20].

Acknowledgements. It was the suggestion of Dusan Repovš that I speak on this topic at Ljubljana that led to the above distillation of my views. So I am grateful to him for more than his warm hospitality. I would also like to thank the numerous people, many of whose names appear below, who kindly offered comments on a draft of this work. Most of all, I thank, and salute, Brian Steer – mathematician, thesis supervisor, colleague, and friend.

References

1. A. Adem: Recent developments in the cohomology of finite groups, *Notices Amer. Math. Soc.* **44** (1997), 806–812.
2. R.C. Alperin and A.J. Berrick: Linear representations of binate groups, *J. Pure Appl. Algebra* **94** (1994), 17–23.
3. R.D. Anderson: The algebraic simplicity of certain groups of homeomorphisms, *Amer. J. Math.* **80** (1958), 955–963.
4. R.D. Anderson: On homeomorphisms as products of conjugates of a given homeomorphism and its inverse, in 'Topology of 3-Manifolds, Proc. U. Georgia Inst. 1961', Prentice-Hall (New Jersey, 1962), 231–234.
5. H. Bass: 'Algebraic K-Theory', Benjamin (New York, 1968).
6. G. Baumslag, E. Dyer and A. Heller: The topology of discrete groups, *J. Pure Appl. Algebra* **16** (1980), 1–47.
7. G. Baumslag, E. Dyer and C.F. Miller III: On the integral homology of finitely presented groups, *Topology* **22** (1983), 27–46.
8. G. Baumslag and K.W. Gruenberg: Some reflections on cohomological dimension and freeness, *J. Algebra* **6** (1967), 394–409.
9. A.J. Berrick: 'An Approach to Algebraic K-Theory', Pitman Research Notes Math. 56 (London, 1982).
10. A.J. Berrick: Two functors from abelian groups to perfect groups, *J. Pure Appl. Algebra* **44** (1987), 35–43.
11. A.J. Berrick: Universal groups, binate groups and acyclicity, in 'Group Theory', Proc. Singapore Conf., de Gruyter (Berlin, 1989), 253 266.
12. A.J. Berrick: Remarks on the structure of acyclic groups, *Bull. London Math. Soc.* **22** (1990), 227–232.
13. A.J. Berrick: Torsion generators for all abelian groups, *J. Algebra* **139** (1991), 190–194.
14. A.J. Berrick: Groups with no nontrivial linear representations, *Bull. Austral. Math. Soc.* **50** (1994), 1–11; Corrigenda, *loc cit* **52** (1995), 345–346.
15. A.J. Berrick: The plus-construction as a localization, in 'Algebraic K-Theory and its Applications' Procs ICTP 1997, World Scientific (Singapore, 1999), 313–336.
16. A.J. Berrick and C. Casacuberta: A universal space for plus-constructions, *Topology* **38** (1999), 467–477.
17. A.J. Berrick and P.H. Kropholler: Groups with infinite homology, in 'Cohomological Methods in Homotopy Theory' Procs BCAT98, Progress in Math. 196, Birkhäuser (Basel, 2001), 27–33.

18. A.J. Berrick and C.F. Miller III: Strongly torsion generated groups, *Math. Proc. Camb. Phil. Soc.* **111** (1992), 219–229.
19. A.J. Berrick and D.J.S. Robinson: Imperfect groups, *J. Pure Appl. Algebra* **88** (1993), 3–22.
20. A.J. Berrick and K. Varadarajan: Binate towers of groups, *Arch. Math.* **62** (1994), 97–111.
21. A.J. Berrick and Y.-L. Wong: Acyclic groups and wild arcs, preprint.
22. F.R. Beyl and J. Tappe: 'Group Extensions, Representations, and the Schur Multiplicator', Springer LNM 958 (Berlin, 1982).
23. D. Blanc and G. Peschke: The plus construction, Postnikov decompositions, and universal central module extensions, preprint.
24. S.G. Brick: A note on coverings and Kervaire complexes, *Bull. Austral. Math. Soc.* **46** (1992), 1–21.
25. J.W. Cannon: ULC properties in neighborhoods of embedded surfaces and curves in E^3, *Canad. J. Math.* **25** (1973), 31–73.
26. J.W. Cannon: The recognition problem for topological manifolds: What is a topological manifold?, *Bull. Amer. Math. Soc.* **84** (1978), 832–866.
27. P. Churchard and D. Spring: Proper knot theory in open 3-manifolds, *Trans. Amer. Math. Soc.* **308** (1988), 133–142.
28. M.M. Cohen: 'A Course in Simple-Homotopy Theory', Springer GTM 10 (New York, 1973).
29. B. Dai and H.-Y. Wang: On minimal symplectic 4-manifolds, preprint.
30. R.K. Dennis and L.N. Vaserstein: On a question of M. Newman on the number of commutators, *J. Algebra* **118** (1988), 150–161.
31. A.N. Dranishnikov and D. Repovš: Cohomological dimension with respect to perfect groups, *Topology and its Appl.* **74** (1996), 123–140.
32. E. Dyer and A.T. Vasquez: Some small aspherical spaces, *J. Austral. Math. Soc.* **16** (1973), 332–352.
33. D.B.A Epstein: Finite presentations of groups and 3-manifolds, *Quart. J. Math.* **12** (1961), 205–212.
34. D.B.A. Epstein: A group with zero homology, *Proc. Camb. Phil. Soc.* **64** (1968), 599–601.
35. D.B.A. Epstein: The simplicity of certain groups of homeomorphisms, *Compositio Math.* **22** (1970), 165–173.
36. G.M. Fisher: On the group of all homeomorphisms of a manifold, *Trans. Amer. Math. Soc.* **97** (1960), 193–212.
37. R.H. Fox: A remarkable simple closed curve, *Ann. of Math.* **50** (1949), 264–265.
38. R.H. Fox and E. Artin: Some wild cells and spheres in three-dimensional space, *Ann. of Math.* **49** (1948), 979–990.
39. B. Freedman and M.H. Freedman: Kneser–Haken finiteness for bounded 3-manifolds, locally free groups, and cyclic covers, *Topology* **37** (1998), 133–147.
40. L. Fuchs: 'Infinite Abelian Groups', Volumes I, II, Academic Press (New York, 1970, 1973).
41. S.M. Gersten: Nonsingular equations of small weight over groups, Combinatorial Group Theory and Topology (Alta, Utah, 1984), Ann. of Math. Stud. 111, Princeton University Press (Princeton, 1987), 121–144.
42. R.E. Gompf: A new construction of symplectic manifolds, *Ann. of Math.* **142** (1995), 527–595.
43. P. Greenberg: Homology of some groups of PL-homeomorphisms of the line, *Topology and its Appl.* **33** (1989), 281–295.

44. P. Greenberg and V. Sergiescu: An acyclic extension of the braid group, *Comment. Math. Helvetici* **66** (1991), 109–138.
45. P. Greenberg and V. Sergiescu: Piecewise projective homeomorphisms and a noncommutative Steinberg extension, *K-Theory* **9** (1995), 529–544.
46. H.B. Griffiths: 'Surfaces', 2nd ed., Cambridge University Press (Cambridge, 1981).
47. K.W. Gruenberg: 'Cohomological Topics in Group Theory', Springer LNM 143 (Berlin, 1970).
48. P. Hall: Some constructions for locally finite groups, *J. London Math. Soc.* **34** (1959), 305–319.
49. P. de la Harpe: Sur le simplicité essentielle du groupe des inversibles et du groupe unitaire dans une C^*-algèbre simple, *Funct. Annal.* **62** (1985), 354–378.
50. P. de la Harpe and D. McDuff: Acyclic groups of automorphisms, *Comment. Math. Helvetici* **58** (1983), 48–71.
51. R. Hartshorne: 'Algebraic Geometry', Springer GTM 52 (New York, 1977).
52. J-C. Hausmann and S. Weinberger: Caractéristiques d'Euler et groupes fondamentaux des variétés de dimension 4, *Comment. Math. Helv.* **60** (1985), 139–144.
53. A. Heller: On the homotopy theory of topogenic groups and groupoids, *Illinois J. Math.* **24** (1980), 576–605.
54. G. Higman: A finitely generated infinite simple group, *J. London Math. Soc.* **26** (1951), 61–64.
55. G. Higman: A note on algebraically closed groups, *J. London Math. Soc.* **27** (1952), 227–242.
56. D. Holt and W. Plesken: 'Perfect Groups', Oxford University Press (Oxford, 1989).
57. J. Howie: Aspherical and acyclic 2-complexes, *J. London Math. Soc.* **20** (1979), 549–558.
58. J. Howie: On pairs of 2-complexes and systems of equations over groups, *J. Reine Angew. Math.* **324** (1981), 165–174.
59. J. Howie and H. Short: The band-sum problem, *J. London Math. Soc.* **31** (1985), 571–576.
60. J. Howie: Nonsingular systems of two length three equations over a group, *Math. Proc. Cambridge Phil. Soc.* **110** (1991), 11–24.
61. D.M. Kan and W.P. Thurston: Every connected space has the homology of a $K(\pi, 1)$, *Topology* **15** (1976), 253–258.
62. O.H. Kegel and B.A.F. Wehrfritz: 'Locally Finite Groups', North Holland (Amsterdam, 1973).
63. M.A. Kervaire: On higher dimensional knots, in 'Differential and Combinatorial Topology' (A Symposium in Honor of Marston Morse) Princeton University Press (Princeton, 1965), 105–119.
64. M.A. Kervaire: Smooth homology spheres and their fundamental groups, *Trans. Amer. Math. Soc.* **144** (1969), 67–72.
65. M.A. Kervaire: Multiplicateurs de Schur et K-théorie, in 'Essays on Topology and Related Topics', Memoires dédiés à Georges de Rham, Springer (Heidelberg, 1970), 212–225.
66. A.A. Klyachko: A funny property of sphere and equations over groups, *Comm. in Alg.* **21** (1993), 2555–2575.
67. R.C. Lyndon: 'Groups and Geometry', LMS Lect. Notes 101, Cambridge University Press (Cambridge, 1985).
68. R.C. Lyndon and P.E. Schupp: 'Combinatorial Group Theory', Springer Ergeb. 89 (Berlin, 1977).

69. J.N. Mather: The vanishing of the homology of certain groups of homeomorphisms, *Topology* **10** (1971), 297–298.
70. J.N. Mather: Commutators of diffeomorphisms, *Comment. Math. Helv.* **49** (1974), 512–528.
71. J.N. Mather: Commutators of diffeomorphisms: II, *Comment. Math. Helv.* **50** (1975), 33–40.
72. D. McDuff: On groups of volume-preserving diffeomorphisms and foliations with transverse volume form, *Proc. London Math. Soc.* **43** (1981), 295–320.
73. D. McDuff: On the tangle complexes and volume-preserving diffeomorphisms of open 3-manifolds, *Proc. London Math. Soc.* **43** (1981), 321–333.
74. J.M. McPherson: On the nullity and enclosure genus of wild knots, *Trans. Amer. Math. Soc.* **144** (1969), 545–555.
75. R.J. Milgram: The cohomology of the Mathieu group M_{23}, *J. Group Theory,* **3** (2000), 7–26.
76. J. Milnor: 'Introduction to Algebraic K-Theory', Ann. Math. Studies 72, Princeton University Press (Princeton, 1971).
77. O. Nanyes: Proper knots in open 3-manifolds have locally unknotted representatives, *Proc. Amer. Math. Soc.* **113** (1991), 563–571.
78. R. Oliver: 'Whitehead Groups of Finite Groups', LMS Lect. Notes 132, Cambridge University Press (Cambridge, 1988).
79. D. Quillen: Cohomology of groups, Actes Congrès Int. Math. Nice, vol. 2, 1970, 47–51.
80. D.J.S. Robinson: 'A Course in the Theory of Groups', Springer GTM 80 (New York, 1982).
81. J. Rosenberg: 'Algebraic K-Theory and its Applications', Springer GTM 147 (New York, 1994).
82. O.S. Rothaus: On the nontriviality of some group extensions given by generators and relations, *Ann. of Math.* **106** (1977), 599–612.
83. P. Sankaran and K. Varadarajan: Acyclicity of certain homeomorphism groups, *Canad. J. Math.* **42** (1990), 80–94.
84. P. Sankaran and K. Varadarajan: On certain homeomorphism groups, *J. Pure Appl. Algebra* **92** (1993), 191–197; Corrections, ibid **114** (1997), 217–219.
85. J. Schreier and S. Ulam: Über topologische Abbildungen der euklidischen Sphären, *Fund. Math.* **23** (1934), 102–118.
86. W.R. Scott: Algebraically closed groups, *Trans. Amer. Math. Soc.* **2** (1951), 118–121.
87. G. Segal: Classifying spaces related to foliations, *Topology* **17** (1978), 367–382.
88. E.H. Spanier: 'Algebraic Topology', McGraw-Hill (New York, 1966).
89. J.R. Stallings: Surfaces in three-manifolds and nonsingular equations in groups, *Math. Z.* **184** (1983), 1–17.
90. M.A. Štan'ko: Embeddings of compacta in Euclidean space, *Mat Sb* **83** (2) (1970), 234–255 (Russian); *Math. USSR-Sb* **10** (1970), 234–254 (English).
91. R. Steinberg: Générateurs, relations et revêtements de groupes algébriques, in 'Colloq. Théorie des Groupes Algébriques', Bruxelles, 1962, Gauthier-Villars (Paris, 1964), 113–127.
92. T. Tsuboi: Small commutators of piecewise linear homeomorphisms of the real line, *Topology* **34** (1995), 815–857.
93. R.C. Thompson: Commutators in the special and general linear groups, *Trans. Amer. Math. Soc.* **101** (1961), 16–33.
94. W. Thurston: Foliations and groups of diffeomorphisms, *Bull. Amer. Math. Soc.* **80** (1974), 304–307.

95. K. Varadarajan: Pseudo-mitotic groups, *J. Pure Appl. Algebra* **37** (1985), 205–213.
96. J.B. Wagoner: Delooping classifying spaces in algebraic K-theory, *Topology* **11** (1972), 349–370.
97. J.H.C. Whitehead: Simplicial spaces, nuclei and m-groups, *Proc. London Math. Soc.* **45** (1939), 243–327.
98. J.H.C. Whitehead: Simple homotopy types, *Amer. J. Math.* **72** (1950), 1–57.
99. J.V. Whittaker: Normal subgroups of some homeomorphism groups, *Pacific J. Math.* **10** (1960), 1469–1478.

2

The geometry of the word problem

MARTIN R. BRIDSON

Introduction

The study of decision problems in group theory is a subject that does not impinge on most geometers' lives – for many it remains an apparently arcane region of mathematics near the borders of group theory and logic, echoing with talk of complexity and undecidability, devoid of the light of geometry. The study of minimal surfaces, on the other hand, is an immediately engaging field that combines the shimmering appeal of soap films with intriguing analytical problems; Plateau's problem has a particularly intuitive appeal. The first purpose of these notes is to explain that despite this sharp contrast in emotions, the study of the large-scale geometry of least-area discs in Riemannian manifolds is intimately connected with the study of the complexity of word problems in finitely presented groups.

Joseph Antoine Ferdinand Plateau was a Belgian physicist who, in 1873, published a stimulating account of his experiments with soap films [90]. The question of whether or not every rectifiable Jordan loop in 3-dimensional Euclidean space bounds a disc of minimal area subsequently became known as Plateau's problem. This problem was solved by Jesse Douglas [37] and Tibor Radó [91] (independently) around 1930. In 1948 C.B. Morrey [75] extended the results of Douglas and Radó to a class of spaces that includes the universal covering of any closed, smooth Riemannian manifold M.

Once one knows that least-area discs exist in this generality, numerous questions come to mind concerning their local and global geometry (cf. [79], [86], and [71]). The questions on which we shall focus in this chapter concern the large-scale geometry of these discs: Can one bound the area of least-area discs in M by a function of the length of their boundaries? If so, what is the least such function? What happens to the asymptotic behaviour of this function when one perturbs the metric or varies M within its homotopy type? What can one say about the diameter of least-area discs? *etc.*

Remarkably, these questions turn out to be intimately connected with the nature of the word problem in the fundamental group of M, i.e. the problem of determining which words in the generators of the group equal the identity. The most important and striking connection of this type is given by the *Filling theorem* in Section 2: the smallest function $\mathrm{Fill}_0^M(l)$ bounding the area of

The author's research is supported by an EPSRC Advanced Fellowship.

least-area discs in terms of their boundary length has qualitatively the same asymptotic behaviour[1] as the *Dehn function* $\delta_{\pi_1 M}(l)$ of the fundamental group of M.

The Dehn function of a finitely presented group $\Gamma = \langle \mathcal{A} \mid \mathcal{R} \rangle$ measures the complexity of the word problem for Γ by giving the least upper bound on the number of defining relations $r \in \mathcal{R}$ that must be applied in order to show that a word w in the letters $\mathcal{A}^{\pm 1}$ is equal to $1 \in \Gamma$; the bound is given as a function of the length of w (see Paragraph 1.2).

The first purpose of these notes is to give a thorough account of the Filling theorem. The second purpose of these notes is to sketch the current state of knowledge concerning Dehn functions. Thus, in Section 3, I shall explain what is known about the set of $\simeq$ classes of Dehn functions (equivalently, isoperimetric functions Fill_0^M of closed Riemannian manifolds), and I shall also describe what is known about the Dehn functions of various groups that are of geometric interest. In later sections we shall see a variety of methods for calculating Dehn functions (some geometric, some algebraic, and some purely combinatorial). Along the way we shall see examples of how the equivalence $\delta_{\pi_1 M} \simeq \text{Fill}_0^M$ can inform in both directions (cf. (2.2) and Section 6).

Historical background

The precise equivalence between filling functions of manifolds and complexity functions for word problems is a modern observation due to Mikhael Gromov, but this connection sits comfortably with the geometric origins of combinatorial group theory.

Topology and combinatorial group theory emerged from the same circle of ideas at the end of the nineteenth century. By 1910 Dehn had realized that the problems with which he was wrestling in his attempts to understand low-dimensional manifolds were instances of more general group-theoretic problems. In 1912 he published the celebrated paper in which he set forth the three basic decision problems that remained the main focus for combinatorial group theory throughout the twentieth century:

> *The general discontinuous group is given by n generators and m relations between them.* [...] *Here there are above all three fundamental problems* [...]
>
> 1: [The Word Problem] *An element of the group is given as a product of generators. One is required to give a method whereby it may be decided in a finite number of steps whether this element is the identity or not.* [2: The Conjugacy Problem. 3: The Isomorphism Problem]

[1] More precisely, Fill_0^M is $\simeq$ equivalent to $\delta_{\pi_1 M}$ in the sense of 1.3.2.

One is already led to them by necessity with work in topology. Each knotted space curve, in order to be completely understood, demands the solution of the three above problems in a special case.[2]

In the present chapter I shall concentrate almost exclusively on the word problem, but in Section 8 I shall explain constructions that translate the complexity of word problems into conjugacy problems and isomorphism problems. These basic decision problems are all unsolvable in the absence of further hypotheses (see [72] for a survey of these matters) and in the spirit of Dehn's comments I should note that this undecidability has consequences for the study of manifolds. For example, the undecidability of the isomorphism problem for groups implies that there is no algorithm to recognize whether or not a closed 4-manifold (given by a finite triangulation, say) is homeomorphic to the 4-sphere [70].

Despite Dehn's early influence, the geometric vein in combinatorial group theory lacked prominence for much of the twentieth century (see [30] for a history up to 1980). A striking example of this neglect concerns a paper [61] written by E.R. van Kampen in 1931 which seems to have gone essentially unnoticed until rediscovered[3] by C. Weinbaum in the 1960s, just after Roger Lyndon [65] rediscovered the paper's main idea. This idea translates many questions concerning word problems into questions concerning the geometry of certain planar 2-complexes called *van Kampen diagrams* (see Section 4). This translation acts as a link between Riemannian filling problems and word problems. The work of Gromov [55], [56] gave full voice to the implications of this link. In the decade since Gromov's foundational work there has been a great deal of activity in this area and I hope that when the reader has finished the present chapter (s)he will have absorbed a sense of this activity and its achievements.

Contents

I have written this chapter with the intention that it should be accessible to graduate students and colleagues working in other areas of mathematics. It is organised as follows. In Section 1 we shall see how a naive head-on approach to the word problem leads to the definition of the Dehn function of a group. In Section 2 we introduce the 2-dimensional, genus-0 isoperimetric function of a closed Riemannian manifold M and state the theorem relating it to the Dehn function of $\pi_1 M$; the proof of this theorem is postponed until Section 5. This theorem is generally regarded as folklore – its validity has been assumed implicitly in many papers, but the absence of a detailed proof in the literature has been the source of comment and disquiet. The proof given here is self-contained. It is based on the notes from my lectures at the conferences in Durham, Lyon,

[2] The special cases referred to here were not resolved fully until the early 1990s, and their ultimate solution rested on some of the deepest geometry and topology of the time, in particular the work of Thurston on the geometric nature of 3-manifolds.

[3] Van Kampen's article was next to the one in which he proved the Seifert–van Kampen Theorem.

and Champoussin in the spring and summer of 1994. José Burillo and Jennifer Taback [26] have suggested an alternative proof, motivated by arguments in [42]. Both proofs rely on van Kampen's lemma, which is proved in complete detail in Section 4.

Section 3 contains a brief survey describing the current state of knowledge about the nature of Dehn functions for groups in general as well as groups that are of particular geometric interest. We shall not prove the results in this section, but several of the key ideas involved are explained in subsequent sections.

Section 6 contains information about the classes of groups whose Dehn functions are linear or quadratic. We shall see that having a linear Dehn function is a manifestation of negative curvature. We shall also see that non-positive curvature is related to having a quadratic Dehn function, although the connection is much weaker than in the linear case.

The final section of this paper contains a brief discussion of different measures of complexity for the word problem, as well as constructions relating the word problem to the other basic decision problems of group theory.

There are three appendices to this paper. The first contains a description of some basic concepts in geometric group theory – this is included to make the arguments in the main body of the paper accessible to a wider audience. The second appendix describes some of the basic vocabulary of length spaces. The third appendix contains the proof of a technical result concerning the geometry of combinatorial discs; this result, which is original, is needed in Section 5.

Exercises are scattered throughout the text, some are routine verifications, some lead the diligent reader through proofs, and others are challenges intended to entice the reader along fruitful tangents.

These notes are dedicated with deep affection to my tutor and friend Brian Steer. Between 1983 and 1986 Brian transformed me into a budding mathematician and thereby determined the course of my adult life.

APPENDIX B: LENGTH SPACES

APPENDIX C: A PROOF OF THE CELLULATION LEMMA

1 The word problem

The purpose of this first section is to indicate why Dehn functions are fundamental to the understanding of discrete groups.

1.1 Presenting groups that arise in nature

Suppose that one wishes to understand a group Γ that arises as a group of transformations of some mathematical object, for example isometries of a metric space. Typically, one might be interested in the group generated by certain basic transformations $\mathcal{A} = \{a_1, \ldots, a_n\}$. One then knows that arbitrary elements of Γ can be expressed as words in these generators and their inverses, but in order to gain a real understanding of the group one needs to know which pairs of words w, w' represent the same element of Γ, i.e. when $w^{-1}w' = 1$ in Γ. Words that represent the identity are called *relations*.

Let us suppose that the context in which our group arose is such that we can identify at least a few relations $\mathcal{R} = \{r_1, \ldots, r_m\}$. How might we use this list to deduce that other words represent the identity?

If a word w contains $r \in \mathcal{R}$ or its inverse as a subword, say[4] $w = w_1 r^{\pm 1} w_2$, then we can replace w by the shorter word $w' = w_1 w_2$, knowing that w' and w represent the same element of Γ. More generally, if r can be broken into (perhaps empty) subwords $r \equiv u_1 u_2 u_3$ and if $w \equiv w_1 u_2^{\pm 1} w_2$, then one knows that $w' \equiv w_1 (u_3 u_1)^{\mp 1} w_2$ equals w in Γ. Under these circumstances[5] one says w' **is obtained from** w **by applying the relator** r.

If we can reduce w to the empty word by applying a sequence of relators $r \in \mathcal{R}$, then we will have deduced that $w = 1$ in Γ. If such a sequence can be found for every word w that represents the identity – in other words, every relation in the group can be deduced from the set $\mathcal{R}$ – then the pair[6] $\langle \mathcal{A} \mid \mathcal{R} \rangle$ is called a *presentation* of Γ, and one writes[7] $\Gamma = \langle \mathcal{A} \mid \mathcal{R} \rangle$.

---

[4] We write $=$ for equality in the free group, and $\equiv$ when words are actually identical.

[5] At this point we are viewing words as elements of the free group $F(\mathcal{A})$, so implicitly we allow the insertion and deletion of subwords of the form aa^{-1}.

[6] If $\mathcal{R} = \{r_1, r_2, \ldots\}$, one often writes $\langle \mathcal{A} \mid r_1 = 1, r_2 = 1, \ldots \rangle$ instead of $\langle \mathcal{A} \mid \mathcal{R} \rangle$, particularly when this creates a desirable emphasis. Likewise, one may write $\langle \mathcal{A} \mid u_1 = v_1, u_2 = v_2, \ldots \rangle$, where $r_i \equiv u_i v_i^{-1}$.

[7] To assign a name to a presentation, P say, one writes $P \equiv \langle \mathcal{A} \mid \mathcal{R} \rangle$.

1.2 Attacking the word problem head-on

A solution to the word problem in Γ is an algorithm that will decide which elements of the group represent the identity and which do not. If one can bound the number of relators that must be applied to a word w in order to show that $w = 1$, and this bound can be expressed as a computable function of the length of w, then one has an effective solution to the word problem. In order to quantify this idea precisely, one works with equalities in the free group $F(\mathcal{A})$.

Suppose that $w' = w_1(u_3u_1)w_2$ has been obtained from $w = w_1u_2^{-1}w_2$ by applying the relator $r \equiv (u_1u_2u_3)^{-1}$. In Γ we have $w = w'$, while in the free group $F(\mathcal{A})$ we have:

$$w \equiv w_1u_2^{-1}w_2 \overset{\text{free}}{=} (x_1^{-1}rx_1)\, w_1u_3u_1w_2 \equiv (x_1^{-1}rx_1)\, w',$$

where $x_1 := u_3^{-1}w_1^{-1}$. If w'' is a word obtained from w' by applying a further relator r', then there is an equality of the form $w \overset{\text{free}}{=} (x_1^{-1}rx_1)\,(x_2^{-1}r'x_2)\, w''$.

Proceeding in this manner, if we can reduce w to the empty word by applying a sequence of N relators from $\mathcal{R}$, then we will have an equality[8]

$$w \overset{\text{free}}{=} \prod_{i=1}^{N} x_i^{-1}r_ix_i, \tag{1.2.1}$$

where $r_i \in \mathcal{R}^{\pm 1}$ and $x_i \in F(\mathcal{A})$.

Thus we see that when one attacks the word problem head-on by simply applying a list of relators to a word w, one is implicitly expressing w as a product of conjugates of those relators. The ease with which one can expect to identify such an expression for w will vary according to the group under consideration, and in particular will depend very much on the number N of factors in a least such expression.

Definition 1.2.2 *Given a finite presentation $P \equiv \langle \mathcal{A} \mid \mathcal{R}\rangle$ defining a group Γ, we say that a word w in the letters $\mathcal{A}^{\pm 1}$ is* null-homotopic *if $w =_\Gamma 1$, i.e. w lies in the normal closure of $\mathcal{R}$ in the free group $F(\mathcal{A})$. We define the algebraic area of such a word to be*

$$\mathrm{Area}_a(w) := \min\Big\{N \mid w \overset{\text{free}}{=} \prod_{i=1}^{N} x_i^{-1}r_ix_i \text{ with } x_i \in F(\mathcal{A}),\ r_i \in \mathcal{R}^{\pm 1}\Big\}.$$

The Dehn function *of P is the function $\delta_P : \mathbb{N} \to \mathbb{N}$ defined by*

$$\delta_P(n) := \max\{\mathrm{Area}_a(w) \mid w =_\Gamma 1, |w| \leqslant n\},$$

where $|w|$ denotes the length of the word w.

---

[8] This equality shows in particular that $\Gamma = \langle \mathcal{A} \mid \mathcal{R}\rangle$ iff the kernel of the natural map $F(\mathcal{A}) \to \Gamma$ is the normal closure of $\mathcal{R}$.

1.3 The Dehn function of a group

Since we are really interested in groups rather than particular finite presentations of them, we would like to talk about the Dehn function of Γ rather than of P. The following exercise illustrates how the Dehn functions of different presentations of a group may vary.

Exercise 1.3.1 Show that the Dehn function of $\langle a \mid \emptyset \rangle$ is $\delta(n) \equiv 0$ and the Dehn function of $\langle a, b \mid b \rangle$ is $\delta(n) = n$. For each positive integer k find a presentation of $\mathbb{Z}$ with Dehn function $\delta(n) = kn$.

Definition 1.3.2 *Two monotone functions* $f, g : [0, \infty) \to [0, \infty)$ *are said to be* $\simeq$ *equivalent if* $f \preceq g$ *and* $g \preceq f$, *where* $f \preceq g$ *means that there exists a constant* $C > 0$ *such that* $f(l) \leqslant C\, g(Cl + C) + Cl + C$ *for all* $l \geqslant 0$.

One extends this equivalence relation to functions $\mathbb{N} \to [0, \infty)$ *by assuming them to be constant on each interval* $[n, n+1)$.

The relation $\simeq$ preserves the asymptotic nature of a function. For example, if $p > 1$ then $n^p \not\simeq n^p \log n$, and $n^p \simeq n^q$ implies $q = p$; likewise, $n^p \not\simeq 2^n$ and $2^{2^n} \not\simeq 2^n$. But $\simeq$ identifies all polynomials of the same degree, and likewise all single exponentials ($k^n \simeq K^n$ for all constants $k, K > 1$).

Proposition 1.3.3 *If the groups defined by two finite presentations are isomorphic, the Dehn functions of those presentations are* $\simeq$ *equivalent.*

Proof First we consider what happens when we add redundant relators $\mathcal{R}'$ to a finite presentation $P \equiv \langle \mathcal{A} \mid \mathcal{R} \rangle$. Let $P' \equiv \langle \mathcal{A} \mid \mathcal{R} \cup \mathcal{R}' \rangle$. To say that the new relators $r \in \mathcal{R}'$ are redundant means that each can be expressed in the free group $F(\mathcal{A})$ as a product Π_r of (say m_r) conjugates of the old relators $\mathcal{R}^{\pm 1}$. Let m be the maximum of the m_r.

If a word $w \in F(\mathcal{A})$ is a product of N conjugates of relators from $\mathcal{R} \cup \mathcal{R}'$ and their inverses, then by substituting Π_r for each occurrence of $r \in \mathcal{R}'$ in this product we can rewrite w (freely) as a product of at most mN conjugates of the relators $\mathcal{R}^{\pm 1}$. Since it is obvious that the area of w with respect to P' is not greater than its area with respect to P, we have $\delta_{P'}(n) \leqslant \delta_P(n) \leqslant m\, \delta_{P'}(n)$ for all $n \in \mathbb{N}$. Hence $\delta_P \simeq \delta_{P'}$.

Next we consider what happens when we add finitely many generators and relators to P. Suppose that we add generators $\mathcal{B}$, and add one relator bu_b^{-1} for each $b \in \mathcal{B}$, where u_b is a word in $F(\mathcal{A})$ that equals b in the group being presented. Let P'' be the resulting presentation. Let M be the maximum of the lengths of the words u_b.

Given a null-homotopic word $w \in F(\mathcal{A} \cup \mathcal{B})$, we first apply the new relators to replace each occurrence of each letter $b \in \mathcal{B}$ with the word u_b. The result is a word in $F(\mathcal{A})$ that has length at most $M|w|$, and this word may be reduced to the empty word by applying at most $\delta_P(M|w|)$ relators from $\mathcal{R}$. Thus $\delta_{P''} \preceq \delta_P$.

We claim that $\delta_P(n) \leqslant \delta_{P''}(n)$ for all $n \in \mathbb{N}$. To prove this claim we must show that if a word $w \in F(\mathcal{A})$ can be expressed in $F(\mathcal{A} \cup \mathcal{B})$ as a product Π of at most N conjugates of the given relators, then it can also be expressed in $F(\mathcal{A})$ as a product of at most N conjugates of the relators $\mathcal{R}^{\pm 1}$. To see that this is the case, one simply looks at the image of Π under the retraction $F(\mathcal{A} \cup \mathcal{B}) \to F(\mathcal{A})$ that sends each $b \in \mathcal{B}$ to u_b.

In general, given two finite presentations $P_1 \equiv \langle \mathcal{A} \mid \mathcal{R} \rangle$ and $P_2 \equiv \langle \mathcal{B} \mid \mathcal{R}' \rangle$ of a group G, one considers the presentation of G that has generators $\mathcal{A} \cup \mathcal{B}$ and relators $\mathcal{R}, \mathcal{R}', \{bu_b^{-1} \mid b \in \mathcal{B}\}$ and $\{av_a^{-1} \mid a \in \mathcal{A}\}$, where u_b (respectively v_a) is a word in $F(\mathcal{A})$ (respectively $F(\mathcal{B})$) that equals b (respectively a) in G. The first two steps of the proof imply that the Dehn function of this presentation is equivalent to that of both P_1 and P_2. □

The first detailed proof of (1.3.3) in the literature is due to Steve Gersten [45]. A more general result given in Appendix A (Proposition A.1.7) lends a geometric perspective to the equivalence in (1.3.3).

Isoperimetric inequalities and δ_Γ. In the light of the preceding proposition we may talk of '*the*' *Dehn function* of a finitely presented group Γ, denoted δ_Γ, with the understanding that this is only defined up to $\simeq$ equivalence.

One says that Γ satisfies a *quadratic isoperimetric inequality* if $\delta_\Gamma(n) \preceq n^2$. Linear (also polynomial, exponential, *etc.*) isoperimetric inequalities are defined similarly.

A finitely generated group is said to have a solvable word problem if there is an algorithm that decides which words in the generators represent the identity and which do not. Readers who are familiar with the rudiments of decidability should treat the following statement as an exercise, and those who are not may treat it as a definition.

Proposition 1.3.4 *A finitely presentable group Γ has a solvable word problem if and only if the Dehn function of every finite presentation of Γ is computable (i.e. is a recursive function).*

Exercise 1.3.5 Two groups are said to be commensurable if they have isomorphic subgroups of finite index. Deduce from the Filling theorem (Section 2) that the Dehn functions of commensurable finitely presented groups are $\simeq$ equivalent. (Hint: Use covering spaces.)

The reader might find it instructive to investigate how awkward it is to prove this fact algebraically.

2 The isoperimetric function Fill_0^M of a manifold

Let M be a closed, smooth, Riemannian manifold. In this section we shall describe the filling function Fill_0^M and its relationship to the Dehn function of the fundamental group of M.

2.1 The Filling theorem

Let D be a 2-dimensional disc and let S^1 be its boundary circle. Let M be a smooth, complete, Riemannian manifold. Let $c : S^1 \to M$ be a null-homotopic, rectifiable loop and define FArea(c) to be the infimum of the areas[9] of all Lipschitz maps $g : D \to X$ such that $g|_{\partial D}$ is a reparameterization[10] of c. If this infimum is attained by a (not necessarily injective) map $f : D \to M$ then, blurring the question of reparametrization, we say that f is a *least-area filling* of the loop $c = f|_{\partial D}$, or simply that f is a *least-area disc.*

If M is the universal covering of a closed manifold, then the existence of least-area discs (for embedded loops) is guaranteed by Morrey's solution to Plateau's problem [75].

Definition 2.1.1 *Let M be a smooth, complete, Riemannian manifold. The* genus zero, 2-dimensional, isoperimetric function *of M is the function* $[0, \infty) \to [0, \infty)$ *defined by*

$$\mathrm{Fill}_0^M(l) := \sup\{\mathrm{FArea}(c) \mid c : S^1 \to M \textit{ null-homotopic, length}(c) \leqslant l\}.$$

One of the main purposes of this chapter is to provide a detailed proof of the following fundamental equivalence:

2.1.2 Filling theorem *The genus zero, 2-dimensional isoperimetric function* Fill_0^M *of any smooth, closed, Riemannian manifold M is $\simeq$ equivalent to the Dehn function $\delta_{\pi_1 M}$ of the fundamental group of M.*

Remark 2.1.3 A similar statement holds with regard to isoperimetric functions of more general classes of spaces with upper curvature bounds (in the sense of Alexandrov [22]) but we shall not dwell on this point as we do not wish to obscure the main ideas with the technicalities required to set-up the required definitions. Nevertheless, in our proof of the filling theorem we shall make a point of isolating the key hypotheses so as to render these generalizations straightforward (cf. 5.2.2). In particular we avoid using any facts concerning the regularity of solutions to Plateau's problem in the Riemannian setting.

We postpone the proof of the Filling theorem to Section 5, but we take a moment now to remove a concern about the definition of Fill_0^M: *a priori* the supremum in the definition of $\mathrm{Fill}_0^M(l)$ could be infinite for certain values of l even if M is compact, but in fact it is not.

---

[9] The situations that we shall be considering are sufficiently regular as to render all standard notions of area equivalent; for definiteness one could take 2-dimensional Hausdorff measure, or the notion of (Lebesgue) area in spaces with upper curvature bounds introduced by Alexandrov [1] and refined by Nikolaev (see [11] and [22] page 425).

[10] When working with filling problems it is usually better to consider loops that are equivalent in the sense of Frechet, but this technicality will have no bearing here.

Lemma 2.1.4 *If M is compact, the* sup *in the definition of* $\mathrm{Fill}_0^M(l)$ *is finite for all* $l \geqslant 0$.

Proof If the sectional curvature of M is bounded above by $k > 0$ then any null-homotopic loop in M of length $l < 2\pi/\sqrt{k}$ bounds a disc whose area is at most the area $A(k,l)$ of the disc enclosed by a circle of length l on the sphere of constant curvature k. Indeed Reshetnyak [93] proved that this bound holds in any complete geodesic space of curvature $\leqslant k$ (cf. appendix to [71]).

Let $\rho > 0$ be less than the injectivity radius of M, fix a finite set S so that every point of M lies in the $\rho/3$-neighbourhood of S and let $e_{x,x'} : [0,1] \to M$ be the constant speed geodesic joining each $x, x' \in S$ with $d(x,x') < \rho$.

Given any constant-speed loop $c : [0,1] \to M$, one can associate to it the concatenation $\hat{c} = e_{x_0,x_1} \dots e_{x_n,x_0}$ where n is the least integer greater than $3l(c)/\rho$ and $x_i \in S$ is such that $d(x_i, c(i/n)) < \rho/3$ (cf. Fig. 5.1.2).

By construction, $|\mathrm{FArea}(c) - \mathrm{FArea}(\hat{c})| \leqslant n\, A(k, 2\rho)$ and $l(\hat{c}) \leqslant 3l(c) + \rho$. It follows that the $\simeq$ class of Fill_0^M remains unchanged if instead of quantifying over all rectifiable loops c one quantifies only over loops that are concatenations of the geodesics $e_{x,x'}$. For all $L > 0$, there are only finitely many such concatenations of length $\leqslant L$, so in particular $\mathrm{Fill}_0^M(l)$ is finite for all l. □

Remark 2.1.5 The reduction to piecewise-geodesic loops in the above proof exemplifies the fact that if one is concerned only with the $\simeq$ class of Fill_0^M then there is no harm in restricting one's attention to well-behaved subclasses of rectifiable loops.

2.2 Filling in Heisenberg groups

The results described in this paragraph are due to Mikhael Gromov. We present them here in order to give an immediate illustration of how one can exploit the equivalence $\mathrm{Fill}_0^M \simeq \delta_{\pi_1 M}$.

Let $n = 2m + 1$. The n-dimensional Heisenberg group $\mathcal{H}_n$ is the group of $(m+1)$-by-$(m+1)$ real matrices of the form:

$$\begin{pmatrix} 1 & x_1 & \dots & x_{m-1} & z \\ 0 & 1 & 0 & 0 & y_1 \\ \vdots & & & \vdots & \vdots \\ 0 & 0 & \dots & 1 & y_{m-1} \\ 0 & 0 & \dots & 0 & 1 \end{pmatrix}.$$

$\mathcal{H}_n$ is a nilpotent Lie group. Its Lie algebra L is generated by $X_1, \dots, X_{m-1}$, $Y_1, \dots, Y_{m-1}, Z = X_m = Y_m$ with relations $[X_i, Y_j] = [X_i, X_j] = [Y_i, Y_j] = 0$ for all $i \neq j$ and $[X_i, Y_i] = Z$ for $i = 1, \dots, m-1$. There is a natural grading $L = L_1 \oplus L_2$, where L_2 is spanned by Z and L_1 is spanned by the remaining X_i and Y_i.

The translates of L_1 by the left action of $\mathcal{H}_n$ form a sub-bundle T_1 of the tangent bundle of $\mathcal{H}_n$. (This codimension-1 sub-bundle gives the standard contact structure on $\mathcal{H}_n$.) A curve or surface mapped to $\mathcal{H}_n$ is said to be *horizontal* if it is differentiable almost everywhere and its tangent vectors lie in T_1. Every smooth curve c in $\mathcal{H}_n$ can be approximated by a horizontal curve whose length is arbitrarily close to that of c. The question of whether every horizontal loop bounds a horizontal disc ('the horizontal filling problem') is delicate, and it is here that we find a connection with Dehn functions.

The following result is an application of the theory developed by Gromov in Section 2.3.8 of his book on partial differential relations [54] and is explained on page 85 of [56].

Proposition 2.2.1 *If every horizontal loop in $\mathcal{H}_n$ can be filled with a horizontal disc, then* $\text{Fill}_0^{\mathcal{H}_n}(l) \simeq l^2$.

The idea of the proof is as follows. First one must argue that there is a constant C such that horizontal loops of length $\leqslant 1$ can be filled with horizontal discs of area at most C. Then one considers the 1-parameter family of maps $h_t = \exp \circ \lambda_t \circ \exp^{-1} : \mathcal{H}_n \to \mathcal{H}_n$, where the Lie-algebra homomorphism $\lambda_t : L \to L$ is multiplication by $t \in [0,1]$ on L_1 and by t^2 on L_2. Note that h_t multiplies the length of horizontal curves by t and the area of horizontal discs by t^2.

Given a horizontal loop $c : S^1 \to \mathcal{H}_n$ of length $l > 1$, we consider $h_{1/l} \circ c$. One can fill this horizontal loop of length 1 with a horizontal disc $f_0 : D \to H_n$ of area at most C and hence obtain a horizontal disc $f := h_{1/l}^{-1} \circ f_0$ of area $\leqslant Cl^2$ that fills c. Since arbitrary loops can be approximated by horizontal loops, it follows that $\mathcal{H}_n$ satisfies a quadratic isoperimetric inequality.

The *integer Heisenberg group* H_n consists of those matrices in $\mathcal{H}_n$ that have integer entries. The subgroup $H_n \subset \mathcal{H}_n$ is discrete, torsion-free and cocompact, hence $M := H_n \backslash \mathcal{H}_n$ is a compact Riemannian manifold with universal covering $\mathcal{H}_n$, and $\delta_{H_n} \simeq \text{Fill}_0^M = \text{Fill}_0^{\mathcal{H}_n}$.

Gromov proves that the horizontal filling problem is solvable in $\mathcal{H}_n$ if and only if $n \geqslant 5$. It therefore follows[11] from the Filling theorem and the above proposition that the integral Heisenberg group H_n has a quadratic Dehn function if $n \geqslant 5$. On the other hand, it is not hard to show by various combinatorial means (see 3.1.4 and 3.3.1 below) that the Dehn function of H_3 is cubic, so from the Filling theorem and the above proposition one gets a proof of the easier 'only if' implication in Gromov's theorem: $\mathcal{H}_3$ contains horizontal loops of finite length that cannot be filled with a horizontal disc.

[11] For a self-contained proof along these lines see Allcock [2]. More recently, a purely combinatorial proof has been discovered by Ol'shanskii and Sapir [83].

3 Which functions are Dehn functions?

The most fundamental question concerning isoperimetric inequalities for finitely presented groups is that of determining which $\simeq$ equivalence classes of functions arise as Dehn functions. The struggle to solve this question was a major theme in geometric group theory in the 1990s. In this section I shall explain why this struggle is almost over. I shall also describe what is known about the Dehn functions of certain groups that are of special interest in geometry and topology.

Section 7 contains a sample of the techniques that were developed to establish the results quoted in the present section.

3.1 The isoperimetric spectrum

The development of knowledge concerning the nature of Dehn functions is best explained in terms of how the set of numbers

$$\mathrm{IP} = \{\rho \in [1, \infty) \mid f(n) = n^\rho \text{ is } \simeq \text{a Dehn function}\}$$

came to be understood. This set is called the *isoperimetric spectrum.*

Since there are only countably many finite presentations of groups, Proposition 1.3.3 implies that there are only countably many $\simeq$ classes of Dehn functions. Thus, intriguingly, IP is a naturally arising countable set of positive numbers.

Integer exponents. In Section 6 we shall discuss the class of groups that have linear Dehn functions. The following exercises describe the simplest examples from this class.

Exercises 3.1.1 (i) Finite groups and free groups have linear Dehn functions.

(ii) Let $\mathbb{H}^2$ denote the hyperbolic plane. There is a constant $C > 0$ such that for all $l > 1$, each loop in $\mathbb{H}^2$ of length $\leqslant l$ bounds a disc of area $\leqslant Cl$.

(iii) Every finitely generated group that acts properly by isometries on $\mathbb{H}^2$ has a linear Dehn function. (Hint: If the action is cocompact you can use (ii). If the action is not cocompact, argue that the group must have a free subgroup of finite index.)

In Section 6 we shall also describe what is known about the class of groups that have quadratic Dehn functions. Finitely generated abelian groups provide the easiest examples in this class.

Example 3.1.2 The Dehn function of $P \equiv \langle a, b \mid [a, b]\rangle$ is quadratic. More precisely, $(l^2 - 2l - 3) \leqslant 16\,\delta_P(l) \leqslant l^2$, the upper bound being attained in the case of words of the form $a^{-n}b^{-n}a^nb^n$.

Exercise 3.1.3 Prove that the inequality in (3.1.2) holds for the natural presentation of any free abelian group $\mathbb{Z}^r$, $r \geqslant 2$, and that it is optimal. (Hint: Given a word w that equals the identity in $\mathbb{Z}^r$, focus on a specific generator a and move all occurrences of $a^{\pm 1}$ to the left in w by applying the relators $[a, b] = 1$, freely reducing the resulting word whenever possible. Repeat for each generator and count the total number of relators applied — cf. Paragraph 1.2. If you have trouble with the lower bound, look at Section 7.)

In about 1988 Bill Thurston [42] and Steve Gersten [45] proved that the 3-dimensional Heisenberg group H_3 has a cubic Dehn function (see paragraph 2.2 and Theorem 3.3.1).

It now seems odd to report that there was a lull of a few years before people discovered sequences of groups $(\Gamma_d)_{d \in \mathbb{N}}$ such that the Dehn function of Γ_d is polynomial of degree d. Such sequences were described by a number of authors at about the same time – Gromov [56], Baumslag, Miller and Short [10], and Bridson-Pittet [23]. The following result, proved by Bridson and Gersten in [21], provides many such sequences, and the literature now contains examples with all manner of additional properties (e.g. having Eilenberg–Maclane spaces of specified dimension [16]).

Theorem 3.1.4 *The Dehn function of each semi-direct product of the form* $\mathbb{Z}^n \rtimes_\phi \mathbb{Z}$ *is* $\simeq$ *either a polynomial or an exponential function. It is polynomial if and only if all of the eigenvalues of* $\phi \in \mathrm{GL}(n, \mathbb{Z})$ *are roots of unity, in which case the degree of the polynomial is* $c + 1$*, where* c *is the size of the largest elementary block in the Jordan form of* ϕ*.*

Notice that groups of the form $\mathbb{Z}^n \rtimes_\phi \mathbb{Z}$ are precisely those that arise as fundamental groups of torus bundles over the circle, and hence the above theorem classifies the isoperimetric functions Fill_0^M of such bundles.

The appearance of the Jordan form in the above theorem is connected to the following facts (cf. 7.1.4).

Exercises 3.1.5 (i) If a matrix $\phi \in \mathrm{GL}(n, \mathbb{Z})$ does not have an eigenvalue of absolute value greater than 1, then all of its eigenvalues are N-th roots of unity, where N depends only on n. (Hint, [21], page 7: Let $P \subset \mathbb{Z}[x]$ be the set of monic polynomials of degree n whose roots all lie on the unit circle. P is finite. If the characteristic polynomial of ϕ lies in P then so does that of each power ϕ^r.)

(ii) Regard $\mathrm{GL}(n, \mathbb{Z})$ as a subset of $\mathbb{R}^{n^2}$ and fix a norm on $\mathbb{R}^{n^2}$. Prove that $m \mapsto \|\phi^m\|$ is $\simeq$ equivalent to an exponential function or a polynomial of degree $c-1$, where c is the size of the largest elementary block in the Jordan form of ϕ.

Filling the gaps in IP. The following theorem is due to Gromov [55]. Detailed proofs were given by Ol'shanskii [81], Bowditch [14] (also [22] page 422) and Papasoglu [87].

Theorem 3.1.6 *If the Dehn function of a group is subquadratic (i.e. $\delta_\Gamma(n) = o(n^2)$) then it is linear ($\delta_\Gamma(n) \simeq n$). Thus* $\mathrm{IP} \cap (1,2)$ *is empty.*

This theorem begs the question of what other gaps there may be in the isoperimetric spectrum, or indeed whether there are any non-integral isoperimetric exponents at all. This last question was settled by the discovery of the *abc* groups [19]. These groups are obtained by taking three torus bundles over the circle (each of a different dimension) and amalgamating their fundamental groups along central cyclic subgroups.

The basic building block is $G_c = \mathbb{Z}^c \rtimes_{\phi_c} \mathbb{Z}$, where $\phi_c \in \mathrm{GL}(c, \mathbb{Z})$ is the unipotent matrix with ones on the diagonal and superdiagonal and zeros elsewhere. G_c has presentation:

$$\langle x_1, \ldots, x_c, t \mid [x_i, x_j] = 1 \text{ for all } i, j,\ [x_c, t] = 1,\ [x_i, t] = x_{i+1} \text{ if } i < c\rangle. \quad (3.1.7)$$

Notice that the centre of G_c is the infinite cyclic subgroup generated by x_c. To emphasize this fact we write z_c in place of x_c.

The *abc* groups $\Gamma(a,b,c)$ are defined as follows: first we amalgamate G_a with $G_b \times \mathbb{Z}$ by identifying the centre of G_a with that of G_b, then we form the amalgamated free product of the resulting group with G_c by identifying the centre of the latter with the right-hand factor of $G_b \times \mathbb{Z}$. In symbols:

$$\Gamma(a,b,c) = G_a *_{z_a = z_b} (G_b \times \langle\, \zeta\,\rangle) *_{\zeta = z_c} G_c.$$

Theorem 3.1.8 *For all integers $1 \leqslant b \leqslant a < c$, the Dehn function of $\Gamma(a,b,c)$ is $\simeq n^{c+(a/b)}$.*

Variations on this construction yield other families of rational exponents [19].

By far the most comprehensive result concerning the structure of Dehn functions is due to Sapir, Birget, and Rips. Their result, which we shall describe in a moment, essentially classifies the Dehn functions $\succeq n^4$. In particular they show that IP is dense in $[4, \infty)$.

Subsequently, Brady and Bridson [15] showed that Gromov's gap $(1,2)$ is the only gap in the isoperimetric spectrum:

Theorem 3.1.9 *For each pair of positive integers $p \geqslant q$, there exist finitely presented groups whose Dehn functions are $\simeq n^{2\alpha}$ where $\alpha = \log_2(2p/q)$.*

Corollary 3.1.10 *The closure of* IP *is* $\{1\} \cup [2, \infty)$.

Note that the exponents described in the above theorem are transcendental if they are not integers [80], Theorem 10.2. The easiest examples of groups as described in the above theorem are

$$G_{p,q} = \langle a, b, s, t \mid [a,b] = 1, sa^q s^{-1} = a^p b,\ ta^q t^{-1} = a^p b^{-1}\rangle,$$

which we shall look at more closely in (7.2.12).

The Sapir–Birget–Rips theorem. In [95] Mark Sapir, Jean-Camille Birget and Eliyahu Rips show that if a number $\alpha > 4$ is such that there is a constant $C > 0$ and a Turing machine that calculates the first m digits of the decimal expansion of α in time $\leqslant C2^{2^{Cm}}$, then $\alpha \in \mathrm{IP}$. Conversely, they show that if $\alpha \in \mathrm{IP}$ then there is a Turing machine that calculates the first m digits of α in time $\leqslant C2^{2^{2^{Cm}}}$. (The discrepancy in the height of the two towers of exponentials is connected to the P = NP problem.) More generally they prove:

Theorem 3.1.11 *Let $\mathcal{D}_4$ be the set of $\simeq$ equivalence classes of Dehn functions $\delta(n) \succeq n^4$. Let $\mathcal{T}_4$ be the set of $\simeq$ classes of time functions $t(n) \succeq n^4$ of arbitrary Turing machines. Let $\mathcal{T}^4$ be the set of $\simeq$ classes of superadditive*[12] *functions that are fourth powers of time functions. Then $\mathcal{T}^4 \subseteq \mathcal{D}_4 \subseteq \mathcal{T}_4$.*

It is unknown whether $\mathcal{T}^4$ coincides with the $\simeq$ classes of all superadditive functions in $\mathcal{T}_4$. If it does, then the above theorem would completely classify Dehn functions $\succeq n^4$. In the light of Theorem 3.1.9, one suspects that Dehn functions $\succeq n^2$ are similarly unrestricted in nature.

As it stands, the above result already implies that any rational or other reasonable number, for example $\pi + e^2$, is the exponent of a Dehn function. Likewise, the following are Dehn functions: $2^{\sqrt{n}}$, e^{n^π}, $n^2 \log_3(\log_7 n), \ldots$

As one might guess from the statement, the theorem is proved by showing that one can encode the workings of a certain class of machines ('S-machines') into group presentations.

3.2 Examples of large Dehn functions

Thus far in this section I have concentrated on IP in order to explain the development of our understanding of Dehn functions. Let me offset this now by pointing out that many naturally occurring groups do not have Dehn functions that are bounded above by a polynomial function. We saw some such examples in (3.1.4). Here are some more simple examples of this type.

Consider the recursively-defined sequence of functions $\varepsilon_i(n) := 2^{\varepsilon_{i-1}(n)}$, where $\varepsilon_0(n) = n$ and $\varepsilon_1(n) = 2^n$. Let

$$B_m = \langle x_0, x_1, \ldots, x_m \mid x_i^{-1} x_{i-1} x_i = x_{i-1}^2 \text{ for } i = 1, \ldots, m \rangle. \tag{3.2.1}$$

The best known of these groups is B_1, which has many manifestations, e.g. as a group of affine transformations of the real line, where x_0 acts as $t \mapsto t+1$ and x_1 as $t \mapsto 2t$.

Proposition 3.2.2 *The Dehn function of B_m is $\simeq \varepsilon_m(n)$.*

For the lower bound, see Exercise 7.2.11. The following exercises explain one method of establishing the upper bound.

---

[12] $f(m+n) \geqslant f(n) + f(m)$ for all $n, m \in \mathbb{N}$.

Exercises 3.2.3 (i) Let w be a word in the generators of B_1. Show that one can transform w into a word of the form $x_1^m x_0^r x_1^{-m'}$ with $m, m' \geqslant 0$ by applying the defining relator $x_1^{-1}x_0x_1x_0^{-2}$ at most 2^n times. (Hint: Move each occurrence of x_1 in w to the left by replacing subwords x_0x_1 with $x_1x_0^2$, and $x_0^{-1}x_1$ with $x_1x_0^{-2}$. Move all occurrences of x_1^{-1} to the right.)

(ii) Prove that $x_0 \in B_1$ has infinite order. (You could consider the representation $B_1 \to \text{Aff}(\mathbb{R})$ described above.[13]) Deduce that $\delta_{B_m}(n) \preceq 2^n$. (Hint: By looking at the map $B_1 \to \langle x_1 \rangle$ that kills x_0 and the map $B_1 \to \mathbb{Z}_q \rtimes \mathbb{Z}$ that kills x_0^q (where q is an arbitrary odd prime), one can see that if $w = 1$ in Γ then the word obtained in (i) has $m = m'$ and $r = 0$.)

A less *ad hoc* proof of (ii) can be based on Britton's lemma (see 7.2.4(ii) or [22], page 498):

(iii) Deduce from Britton's lemma that if a word in the generators of B_m represents the identity and contains at least one occurrence of $x_m^{\pm 1}$ then it contains a subword of the form $w_0 = x_m^e w_1 x_m^{-e}$, where $e = \pm 1$ and w_1 is a word in the letters $\{x_i \mid i < m\}$ with $w_1 = x_{m-1}^p$ in B_{m-1}, where p is even if $e = 1$.

Arguing by induction on m, and a secondary induction on the number of occurrences of $x_m^{\pm 1}$ in w_0, show that one can replace w_0 by $x_{m-1}^{p/2}$ or x_{m-1}^{2p} by applying at most $\varepsilon_{m-1}(2p)$ relators from the presentation of B_{m-1}. Deduce that $\delta_{B_m}(n) \leqslant \varepsilon_m(n)$.

Example 3.2.4 Steve Gersten [45] showed that the Dehn function of the group

$$S = \langle x, y \mid (yxy^{-1})^{-1}x(yxy^{-1}) = x^2 \rangle$$

grows faster than any iterated exponential. Specifically, $\delta_S(n) \simeq \varepsilon_n(n)$. A classical theorem of Magnus states that all 1-relator groups have a solvable word problem. It is conjectured that $\varepsilon_n(n)$ is an upper bound on the Dehn functions of all 1-relator groups; in [46] Gersten established a weaker upper bound.

Exercise 3.2.5 Show that for every $m > 0$ there exists a monomorphism $B_m \to S$. (Hint: Conjugate $(y^i x y^{-i})$ by $(y^{i+1} x y^{-(i+1)})$.)

3.3 Groups of classical interest

In this subsection I shall describe what is known about the Dehn functions of various groups that are of interest for geometric reasons.

---

[13] More ambitiously, you could try to prove the following result of Higman, Neumann, and Neumann (see [97] for a geometric treatment). Given a group $\Gamma = \langle \mathcal{A} \mid \mathcal{R} \rangle$ and an isomorphism $\phi : S_1 \to S_2$ between subgroups of Γ, one can form the HNN extension $\Gamma *_\phi = \langle \mathcal{A}, t \mid \mathcal{R}, \phi'(s) = t^{-1}st, \forall s' \in S \rangle$, where $t \notin \mathcal{A}$, $S \subset F(\mathcal{A})$ is a set of words that maps bijectively to S_1, and for each $s \in S$ the word $\phi'(s) \in F(\mathcal{A})$ maps to $\phi(s) \in S_2$. Show that the map $\Gamma \to \Gamma *_\phi$ induced by $\text{id}_\mathcal{A}$ is an injection.

Low-dimensional topology. If S is a compact 2-manifold, then $\pi_1 S$ has a linear Dehn function unless S is a torus or a Klein bottle, in which case $\pi_1 S$ has a quadratic Dehn function (see 3.1.1, 3.1.2, 1.3.5). The following theorem describes the situation for 3-dimensional manifolds – it follows easily from results of Epstein and Thurston [42] (cf. [17] and 7.1.4 below). Since all finitely presented groups arise as fundamental groups of closed n-manifolds for each $n \geqslant 4$ (see A.3.1), there can be no such general statement in higher dimensions.

Theorem 3.3.1 *Let M be a compact 3-manifold. Suppose that M satisfies Thurston's geometrization conjecture.*[14]

The Dehn function of $\pi_1 M$ is linear, quadratic, cubic, or exponential. It is linear if and only if $\pi_1 M$ does not contain $\mathbb{Z}^2$. It is quadratic if and only if $\pi_1 M$ contains $\mathbb{Z}^2$ but does not contain a subgroup $\mathbb{Z}^2 \rtimes_\phi \mathbb{Z}$ with $\phi \in \mathrm{GL}(2, \mathbb{Z})$ of infinite order. Subgroups $\mathbb{Z}^2 \rtimes_\phi \mathbb{Z}$ arise only if a finite-sheeted covering of M has a connected summand that is a torus bundle over the circle, and the Dehn function of $\pi_1 M$ is cubic only if each such summand is a quotient of the Heisenberg group (in which case ϕ is unipotent).[15]

Remark 3.3.2 [Free-by-cyclic groups] If a 3-manifold M fibres over the circle then one sees from the long exact sequence in homotopy that $\pi_2 M = 0$ and that $\pi_1 M$ is a semi-direct product $\Sigma \rtimes_\phi \mathbb{Z}$, where Σ is the fundamental group of the surface fibre. Since $\pi_2 M = 0$, one knows that M does not split as a non-trivial connected sum, so the above theorem implies that if $\mathbb{Z}^2 \not\subseteq \Sigma$, then the Dehn function of $\pi_1 M$ is either linear or quadratic.

If M has boundary then Σ will be a finitely generated free group. Not all free-group automorphisms arise from fibrations of 3-manifolds, and it is has yet to be proved that the Dehn functions of arbitrary semi-direct products of the form $\Gamma = \Sigma \rtimes_\phi \mathbb{Z}$, with Σ free, are at most quadratic, cf. [69]. In [12] Bestvina and Feighn show that the Dehn function of Γ is linear if and only if $\mathbb{Z}^2 \not\subseteq \Gamma$.

There are strong analogies between mapping class groups of surfaces, Braid groups (more generally, Artin groups), and automorphism groups of free groups. These groups play important roles in low-dimensional topology. Bill Thurston proved that the Braid groups are automatic, [42] Chapter 9 (see also Charney [31]), and Lee Mosher proved that the mapping class groups of all surfaces of finite type are automatic [76]. As a consequence (see 6.3.2) we obtain:

Theorem 3.3.3 *The mapping class group of any surface of finite type satisfies a quadratic isoperimetric inequality.*

Hatcher and Vogtmann [58] and Gersten (unpublished) proved that the Dehn function of the group of (outer) automorphisms of any finitely generated free

[14] In the absence of this assumption it remains unknown whether every compact 3-manifold has a solvable word problem.

[15] $\pi_1 M$ has an exponential Dehn function if and only if M has a connected summand that is modelled on the geometry Sol – cf. 3.1.4

group is $\preceq 2^n$. Bridson and Vogtmann [24] proved that this bound is sharp in rank 3, and special considerations apply in rank 2.

Theorem 3.3.4 *Let F_r denote a free group of rank r. The Dehn function of* $\mathrm{Out}(F_2)$ *is linear. The Dehn function of* $\mathrm{Aut}(F_2)$ *is quadratic. The Dehn functions of* $\mathrm{Aut}(F_3)$ *and* $\mathrm{Out}(F_3)$ *are exponential. In general the Dehn functions of* $\mathrm{Aut}(F_r)$ *and* $\mathrm{Out}(F_r)$ *are* $\preceq 2^n$.

Lattices in semisimple Lie groups. Let G be a connected semisimple Lie group with finite centre and no compact factors. Associated to G one has a Riemannian symmetric space $X = G/K$, where $K \subseteq G$ is a maximal compact subgroup. A discrete subgroup $\Gamma \subset G$ is called a *lattice* if the quotient $\Gamma\backslash X$ has finite volume; the lattice is called *uniform* (or cocompact) if $\Gamma\backslash X$ is compact. The *rank* of G is the dimension of the maximal isometrically embedded flats $\mathbb{E}^r \hookrightarrow X$.

If G has rank 1 then X has strictly negative curvature (e.g. $G = \mathrm{SO}(n,1)$ and $X = \mathbb{H}^n$) and in general (e.g. $G = \mathrm{SL}(n,\mathbb{R})$) X has non-positive curvature (see, for example, [22] Chapter II.10). It follows that the Dehn functions of uniform lattices are linear (in the rank 1 case) or quadratic (the higher rank case) – see Section 6.

Each non-uniform lattice in a rank 1 group contains non-trivial subgroups that stabilize points at infinity in the symmetric space X; these subgroups leave invariant the horospheres centred at the fixed points at infinity. We use the term *horospherical* to describe these subgroups. An example of a horospherical subgroup is the fundamental group of the boundary torus in a hyperbolic knot complement. Each maximal horospherical subgroup contains a nilpotent subgroup of finite index: in the case $G = \mathrm{SO}(n,1)$, this nilpotent subgroup is isomorphic to $\mathbb{Z}^{n-1}$, and in the case $G = \mathrm{SU}(n,1)$ it is isomorphic to H_{2n-1}, the integer Heisenberg group.

Theorem 3.3.5 *Let G be a semisimple Lie group of rank 1 and let $\Gamma \subset G$ be a lattice. If Γ is uniform then its Dehn function is linear. If Γ is non-uniform then its Dehn function is equal to that of each of its maximal horospherical subgroups.*

This result is due to Gromov [56].

Example 3.3.6 It follows from our discussion in 2.2 that non-uniform lattices in $\mathrm{SU}(2,1)$ have cubic Dehn functions, whereas those in $\mathrm{SU}(n,1)$ with $n > 2$ have quadratic Dehn functions. More generally, it follows from the above theorem that a non-uniform lattice in a rank 1 group G will have a quadratic Dehn function unless the symmetric space for G is the hyperbolic plane over the real, complex, quaternionic, or Cayley numbers. For the real hyperbolic plane the Dehn function of non-uniform lattices is linear (3.1.1), in the complex case ($G = \mathrm{SU}(2,1)$) it is cubic, and it is also believed to be cubic in the remaining cases.

The following theorem of Leuzinger and Pittet [62], which builds on the work of Gromov on solvable groups [56], completes the picture of Dehn functions for lattices in rank 2.

Theorem 3.3.7 *If G is a connected semisimple Lie group with finite centre and rank 2, then the Dehn function of any irreducible, non-uniform lattice in G is $\simeq 2^n$.*

The situation for non-uniform lattices in rank $\geqslant 3$ is more complicated and is the subject of active research. We refer the reader to Gromov [56] for an exciting glimpse of some of the issues that arise and to Druţu [38] and Leuzinger-Pittet [63] for significant recent progress in this direction. The following assertion of Bill Thurston illustrates some of the subtleties involved in higher rank: *the Dehn function of* $\mathrm{SL}(3,\mathbb{Z})$ *is exponential, but the Dehn function of* $\mathrm{SL}(n,\mathbb{Z})$ *is quadratic if* $n > 3$.

See [42] page 230 for a proof of this statement in the case $n = 3$ (cf. [38] and [56] page 91). A complete proof is not available in the case $n > 3$. Druţu's recent work has helped to clarify the situation, but there remains much work to be done in this direction.

Nilpotent groups. We saw in (3.1.2) that abelian groups satisfy a quadratic isoperimetric inequality. Using a modest amount of knowledge about the structure of nilpotent groups, it is not hard to show that all finitely generated nilpotent groups satisfy a polynomial isoperimetric inequality (see [56] for example). But determining the degree of the optimal bound on the Dehn function, both in general and for specific examples, is a more delicate matter, as our earlier discussion of the Heisenberg groups illustrates.

Gromov, [56] Chapter 5, gives an enticing overview of this area. In particular he sketches a reason why nilpotent groups of class c should have Dehn functions that are polynomial of degree $\leqslant c+1$ and gives a proof of this inequality for groups where the Lie algebra of the Malcev completion is graded. (For a detailed account of this last result, and extensions, see Pittet [89].) A number of other researchers have obtained related results using both geometric and combinatorial methods. In particular, Hidber [59] gives a purely algebraic proof that the Dehn function of a nilpotent group of class c is bounded above by a polynomial of degree $2c$.

Finally, I should mention that the study of Dehn functions of non-nilpotent solvable groups is also an active area of research. Indeed this is closely connected to the study of Dehn functions for higher-rank lattices.

Let me end this brief survey of our knowledge of Dehn functions for specific groups by making it clear that I have omitted far more than I have included. I apologise to the many colleagues whose excellent work I have been forced to ignore by reason of space and time.

3.4 Dehn functions of products

The following exercises describe how Dehn functions behave under the formation of products. Their behaviour under more complicated operations such as amalgamated free products, HNN extensions, and central extensions is less straightforward.

Exercises 3.4.1 (i) A subgroup H of a group G is called a *retract* if there is a homomorphism $G \to H$ whose restriction to H is the identity. Show that if H is a retract of the finitely presented group G, then H is finitely presented and $\delta_H(n) \preceq \delta_G(n)$. (Hint: First note that H is finitely generated. Take a finite subset that generates H and argue that it can be extended to a finite generating set for G by adding elements k of the kernel of $G \to H$. Argue that one can take a finite presentation for G with this generating set. Add the relations $k = 1$.)

(ii) Let G_1 and G_2 be infinite, finitely presented groups. Show that the Dehn function of $G_1 \times G_2$ is $\simeq \max\{n^2, \delta_{G_1}(n), \delta_{G_2}(n)\}$, and that that of the free product $G_1 * G_2$ is $\simeq \max\{\delta_{G_1}(n), \delta_{G_2}(n)\}$. (Use (i) for the bounds $\succeq$.)

4 Van Kampen diagrams

Let $\langle \mathcal{A} \mid \mathcal{R} \rangle$ be a finite presentation of a group Γ and let w be a word in the letters $\mathcal{A}^{\pm 1}$. Suppose that $w = 1$ in Γ. Roughly speaking, a van Kampen diagram for w is a planar CW complex that portrays a scheme for reducing w to the empty word by applying a sequence of relations $r \in \mathcal{R}$; the number of 2-cells in the diagram is the number of relations that one applies and is therefore at least as great as $\text{Area}_a(w)$, as defined in (1.2.2). Conversely, we shall see that one can always construct a van Kampen diagram for w that has $\text{Area}_a(w)$ 2-cells. It follows that the Dehn function of $\langle \mathcal{A} \mid \mathcal{R} \rangle$ can be interpreted in terms of isoperimetric inequalities for planar diagrams.

Max Dehn was the first to use planar diagrams in order to study word problems [34], but his diagrams arose in concrete settings (primarily as regions in a tessellated hyperbolic plane). The idea of using diagrams to study relations in arbitrary finitely presented groups is due to E. van Kampen [61]. The idea was rediscovered by Roger Lyndon in the 1960s. At about the same time C. Weinbaum brought van Kampen's original paper to light and made interesting applications of it.

There are a number of correct proofs of the celebrated van Kampen lemma in the literature. The use of pictures in these proofs causes disquiet in some circles, so I have tried to fashion the following proof in a manner that will allay such misgivings.

4.1 Singular disc diagrams

Fix an orientation on $\mathbb{R}^2$. A *singular disc diagram* D is a compact, contractible subset of the plane endowed with the structure of a finite combinatorial 2-complex. (See Appendix A for basic definitions concerning combinatorial complexes.)

We write $\text{Area}_c\, D$ to denote the number of 2-cells in D. And given a vertex $p \in D$ we write $\text{Diam}_p\, D$ to denote the maximum of the distance from p to the other vertices $v \in D$, where 'distance' is the number of 1-cells traversed by a shortest path joining p to v in the 1-skeleton of D.

To avoid pathologies, we assume the 1-cells $e : [0,1] \to D \hookrightarrow \mathbb{R}^2$ are smoothly embedded. Associated to each 1-cell one has two *directed edges* $\varepsilon(t) = e(t)$ and $\bar{\varepsilon}(t) = e(1-t)$. Let $\mathcal{A}_D$ denote the set of directed edges. (By definition $\bar{\bar{\varepsilon}} = \varepsilon$.)

The *boundary cycle* of D is the loop of directed edges describing the frontier of the metric completion of $\mathbb{R}^2 \smallsetminus D$ in the positive (anticlockwise) direction – it consists of a *thin part*, where the underlying 1-cells do not lie in the boundary of any 2-cell, and a *thick part*; the boundary cycle traverses each 1-cell in the thick part once and each 1-cell in the thin part twice.

Definition 4.1.1 [Labelled diagrams] *Let $\mathcal{A}$ be a set and let $\mathcal{A}^{-1}$ be the set of symbols $\{a^{-1} \mid a \in \mathcal{A}\}$. A diagram over $\mathcal{A}$ consists of a singular disc diagram D and a (labelling) map $\lambda : \mathcal{A}_D \to \mathcal{A} \cup \mathcal{A}^{-1}$ such that $\lambda(\bar{\varepsilon}) = \lambda(\varepsilon)^{-1}$ for all $\varepsilon \in \mathcal{A}_D$.*

λ extends to a map from the set of directed edge-paths in D to the set of words in the letters $\mathcal{A} \cup \mathcal{A}^{-1}$. The face labels of D are the words that this map assigns to the attaching loops of the 2-cells of D (beginning at any vertex and proceeding with either orientation).

Proposition 4.1.2 *Let $\mathcal{A}$ be a set, let D be a diagram over $\mathcal{A}$ and let $\mathcal{R}_*$ be a set of words that contains the face labels of D. If a word w occurs as the label on the boundary cycle of D, read from some vertex p in the boundary of D, then in the free group $F(\mathcal{A})$*

$$w = \prod_{i=1}^{\alpha} x_i^{-1} r_i x_i,$$

where $\alpha = \text{Area}_c\, D$, the words x_i have length $|x_i| \leqslant \text{Diam}_p\, D$, and $r_i \in \mathcal{R}_*$. In particular $w = 1$ in the group $\langle \mathcal{A} \mid \mathcal{R}_* \rangle$.

Proof Fix D and p. In the 1-skeleton of D we choose a geodesic spanning tree T rooted at p (see Exercise 4.1.3).

Arguing by induction (the base step is trivial) we may assume that the proposition has been proved for diagrams D' with $\text{Area}_c\, D' < \text{Area}_c\, D$ and for diagrams with $\text{Area}_c\, D' = \text{Area}_c\, D$ where D' has fewer 1-cells than D.

We say that D has a dangling edge if it has a vertex other than p that has only one edge incident at it. If D has such an edge then we may apply our inductive hypothesis to the diagram obtained by removing it – the resulting diagram has the same area as D, its diameter is no greater than that of D, and its boundary label is obtained from that of D by free reduction. Thus we may assume that D has no dangling edges.

If D were a tree it would have dangling edges (or be a single point). Thus $D \neq T$. We follow the boundary cycle of D from p until we encounter the first directed edge ε that is not in T; let a be the label on ε, let w_1 be the label on the segment of the boundary cycle that precedes ε and let w_2 be the label on the segment that follows it. The part of the boundary cycle labelled w_1 is an injective path, because it lies entirely in the tree T and must be locally injective since a backtracking would imply that D had a dangling edge. In particular w_1 has length at most $\text{Diam}_p D$.

Since T contains all of the vertices of D, we do not disconnect D by removing the open 1-cell underlying ε, and hence this 1-cell must lie in the boundary of some 2-cell E. Suppose that the attaching loop of E (read in the positive direction from the initial vertex of ε) has label $r^{-1} := au^{-1}$.

Consider the subcomplex D' obtained from D by deleting the open 1-cell labelled ε and the interior of E. Note that D' is again a diagram over $\mathcal{A}$ (its labelling map is just the restriction of the labelling map of D), its set of face labels is a subset of the face labels of D, its diameter is the same as that of D (because the geodesic spanning tree T is entirely contained in D') and its boundary cycle, read from p, is $w' := w_1uw_2$. In the free group $F(\mathcal{A})$ we have

$$w' = (w_1rw_1^{-1})(w_1aw_2) = (w_1rw_1^{-1})w.$$

We have argued that $|w_1| \leqslant \text{Diam}_p D = \text{Diam}_p D'$. And by induction we may assume that w' can be expressed as a product of conjugates of at most $\text{Area}_c D' = \text{Area}_c D - 1$ face labels, with conjugating elements of length at most $\text{Diam}_p D' = \text{Diam}_p D$. This completes the induction. □

One can give a shorter proof of the above proposition if one ignores the length of the conjugating elements x_i; this weaker form of the result is more standard, e.g. [66].

Exercise 4.1.3 Let $\mathcal{G}$ be a connected graph (1-dimensional CW complex). Let d be a length metric in which each edge has length 1. Fix a vertex $p \in \mathcal{G}$. Prove that $\mathcal{G}$ contains a geodesic spanning tree rooted at p, i.e. a 1-connected subgraph T that contains a path of length $d(p, v)$ from p to each vertex $v \in \mathcal{G}$.

4.2 Van Kampen's lemma

Definition 4.2.1 [Van Kampen diagrams] *Let $\mathcal{A}$ be a set, let $\mathcal{R}$ be a set of words in the letters $\mathcal{A}^{\pm 1}$ and let $\mathcal{R}_*$ be the smallest set of words that contains $\mathcal{R}$ and is closed under the operations of taking cyclic permutations and inverses of words. (Note that $\langle \mathcal{A} \mid \mathcal{R} \rangle \cong \langle \mathcal{A} \mid \mathcal{R}_* \rangle$.)*

If w, D, and p are as in the above proposition, then D is called a van Kampen diagram *for w over $\langle \mathcal{A} \mid \mathcal{R} \rangle$ with basepoint p.*

Theorem 4.2.2 (Van Kampen's lemma) *Let $\mathcal{A}$ be a set, let w be a word in the letters $\mathcal{A} \cup \mathcal{A}^{-1}$, and let $\mathcal{R}$ be a set of words in these letters.*

(1) *$w = 1$ in the group $\Gamma = \langle \mathcal{A} \mid \mathcal{R} \rangle$ if and only if there exists a van Kampen diagram for w over $\langle \mathcal{A} \mid \mathcal{R} \rangle$.*
(2) *If $w = 1$ in Γ then*

$$\mathrm{Area}_a(w) = \min\{\mathrm{Area}_c\, D \mid D \text{ a van Kampen diagram for } w \text{ over } \langle \mathcal{A} \mid \mathcal{R} \rangle\}.$$

In order to complete the proof of this theorem we shall need two lemmas. In the first we consider the following ordering on diagrams over $\mathcal{A}$ that have an initial vertex[16] specified in the boundary cycle: $D' \prec D$ if D' has fewer 1-cells than D and the words labelling the boundary cycles of D and D', read from their initial vertices, are equal as elements of the free group $F(\mathcal{A})$.

Lemma 4.2.3 *If D, with initial point p, is minimal in the ordering $\prec$, then the boundary label of D is a freely reduced word.*

Proof We shall assume that D is a diagram whose boundary label w is not freely reduced and construct a diagram $\prec D$.

Since w is not reduced, there is a pair of successive directed edges $\varepsilon, \varepsilon'$ in the boundary cycle that are labelled a, a^{-1} respectively, where $a \in \mathcal{A} \cup \mathcal{A}^{-1}$. If the initial vertex of ε is equal to the terminal vertex of ε' then we can delete from D these edges together with the contractible region that they enclose, thus obtaining a diagram $\prec D$.

If the initial vertex of ε is not equal to the terminal vertex of ε' then[17] we can connect the latter vertex to the former by a smooth arc $c : [-1, 1] \to \mathbb{R}^2$ that intersects D only at its endpoints. Let $T \subset \mathbb{R}^2$ be the open disc enclosed by the loop $\varepsilon\varepsilon' c$; we shall collapse T in a controlled manner. Let $\Delta = \{(x, y) \mid 1 > y > |x|, |x| < 1\} \subset \mathbb{R}^2$ and fix a diffeomorphism $\phi : \Delta \to T$ that has a continuous extension to $\overline{\Delta}$ with $\phi|_{[-1,1]\times\{1\}} = c$ and $\phi(-t, t) = \varepsilon'(t)$ and $\phi(t, t) = \varepsilon(1 - t)$ for all $t \in [0, 1]$. The map $T \to \{0\} \times \mathbb{R}$ that sends $z = \phi(x, y)$

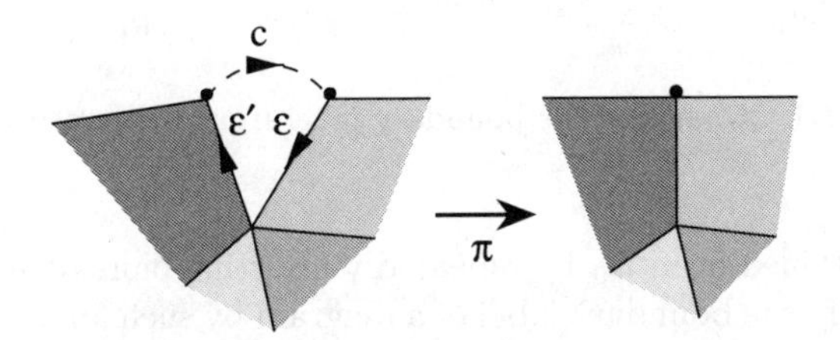

Fig. 4.2.4 Reducing the boundary label

[16] A choice of 'initial vertex' includes the specification of which edge of the boundary cycle is to be traversed first. Nevertheless, when no confusion is threatened, one talks as if the 'initial vertex' is simply a vertex of D.

[17] There are no hidden assumptions here: ε and ε' may be in the thin part of the boundary or in the thick part, and one of them might be a loop.

to y has a continuous extension $\pi : \mathbb{R}^2 \to \mathbb{R}^2$ that is a diffeomorphism on the complement of the closure of T.

$\overline{D} = \pi(D)$ inherits a combinatorial structure from D as well as a choice of initial point for its boundary cycle. $\overline{D}$ has fewer 1-cells than D because $\pi \circ \varepsilon^{-1} = \pi \circ \varepsilon'$. The directed edges $\pi \circ \varepsilon_i$ of $\overline{D}$ inherit the labelling $\lambda(\varepsilon_i)$ from D, and the label on the boundary cycle of $\overline{D}$, read from its initial point, is obtained from w by deleting the subword aa^{-1} corresponding to $\varepsilon\varepsilon'$. Thus $\overline{D} \prec D$. □

Remark 4.2.5 If one employs a suitably natural procedure for choosing the edge ε, then the proof given above actually constitutes an algorithm for transforming a diagram D whose boundary label is not freely reduced into a diagram $D' \prec D$. By repeated application of this algorithm one obtains a diagram $D_0 \prec D$ whose boundary label is reduced. Moreover, the set of face labels of D_0 is contained in the set of face labels of D, and $\mathrm{Area}_c\, D_0 \leqslant \mathrm{Area}_c\, D$.

The following lemma is used to pass from diagrams whose boundary labels are reduced to those whose labels are not.

Lemma 4.2.6 *Let $\mathcal{A}$ be a set, let w be a word in the letters $\mathcal{A} \cup \mathcal{A}^{\pm 1}$ and let w_0 be the reduced word that is equal to w in $F(\mathcal{A})$. Given a diagram D_0 for w_0 over $\mathcal{A}$, one can construct a diagram D for w with* $\mathrm{Area}_c\, D_0 = \mathrm{Area}_c\, D$ *so that the set of face labels of D is the same as that of D_0.*

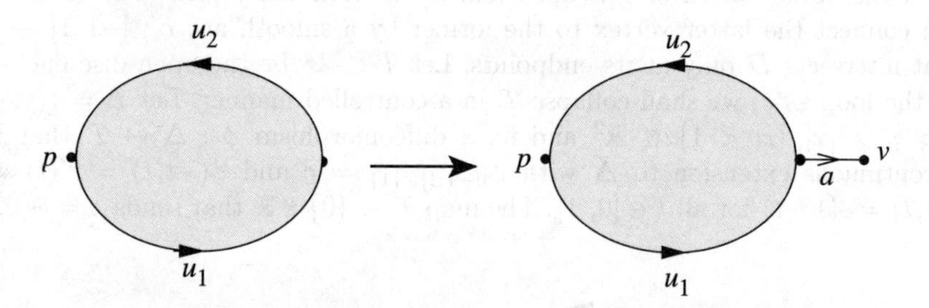

Fig. 4.2.7 Changing the boundary label from u_1u_2 to $u_1aa^{-1}u_2$.

Proof w is obtained from w_0 by repeatedly inserting pairs of letters aa^{-1} with $a \in \mathcal{A}$. To modify the boundary label of a diagram by such an insertion, one adds a new vertex v of valence 1 and an edge labelled a to v from the appropriate vertex of the boundary cycle (Fig. 4.2.7).

The proof of Van Kampen's lemma. If $w = 1$ in $\Gamma = \langle \mathcal{A} \mid \mathcal{R} \rangle$ then in the free group $F(\mathcal{A})$ we have:

$$w \overset{\text{free}}{=} \prod_{i=1}^{N} x_i r_i x_i^{-1}$$

where $r_i \in \mathcal{R}^{\pm 1}$ and $N = \mathrm{Area}_a\, w$. The word W on the right of this equality is the boundary label on the 'lollipop' diagram D_1 shown in Fig. 4.2.8; note that $\mathrm{Area}_c\, D_1 = N$.

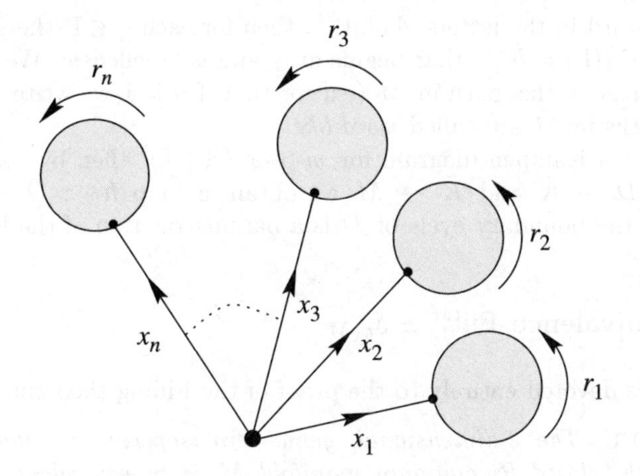

Fig. 4.2.8 The lollipop diagram

Let $D_0 \preceq D_1$ be a $\prec$-minimal diagram. The boundary label of D_0 is the freely reduced word w_0 that is equal to w in $F(\mathcal{A})$, the face labels of D_0 are a subset of those of D_1, and $\mathrm{Area}_c\, D_0 \leqslant \mathrm{Area}_c\, D_1 = \mathrm{Area}_a\, w$ (Lemma 4.2.3 and (4.2.5)). By applying Lemma 4.2.6 to D_0 we obtain a van Kampen diagram D of area $\leqslant N$ for w over $\langle \mathcal{A} \mid \mathcal{R} \rangle$. This proves the implication 'only if' in (1) and the inequality $\geqslant$ in (2). Proposition 4.1.2 provides the complementary 'if' implication and $\leqslant$ inequality. □

4.3 Words and Van Kampen diagrams as maps

In this subsection I shall assume that the reader is familiar with the material in Section A.2 (Appendix A).

If D is a van Kampen diagram over $\langle \mathcal{A} \mid \mathcal{R} \rangle$ with basepoint p, then there is a unique label-preserving combinatorial map from the 1-skeleton of D to the Cayley graph $\mathcal{C}_{\mathcal{A}}(\Gamma)$ that sends p to the vertex $1 \in \Gamma$. This extends to a combinatorial map from D to the universal covering $\tilde{K}$ of the standard 2-complex $K(\mathcal{A} : \mathcal{R})$.

Let M be a closed, smooth Riemannian manifold and let $\Gamma = \langle \mathcal{A} \mid \mathcal{R} \rangle$ be a finite presentation of the fundamental group of M. Let $\tilde{K}$ be the universal covering of $K(\mathcal{A} : \mathcal{R})$. We fix a basepoint $p \in \tilde{M}$, and for every $a \in \mathcal{A}$ we choose a geodesic c_a joining p to $a \cdot p$. These choices give rise to a Γ-equivariant map from $\mathcal{C}_{\mathcal{A}}(\Gamma) = \tilde{K}^{(1)}$ to $\tilde{M}$: this map sends the 1-cell labelled a emanating from

γ homeomorphically onto the segment $\gamma \cdot c_a$. Since $\tilde{M}$ is simply-connected, we may extend this map across the 2-cells of $\tilde{K}$ in a Γ-equivariant manner. We choose this extension so that on each 2-cell it is smooth almost everywhere and has finite area.

If w is a word in the letters $\mathcal{A} \cup \mathcal{A}^{-1}$, then for each $\gamma \in \Gamma$ there is a unique edge-path in $\mathcal{C}_{\mathcal{A}}(\Gamma) = \tilde{K}^{(1)}$ that begins at γ and is labelled w. We write $\hat{w}^{\gamma}$ to denote the image of this path in $\tilde{M}$ (except that if $\gamma = 1$ we write $\hat{w}$ instead of $\hat{w}^1$). Such paths in $\tilde{M}$ are called *word-like.*

If D is a van Kampen diagram for w over $\langle \mathcal{A} \mid \mathcal{R} \rangle$, then by composing the above maps $D \to \tilde{K}$ and $\tilde{K} \to \tilde{M}$ we obtain a map $h_D : D \to \tilde{M}$ whose restriction to the boundary cycle of D is a parametrization of the loop $\hat{w}$.

5 The equivalence $\mathrm{Fill}_0^M \simeq \delta_{\pi_1 M}$

This section is devoted entirely to the proof of the Filling theorem:

Theorem 5.0.1 *The 2-dimensional, genus-zero isoperimetric function* Fill_0^M *of any smooth, closed Riemannian manifold* M *is* $\simeq$ *equivalent to the Dehn function* $\delta_{\pi_1 M}$ *of the fundamental group of* M*.*

5.1 The bound $\mathrm{Fill}_0^M \preceq \delta_{\pi_1 M}$

This direction of the proof is substantially easier than the other. In order to understand the proof, the reader will need to have absorbed the definition of a van Kampen diagram.

Proposition 5.1.1 *If* M *is a smooth, closed Riemannian manifold then* $\Gamma := \pi_1 M$ *is finitely presented and* $\mathrm{Fill}_0^M \preceq \delta_\Gamma$*.*

Proof Corollary A.4.2 of Appendix A shows that Γ is finitely presented. We fix a finite presentation for Γ and assume that the universal cover $\tilde{K}$ of the standard 2-complex of this presentation has been mapped to $\tilde{M}$ as explained in the preceding subsection. We identify Γ (the 0-skeleton of $\tilde{K}$) with its image in $\tilde{M}$. We define λ to be the maximum distance of any point of $\tilde{M}$ from Γ, we define μ to be the maximum of the lengths of the 1-cells of $\tilde{K}$, as measured in $\tilde{M}$, and we define $m = \max\{d_\Gamma(\gamma,\gamma') \mid d_{\tilde{M}}(\gamma,\gamma') \leqslant 2\lambda + 1\}$, where d_Γ is the word metric associated to our chosen generators for Γ.

The images in $\tilde{M}$ of the 2-cells of $\tilde{K}$ are discs of finite area; let α be the maximum of these areas.

Let w be a word in the given generators that equals $1 \in \Gamma$ and consider the corresponding piecewise-geodesic loop $\hat{w}$ in $\tilde{M}$. Choose a van Kampen diagram D for w with $\mathrm{Area}_a(w)$ 2-cells, and consider the associated map $h_D : D \to \tilde{M}$, which fills $\hat{w}$. The area of this map is at most α times the number of 2-cells in D, hence

$$\mathrm{FArea}(\hat{w}) \leqslant \alpha \mathrm{Area}_a(w) \leqslant \alpha\, \delta_\Gamma(|w|).$$

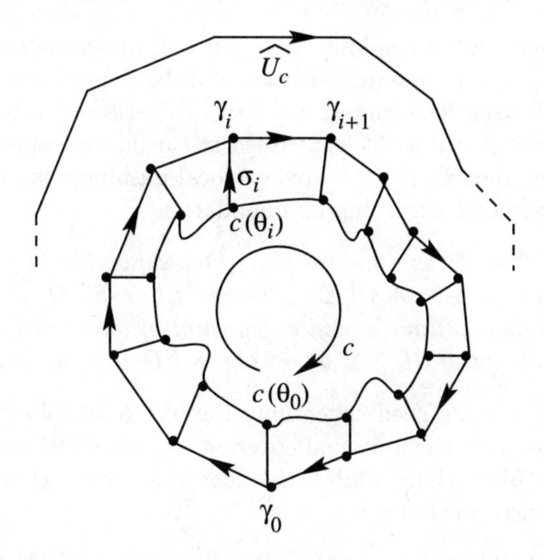

Fig. 5.1.2 Approximating c by the word-like loop $\hat{U}_c$

Given a loop $c : \mathbb{S}^1 \to \tilde{M}$ of finite length $l(c)$, parametrized by arc length, we choose a set of n equally-spaced points $\theta_0, \dots, \theta_{n-1} \in \mathbb{S}^1$, where n is the least integer greater than $l(c)$. We then choose a geodesic segment σ_i from each $c(\theta_i)$ to a nearest point $\gamma_i \in \Gamma \subset \tilde{M}$. The distance in $\tilde{M}$ between successive γ_i (indices mod n) is at most $2\lambda + 1$ and hence γ_i can be connected to γ_{i+1} by a word-like path $\hat{u}_i^{\gamma_i}$ of length at most $m\mu$, where u_i is a word of (algebraic) length m. Since each of the loops[18] $\sigma_i \hat{u}_i^{\gamma_i} \bar{\sigma}_{i+1} \bar{c}|_{[\theta_i, \theta_{i+1}]}$ has length at most $L := m\mu + 1 + 2\lambda$, we have

$$\mathrm{FArea}(c) \leqslant \mathrm{FArea}(\hat{U}_c) + n\,\mathrm{Fill}_0^M(L),$$

where U_c is the concatenation of the words u_i (see Fig. 5.1.2).

The loop c is arbitrary, the word U_c has (algebraic) length at most $nm\mu$, and $n \leqslant l(c) + 1$. Thus, the two inequalities displayed above imply that

$$\mathrm{Fill}_0^M(l) \leqslant \alpha\delta_\Gamma\big(m(l+1)\big) + (l+1)\,\mathrm{Fill}_0^M(L)$$

for all $l > 0$. In particular, since $\mathrm{Fill}_0^M(L)$ is a constant, $\mathrm{Fill}_0^M \preceq \delta_\Gamma$.

5.2 The bound $\delta_{\pi_1 M} \preceq \mathrm{Fill}_0^M$

There are many subtleties concerning the nature of solutions to Plateau's problem in Riemannian manifolds, but the existence of least-area discs (although highly non-trivial) has little to do with the fine structure of the spaces concerned. Indeed Igor Nikolaev [78] showed that one can solve Plateau's problem in any

[18] An overbar denotes reversed orientation.

complete simply-connected geodesic space with an upper curvature[19] bound k. In this generality, when endowed with the pull-back path metric, a least-area spanning disc will itself have curvature $\leqslant k$. (To get some intuition about why this is true, observe that if a disc embedded in Euclidean 3-space has a point of positive curvature, then there is an obvious local pushing move that reduces the area of the disc without disturbing its boundary.)

Definition 5.2.1 *Let D be a metric space homeomorphic to a (perhaps singular[20]) 2-disc and fix $\varepsilon > 0$. A set $\Sigma \subset D$ is said to ε-fill D if every point of D is a distance less than ε from Σ and every point of the boundary cycle ∂D can be connected to a point of $\partial D \cap \Sigma$ by an arc in ∂D that has length at most ε.*

The only fact that we need concerning the nature of solutions to Plateau's problem is that loops in the universal covering of a closed Riemannian manifold can be filled by discs that exhibit the following crude consequence of the curvature bound described above.

Proposition 5.2.2 *If M is a complete Riemannian manifold of curvature $\leqslant k$, then the induced metric on every least-area disc $D \to \tilde{M}$ is such that D can be ρ_k-filled by a set of cardinality less than $\lambda_k(\text{Area}(D) + |\partial D| + 1)$, where $|\partial D|$ denotes the length of the boundary of D and the constants λ_k and ρ_k depend only on k.*

Proof When equipped with the pull-back metric D has curvature $\leqslant k$. In the Riemannian setting this means that if a metric ball of radius $r < \pi/(2\sqrt{k})$ is contained in the interior of D, then the area of that ball is at least as great as the area of a disc of radius r in M_k^2. So if $a(k, r)$ denotes the area of such a disc, then there can be at most $\text{Area}(D)/a(k, r)$ disjoint balls of radius r in the interior of D.

We set $r = r_k := \pi/(4\sqrt{k})$ and choose a maximal collection of disjoint balls of radius r in the interior of D. Let Σ_0 denote the set of centres of these balls. We also choose a collection Σ_1 of no more than $(1/r_k)|\partial D| + 1$ points along ∂D so that every point of ∂D can be connected to a point in Σ_1 by an arc of length less than r_k. By construction, the balls of radius $2r_k$ centred at the points $\Sigma_0 \cup \Sigma_1$ cover D and the cardinality of $\Sigma_0 \cup \Sigma_1$ is bounded above by $(1/a(k, r_k))\text{Area}(D) + (1/r_k)|\partial D| + 1$. Set $\rho_k = 2r_k$ and $\lambda_k = \max\{(1/a(k, r_k)), (1/r_k)\}$. □

In order to establish the reverse inequality in the Filling theorem we shall use the following technical tool for manufacturing combinatorial discs out of ε-filling sets.

---

[19] In the sense of A.D. Alexandrov; see Appendix B for the definition. In Nikolaev's theorem there is a natural restriction on the length of the loops being filled if $k > 0$.

[20] A singular disc is a space homeomorphic to the underlying space of a singular disc diagram, as defined in (4.1). In the Riemannian setting (dimension $\geqslant 3$) one can avoid the need to discuss (topologically) singular discs by considering fillings of embedded loops only (cf. 2.1.5).

5.2.3 Cellulation lemma *Let D be a length space homeomorphic to a (perhaps singular) 2-disc, and suppose that D is ε-filled by a set Σ of cardinality N. Then there exists a combinatorial 2-complex Φ, homeomorphic to the standard 2-disc, and a continuous map $\phi : \Phi \to D$ such that:*

(1) *Φ has less than $8N$ faces (2-cells) and each is a k-gon with $k \leqslant 12$;*
(2) *the restriction of ϕ to each 1-cell in Φ is a path of length at most 2ε;*
(3) *$\phi|_{\partial\Phi}$ is a monotone parametrization of ∂D and $\Sigma \cap \partial D$ lies in the image of the 0-skeleton of $\partial\Phi$.*

In the case of the ε-fillings yielded by Proposition 5.2.2 (which are the focus of our concern), instead of using the decomposition of D furnished by the Cellulation lemma, one might use the dual to the Voronoi decomposition for the given filling — this dual will generically be a triangulation (cf. [99] 5.58). Some care is needed in pursuing this remark, but nevertheless we use it as a pretext[21] for relegating the proof of the Cellulation lemma to Appendix C.

The remainder of the proof of the Filling theorem. It remains to show that $\delta_\Gamma \preceq \mathrm{Fill}_0^M$, where $\Gamma = \pi_1 M$. Let $k > 0$ be an upper bound on the sectional curvature of M.

We fix a basepoint $p \in \tilde{M}$ and choose a number $\rho > 0$ sufficiently large to ensure that the balls of radius $\rho/8$ about $\{\gamma \cdot p \mid \gamma \in \Gamma\}$ cover $\tilde{M}$ and that $\rho > 8\rho_k$ (notation of 5.2.2). Let $\mathcal{A}$ be the set of $a \in \Gamma$ such that $d(a \cdot p, p) < \rho$ and let $\mathcal{R}$ be the set of words in the symbols $\mathcal{A} \cup \mathcal{A}^{-1}$ that have length $\leqslant 12$ and equal the identity in Γ. (Note that $\mathcal{A}$ contains a letter that represents $1 \in \Gamma$.) Corollary A.4.2 shows that $\langle \mathcal{A} \mid \mathcal{R} \rangle$ is a presentation of Γ. We shall show that every null-homotopic word w over this presentation satisfies

$$\mathrm{Area}_a(w) \leqslant 4\lambda_k \left(\mathrm{Fill}_0^M(\rho\,|w|) + \rho\,|w| + 1\right).$$

Given a word w with $w = 1$ in Γ we consider the piecewise geodesic loop $\hat{w}$ in $\tilde{M}$ (notation of 4.3). This loop has length less than $\rho|w|$ and hence[22] can be filled with a least-area disc $f : D \to \tilde{M}$ of area at most $\mathrm{Fill}_0^M(\rho\,|w|)$. Using Proposition 5.2.2 we can ρ_k-fill D with a set Σ of cardinality less than $N := \lambda_k(\mathrm{Fill}_0^M(\rho\,|w|) + \rho\,|w| + 1)$. Increasing the cardinality of Σ by at most $|w|$, we may assume that it contains the vertices of $\hat{w}$.

Consider a combinatorial 2-disc Φ and a map $\phi : \Phi \to D$ as furnished by the Cellulation lemma. Our aim is to label Φ so that it becomes a van Kampen diagram for w over $\langle \mathcal{A} \mid \mathcal{R} \rangle$. The composition $f \circ \phi : \Phi \to \tilde{M}$ will guide us in this construction. Note that the restriction of $f \circ \phi$ to $\partial\Phi$ is a monotone parametrization of $\hat{w}$. The initial point of $\hat{w}$ determines a basepoint for Φ.

---

[21] The honest reason for this deferral is that the proof is lengthy and inelegant.

[22] If one wants to quote Morrey directly here one should perturb $\hat{w}$ to ensure that it is embedded.

For each vertex v in the interior of Φ we choose a point $\gamma_v \cdot p$ in the Γ orbit of p that is closest to $f \circ \phi(v)$. If v and v' are the vertices of a 1-cell in Φ, then $f \circ \phi(v)$ and $f \circ \phi(v')$ are a distance at most $2\rho_k$ apart in $\tilde{M}$ (the second property of Φ in the Cellulation lemma). It follows that $d(\gamma_v \cdot p, \gamma_{v'} \cdot p) \leqslant 2\rho_k + \rho/4$, which is less than $\rho/2$. Hence there exists a generator $a \in \mathcal{A}$ such that $a = \gamma_v^{-1}\gamma_{v'}$ in Γ. We introduce the label a on the edge in Φ joining v to v'.

Among the vertices of $\partial\Phi$ we have a set of distinguished vertices, namely those mapping to the vertices of $\hat{w}$. Call these $x_0, \dots, x_{n-1}$, corresponding to the vertices $w_i \cdot p$ on $\hat{w}$, where w_i is the i-th prefix of w.

If $v \in \Phi \smallsetminus \partial\Phi$ is the initial point of an edge whose endpoint v' lies on the arc joining x_{i-1} to x_i in $\partial\Phi$, then γ_v is a distance less than $\rho/8 + 2\rho_k + \rho/2$ from either $w_{i-1} \cdot p$ or $w_i \cdot p$, depending on which side of the midpoint of the arc v' lies (where 'midpoint' is measured in the arc length pulled back from $\tilde{M}$).

For each $i = 1, \dots, n$ we collapse all but one of the edges along the arc of $\partial\Phi$ joining x_{i-1} to x_i; the edge containing the midpoint is not collapsed,[23] and its image in the quotient disc $\overline{\Phi}$ is labelled with the i-th letter of w. The image in $\overline{\Phi}$ of the quotient of the edge $[v, v']$ discussed in the previous paragraph is labelled either $\gamma_v^{-1} w_{i-1}$ or $\gamma_v^{-1} w_i$, according to the side of the midpoint on which v' lies. (This label will be an element of $\mathcal{A}$ because $\rho/8 + 2\rho_k + \rho/2 < \rho$.)

At this stage we have constructed a combinatorial 2-disc $\overline{\Phi}$ with a label from $\mathcal{A}$ on each directed 1-cell. The label on the boundary circle $\partial\overline{\Phi}$ is our original null-homotopic word w. The label on the boundary cycle of each 2-cell is, by construction, a word of length at most 12 in the letters $\mathcal{A}$ that represents the identity in Γ, because the faces of Φ, and hence $\overline{\Phi}$, are k-gons with $k \leqslant 12$. Thus $\overline{\Phi}$ is a van Kampen diagram for w over our chosen presentation of $\Gamma = \pi_1 M$.

The Cellulation lemma gave us Φ and told us that it had at most $8N$ faces, where $N = \lambda_k(\mathrm{Fill}_0^M(\rho\,|w|) + \rho\,|w| + 1)$. And $\overline{\Phi}$ has the same number of faces as Φ. Thus we have established the desired upper bound on the algebraic area of the arbitrary null-homotopic word w, and we deduce that $\delta_\Gamma \preceq \mathrm{Fill}_0^M$. □

6 Linear and quadratic Dehn functions

In this section we shall see that the groups that have linear Dehn functions are precisely those that are negatively curved on the large scale, i.e. *hyperbolic* in the sense of 6.1.3. This fundamental insight is due to Misha Gromov [55].

We shall also discuss the weaker link between non-positive curvature and the class of groups that have a quadratic Dehn function.

6.1 Hyperbolicity: from Dehn to Gromov

Given a finite set of generators $\mathcal{A}$ for a group Γ, one would have a particularly efficient algorithm for solving the word problem if one could construct a finite

[23] This involves a choice if the midpoint is a vertex.

list of words $u_1, v_1, u_2, v_2, \ldots, u_n, v_n$, with $u_i =_\Gamma v_i$ and $|v_i| < |u_i|$, such that every freely-reduced word in the letters $\mathcal{A}^{\pm 1}$ that represents $1 \in \Gamma$ contains at least one of the u_i as a subword.

If such a list of words exists then one proceeds as follows: given an arbitrary reduced word w, look for subwords of the form u_i; if there is no such subword, stop and declare that w does not represent $1 \in \Gamma$; if u_i occurs as a subword, replace u_i with v_i, freely reduce the resulting word w' and then repeat the search for subwords of the form u_j (noting that $w = w'$ in Γ). Proceeding in this way, after at most $|w|$ steps one will have either reduced w to the empty word (in which case $w = 1$ in Γ) or else verified that $w \neq 1$ in Γ.

Definition 6.1.1 *When it exists, the above procedure for solving the word problem is called a* Dehn algorithm *for* Γ*; it is encoded in* $\langle \mathcal{A} \mid u_1 v_1^{-1}, \ldots, u_n v_n^{-1} \rangle$*, which we call a* Dehn presentation.

Max Dehn proved that Fuchsian groups admit Dehn presentations [35]. Jim Cannon proved that the fundamental groups of all closed negatively curved manifolds admit Dehn presentations [27]. The following *small cancellation* condition provides many other examples (see [66] Chapter V).

Example 6.1.2 Let $\langle \mathcal{A} \mid \mathcal{R} \rangle$ be a finite presentation in which each relator is freely reduced. Assume that if $r \in \mathcal{R}$ then r^{-1} and every cyclic permutation of r is in $\mathcal{R}$. And suppose that whenever there exist distinct $r, r' \in \mathcal{R}$ with a common prefix u (i.e. $r \equiv uv$ and $r' \equiv uv'$), the inequality $|u| < |r|/6$ holds. Then $\langle \mathcal{A} \mid \mathcal{R} \rangle$ is a Dehn presentation.

It requires only a moment's clear thought to see that the existence of a Dehn algorithm for a group Γ implies that Γ has a linear Dehn function (cf. paragraph 1.2). A more profound observation is that the converse is also true. The proof of this fact is indirect, proceeding via Gromov's notion of a hyperbolic group [55].

Gromov made the following remarkable discovery: the simple geometric condition given in (6.1.3) forces a geodesic metric space, regardless of its local structure, to exhibit many of the large-scale features that one associates with simply-connected manifolds of negative curvature. Thus he was able to extend the power of negative curvature well beyond its traditional realm[24] in Riemannian geometry. This stripping away of extraneous structure leads to a deeper understanding of the fundamental groups of closed negatively curved manifolds, and extends such an understanding to much wider classes of groups.

Definition 6.1.3 *A geodesic metric space X is* hyperbolic *(in the sense of Gromov) if there exists a constant $\eta > 0$ such that for every geodesic triangle*[25]

---

[24] The work of H. Busemann and, more particularly, A.D. Alexandrov, had already expanded the range of spaces in which one can discuss negative and non-positive curvature (see [22]), but that work was based on local definitions of curvature, whereas in Gromov's approach one ignores the local structure of the space.

[25] See Appendix B for definitions such as that of a triangle in an arbitrary metric space.

$\Delta \subseteq X$, each edge of Δ lies in the η-neighbourhood of the union of the other two edges. (One writes "X is η-hyperbolic" when it is useful to specify the constant.)

A finitely generated group Γ is said to be *hyperbolic* if its Cayley graph[26] is η-hyperbolic for some $\eta > 0$.

Exercises 6.1.4 (i) Prove that real hyperbolic space $\mathbb{H}^n$ is hyperbolic in the above sense and find the optimal η. (Hint: There is a bound on the area of semicircular discs that can be inscribed in geodesic triangles in $\mathbb{H}^2$.)

(ii) Deduce that the universal covering X of any closed manifold of negative sectional curvature is hyperbolic in the sense of Gromov. (Hint: If one scales the metric so that the curvature of X is bounded above by -1, then every geodesic triangle $\Delta \subseteq X$ is the image of a non-expanding map $\phi : \overline{\Delta} \to \Delta$, where $\overline{\Delta}$ is a triangle in $\mathbb{H}^2$ and the restriction of ϕ to each edge of $\overline{\Delta}$ is an isometry. This is called the CAT(-1) inequality [22].)

The following results are due to Gromov [55] (see also Cannon [28]). Detailed references and proofs can be found in Chapter III.Γ of [22].

Theorem 6.1.5 *The following statements are equivalent for finitely presented groups Γ:*

(1) *Γ is a hyperbolic group.*
(2) *Γ has a finite Dehn presentation.*
(3) *Γ has a linear Dehn function.*
(4) *The Dehn function of Γ is subquadratic (i.e. $\delta_\Gamma(n) = o(n^2)$).*

Proceeding in cyclic order, the only non-trivial implications are (4) $\Rightarrow$ (1) and (1) $\Rightarrow$ (2). We shall not discuss (4) $\Rightarrow$ (1) except to say that Cornelia Druţu [38] recently discovered an elegant proof that uses asymptotic cones (cf. 3.1.6).

The proof that (1) $\Rightarrow$ (2) requires an understanding of the following types of locally-efficient paths. Let $I \subset \mathbb{R}$ be an interval and let X be a metric space. A map $c : I \to X$ is called a *k-local* geodesic if $d(c(t), c(t')) = |t - t'|$ for all $t, t' \in I$ with $|t - t'| \leqslant k$. And c is called a *(λ, ε)-quasi-geodesic* if

$$\frac{1}{\lambda}\,|t - t'| - \varepsilon \leqslant d(c(t), c(t')) \leqslant \lambda\,|t - t'| + \varepsilon$$

for all $t, t' \in I$.

In hyperbolic spaces one has the following local criterion for recognizing certain quasi-geodesics (see [22] page 405).

Lemma 6.1.6 *If X is η-hyperbolic then every 8η-local geodesic in X is a (λ, ε)-quasi-geodesic, where the constant $\lambda > 0$ depends only on η, and ε is less than 8η.*

---

[26] The ambiguity that arises from the fact that we have not specified a generating set is removed by Exercise 6.1.9(ii).

The implication (1) $\Rightarrow$ (2) in Theorem 6.1.5 follows easily from this lemma:

Exercise 6.1.7 Suppose that the Cayley graph of Γ with respect to the finite generating set $\mathcal{A}$ is η-hyperbolic. Let $\mathcal{R}$ be the set of words $u_i v_i^{-1}$, where u_i runs over all words of length $\leqslant 8\eta$ in the letters $\mathcal{A}^{\pm 1}$ for which there exists a word v_i with $|v_i| < |u_i|$ and $u_i = v_i$ in Γ. Show that $\langle \mathcal{A} \mid \mathcal{R} \rangle$ is a Dehn presentation.

The following stability property of quasi-geodesics marks an important difference between spaces of non-positive curvature and spaces of strictly negative curvature (see [22] page 401).

Proposition 6.1.8 *For all $\eta, \lambda, \varepsilon > 0$ there exists $R(\eta, \lambda, \varepsilon) > 0$ such that: if X is η-hyperbolic and $c : [a, b] \to X$ is (λ, ε)-quasi-geodesic with endpoints p and q, then the Hausdorff distance between the image of c and each geodesic segment joining p to q is less than $R(\eta, \lambda, \varepsilon)$.*

This proposition provides a proof (independent of the Filling theorem) that the fundamental groups of closed negatively curved manifolds have linear Dehn functions – see 6.1.4(ii) and 6.1.9(iii).

The following exercises require the reader to understand certain items from Appendix A, namely the definition of quasi-isometry, the Švarc–Milnor lemma and A.1.3(ii).

Exercise 6.1.9 (i) Let X be a geodesic space. If X is quasi-isometric to a η-hyperbolic space, then X is η'-hyperbolic for some $\eta' > 0$. (Hint: Consider quasi-geodesic triangles.)

(ii) If the Cayley graph of a group with respect to one finite generating set is hyperbolic, then so is the Cayley graph of that group with respect to any other finite generating set.

(iii) If a group acts properly and cocompactly by isometries on a hyperbolic geodesic space, then that group has a linear Dehn function.

We refer the reader to Chapter III.Γ of [22] for an introduction to the rich theory of hyperbolic metric spaces (the references given therein will also point the reader to recent developments in this active field). Here are a few of the basic properties of hyperbolic groups.

Theorem 6.1.10 *If a group Γ has a linear Dehn function then:*

(1) Γ *does not contain* $\mathbb{Z}^2$;
(2) Γ *has a solvable conjugacy problem;*
(3) Γ *has only finitely many conjugacy classes of finite subgroups;*
(4) Γ *acts on a contractible simplicial complex with compact quotient and finite stabilizers.*

(5) *Let $\mathcal{A}$ be a finite generating set for Γ and let d be the associated word metric. Define $\tau(\gamma) = \lim_{n\to\infty} d(1,\gamma^n)/n$. Then there is an integer N such that $\{N\,\tau(\gamma) \mid \gamma \in \Gamma \smallsetminus \{1\}\}$ is a set of positive integers.*

6.2 Quadratic Dehn functions and non-positive curvature

If a geodesic metric space X is complete, 1-connected and non-positively curved in the sense of A.D. Alexandrov (see Appendix B), then its metric is *convex* in the sense that $d(c(t), c'(t)) \leqslant t\, d(c(1), c'(1)) + (1-t)\, d(c(0), c'(0))$ for all geodesics $c, c' : [0,1] \to X$ parametrized proportional to arc length. This class of spaces includes the universal covering $\tilde{M}$ of any compact Riemannian manifold whose sectional curvatures are non-positive, and hence the following result applies to the fundamental groups of such manifolds (acting by deck transformation on $\tilde{M}$). It also applies to cocompact lattices in semisimple Lie groups (cf. 3.3.7).

Theorem 6.2.1 *Let X be a complete geodesic space whose metric is convex. If the group Γ acts properly by isometries on X and the quotient of this action is compact, then Γ is finitely presented and its Dehn function is either linear or quadratic.*

The following proof is adapted from [5] and [22], and has earlier origins, e.g. [42].

Proof The point of the proof is to construct the diagram shown in Fig. 6.2.3. Let d be the metric on X. Fix $p \in X$ and let $\rho \geqslant 1$ be such that the balls of radius ρ about the Γ-orbit of p cover X. Let c_γ be the arc-length parametrization of the unique geodesic segment joining p to $\gamma \cdot p$. Let $\mathcal{A} \subset \Gamma$ be the set of $\gamma \in \Gamma$ such that $d(p, \gamma \cdot p) \leqslant 3\rho$. Given $\gamma \in \Gamma$, let m be the least integer greater than $d(p, \gamma \cdot p)/\rho$ and for each positive integer $t < m$ choose $\gamma_t \in \Gamma$ with $d(c_\gamma(\rho t), \gamma_t \cdot p) \leqslant \rho$. Define $\gamma_0 = 1$ and $\gamma_m = \gamma$.

Consider the word $\sigma_\gamma := a_1 \dots a_m$ where $a_i := \gamma_{i-1}^{-1}\gamma_i \in \mathcal{A}$ for $i = 1, \dots, m$. With an eye on future generalizations, we write $\sigma_\gamma(i)$ instead of γ_i to denote the image in Γ of the i-th prefix of σ_γ; by definition $\sigma_\gamma(i) = \gamma$ if $i > m$. (In general we write $w(i)$ for the image in Γ of the i-th prefix of any word w.)

It follows from the convexity of the metric on X that in the word metric $d_{\mathcal{A}}$ on Γ one has

$$d_{\mathcal{A}}(\sigma_\gamma(i), \sigma_{\gamma'}(i)) \leqslant 3\, d_{\mathcal{A}}(\gamma, \gamma') \tag{6.2.2}$$

for all $\gamma, \gamma' \in \Gamma$ and all integers $i > 0$ (see Exercise 6.2.4(i)). We shall use this inequality to construct efficient diagrams for null-homotopic words.

Let w be a null-homotopic word, of length n say. We draw an oriented circle in $\mathbb{R}^2$, mark vertices $v_0, \dots, v_{n-1}$ (in cyclic order) on the circle and label the oriented arc (v_{i-1}, v_i) with the i-th letter of w (indices mod n). We then connect v_0 to each of the vertices v_i with a line segment $[v_0, v_i]$ divided into $|\sigma_{w(i)}|$ 1-cells; these 1-cells are oriented and labelled by the letters of $\sigma_{w(i)}$ in the obvious manner. Define $\sigma_{w(0)} = \sigma_{w(n)}$ to be the empty word, and for $j \geqslant |\sigma_{w(i)}|$ define 'the j-th vertex of $[v_0, v_i]$' to be v_i. Let $J(i) = \max\{|\sigma_{w(i)}|, |\sigma_{w(i+1)}|\}$.

We complete the construction of our diagram for w by introducing an edge from the j-th vertex of $[v_0, v_i]$ to the j-th vertex of $[v_0, v_{i+1}]$ for $i = 0, \ldots, n-1$ and $j = 1, \ldots, J(i)$; this edge is labelled by a word of minimal length that equals $\sigma_{w(i)}(j)^{-1}\sigma_{w(i+1)}(j) \in \Gamma$; according to (6.2.2) this word has length at most 3.

We have constructed a diagram over $\mathcal{A}$ with boundary label w, where w is an arbitrary null-homotopic word. The face labels are null-homotopic words of length $\leqslant 8$; let $\mathcal{R}$ be the set of all such words. Proposition 4.1.2 tells us that $\Gamma = \langle \mathcal{A} \mid \mathcal{R} \rangle$ and that $\text{Area}_a(w)$ is at most the number of faces in the diagram. Thus $\text{Area}_a(w) \leqslant |w| \max\{|\sigma_{w(i)}| \colon i \leqslant |w|\}$. And since $d_{\mathcal{A}}(1, w(i)) \leqslant |w|/2$ for all i, Exercise 6.2.4(ii) tells us that $\text{Area}_a(w) \leqslant (3/2)|w|^2$. ☐

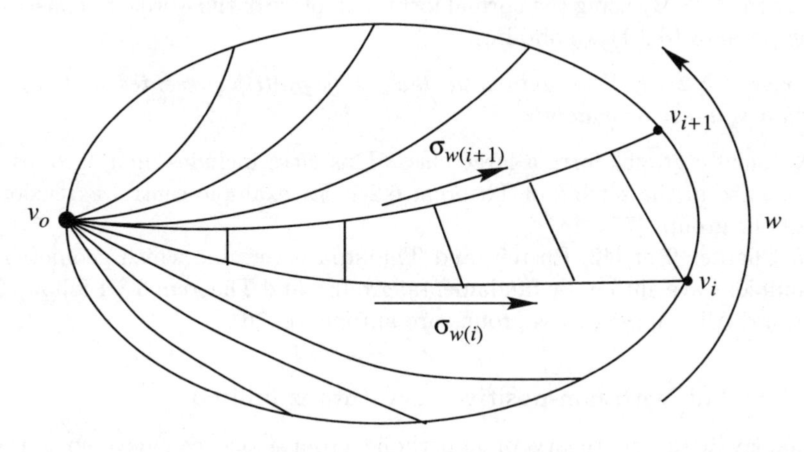

Fig. 6.2.3 Using the combing σ_γ to construct a van Kampen diagram

Exercises 6.2.4 (i) Establish the inequality 6.2.2. (Hint: If $m = d_{\mathcal{A}}(\gamma, \gamma')$ then $d(\gamma \cdot p, \gamma' \cdot p) \leqslant 3m\rho$. Hence, by the convexity of the metric, $d(c_\gamma(\rho t), c_{\gamma'}(\rho t) \leqslant 3m\rho$ for all $t > 0$. Recall that, by definition, $\sigma_g(t) = g_t$. Divide the geodesic $[c_\gamma(\rho t), c_{\gamma'}(\rho t)]$ into $3m$ segments of equal length, and associate to each division point a closest point of $\Gamma \cdot p$, with γ_t and γ'_t associated to the endpoints.)

(ii) Deduce that for all $\gamma \in \Gamma$ the length of the word σ_γ in the above proof is at most $3\, d_{\mathcal{A}}(1, \gamma)$.

6.3 Automatic groups

The intensive study of isoperimetric inequalities for finitely presented groups began in the late 1980s. It emerged primarily from the work of Gromov [55], but a certain impetus also came from the theory of automatic groups. This theory sprang from conversations between Jim Cannon and Bill Thurston and grew into a rich theory due to a team effort orchestrated by David Epstein – see [42].

Roughly speaking, a group Γ with finite generating set $\mathcal{A}$ is *automatic* if one can construct its Cayley graph by computations on finite state automata: there must exist a set of words $\mathcal{L} = \{\sigma_\gamma \mid \gamma \in \Gamma\}$ in the letters $\mathcal{A}^{\pm 1}$, with $\sigma_\gamma = \gamma$ in Γ, such that membership of $\mathcal{L}$ can be determined by a finite state automaton (FSA); and for each $a \in \mathcal{A}$ there must exist a FSA that recognizes those pairs of words $(\sigma_\gamma, \sigma_{\gamma'})$ for which $\gamma' = \gamma a$.

The finiteness of these FSA forces the existence of constants $k, K > 0$ such that $|\sigma_\gamma| \leqslant k\, d_{\mathcal{A}}(1,\gamma)$ and

$$d_{\mathcal{A}}(\sigma_\gamma(i), \sigma_{\gamma'}(i)) \leqslant K\, d_{\mathcal{A}}(\gamma, \gamma') \tag{6.3.1}$$

for all $\gamma, \gamma' \in \Gamma$. By using the normal form $\mathcal{L}$ in place of the words σ_γ constructed in the proof of (6.2.1) we obtain:

Theorem 6.3.2 *If Γ is automatic then it is finitely presented and its Dehn function is linear or quadratic.*

Automatic groups form a large class. This class includes many groups that do not arise in the setting of Theorem 6.2.1, for example central extensions of hyperbolic groups [77].

In Chapter 9 of [42] Epstein and Thurston determine which geometrizable 3-manifolds have automatic fundamental groups, and Theorem 3.3.1 follows from this work. All mapping class groups are automatic [76].

6.4 The link with non-positive curvature is limited

In analogy with the theory of hyperbolic groups, one can develop a theory of *semihyperbolic groups*, defined by a coarse geometric constraint that forces such groups to satisfy most of the useful properties enjoyed by the fundamental groups of compact non-positively curved manifolds (cf. Alonso and Bridson [5] and Gromov [56]).

With Theorems 3.1.6 and 6.1.10 in mind, one might hope that requiring a group to satisfy a quadratic isoperimetric inequality would force it to behave in a 'semihyperbolic' manner, satisfying a list of properties analogous to (6.1.10). The examples that we have seen thus far support this hope to some extent — abelian groups, hyperbolic groups, automatic groups, fundamental groups of compact non-positively curved spaces, $\mathrm{SL}(n,\mathbb{Z})$ for $n \geqslant 4$, various nilpotent groups N, and those non-uniform lattices in rank 1 Lie groups that have these N as cusp groups. But the examples discovered more recently indicate that the class of groups that have a quadratic Dehn function is wilder than this list would suggest, but quite how wild is not clear. For example, it is unknown if a group Γ that has a quadratic[27] Dehn function can have an unsolvable conjugacy problem (it is conjectured that if such a group exists, it should not have a 2-dimensional $K(\Gamma,1)$).

[27] There do exist examples with cubic Dehn functions, [20] Example 2.9.

Besides the property that defines them, the most significant property that is known to be enjoyed by groups with quadratic Dehn functions is the fact that their asymptotic cones are all simply-connected [88]. This property is not enjoyed by all groups with polynomial Dehn functions [18].

7 Techniques for estimating isoperimetric functions

This section contains a sample of the methods that have been developed to calculate Dehn functions. The techniques that I shall describe have been used widely, but I must emphasize that this is only a sample, not a thorough survey. This sample is biased in favour of the methods that I have found most useful in my own work.

7.1 Upper bounds

In general it is easier to obtain upper bounds on Dehn functions than it is to obtain lower bounds. Indeed whenever one has an explicit solution to the word problem in a finitely presented group, one can look for an upper bound on the Dehn function by analysing the use of relations in that solution (cf. paragraph 1.1). Thus there are many *direct methods* for obtaining upper bounds, each adapted to the groups at hand. We have already seen examples of such methods in (3.1.2), (2.2), (6.1.1), and (3.2.3). One might also think of results such as 3.4.1 in this light. Direct methods of a geometric nature are to be found in many of the papers listed in the bibliography, e.g. [19], [15], [95], and [18].

The following general method for obtaining upper bounds on Dehn functions has been used in many contexts.

Using combings to get upper bounds. Let Γ be a group with finite generating set $\mathcal{A}$ and let d be the associated word metric. A *combing* (normal form) for Γ is a set of words $\{\sigma_\gamma \mid \gamma \in \Gamma\}$ in the letters $\mathcal{A}^{\pm 1}$ such that $\sigma_\gamma = \gamma$ in Γ. Whenever one can find a geometrically-efficient combing for a group Γ one can estimate the Dehn function δ_Γ by modifying the proof of Theorem 6.2.1. The control that one needs in order to get non-trivial bounds is remarkably weak [16]. We content ourselves with one of the simplest and most widely used methods of control, wherein one weakens *the fellow-traveller property* (6.3.1) by allowing reparametrizations of the words σ_γ (thought of as paths in the Cayley graph of Γ).

Definition 7.1.1 *Let*

$$R = \{\rho : \mathbb{N} \to \mathbb{N} \mid \rho(0) = 0;\ \rho(n+1) \in \{\rho(n), \rho(n)+1\}\ \forall n;\ \rho \text{ unbounded}\}.$$

Given words w_1, w_2 in the letters $\mathcal{A}^{\pm 1}$, define

$$D(w_1, w_2) = \min_{\rho, \rho' \in R} \Big\{ \max_{t \in \mathbb{N}} \{d(w_1(\rho(t)), w_2(\rho'(t))\} \Big\}.$$

A combing $\gamma \mapsto \sigma_\gamma$ is said to satisfy the asynchronous fellow-traveller property if there is a constant $K > 0$ such that

$$D(\sigma_\gamma, \sigma_{\gamma'}) \leqslant K\, d_{\mathcal{A}}(\gamma, \gamma')$$

for all $\gamma, \gamma' \in \Gamma$. The length *of σ is a function $\mathbb{N} \to \mathbb{N}$:*

$$L_\sigma(n) := \max\{|\sigma_\gamma| \mid d_{\mathcal{A}}(1,\gamma) \leq n\}.$$

Proposition 7.1.2 *If a finitely generated group Γ admits a combing σ that satisfies the asynchronous fellow-traveller property, then Γ is finitely presented and its Dehn function satisfies $\delta_\Gamma(n) \preceq nL_\sigma(n)$. And regardless of the length of the combing, $\delta_\Gamma(n) \preceq 2^n$.*

Exercise 7.1.3 Prove the assertions in the first sentence of the above proposition. (Hint: Follow the construction of Fig. 6.2.3 in the proof of Theorem 6.2.1, but instead of connecting $\sigma_{w(i)}(j)$ to $\sigma_{w(i+1)}(j)$ with a 1-cell, connect $\sigma_{w(i)}(\rho(j))$ to $\sigma_{w(i+1)}(\rho'(j))$, where ρ and ρ' are reparametrizations as in the definition of the asynchronous fellow-traveller property.)

Examples 7.1.4 (i) The upper bound described in Theorem 3.1.4 was established in [23] using the combings constructed in [17]. Given $\Gamma = \mathbb{Z}^m \rtimes_\phi \langle t\rangle$ one can write each $\gamma \in \Gamma$ uniquely in the form $t^n x$ with $x \in \mathbb{Z}^m$. One fixes a basis for $\mathbb{Z}^m$ and represents x by a word l_x that (viewed as a path in the lattice $\mathbb{Z}^m$) stays closest to the Euclidean segment $[0, x]$ in $\mathbb{R}^m = \mathbb{Z}^m \otimes \mathbb{R}$. One then defines $\sigma_\gamma = t^n l_x$, checks that σ satisfies the asynchronous fellow-traveller property and calculates that $L_\sigma(n) \simeq n\,\|\phi^n\|$ (see [23] page 215).

(ii) I proved in [17] that if a compact 3-manifold M satisfies the geometrization conjecture, then $\pi_1 M$ admits a combing that satisfies the asynchronous fellow-traveller property, whence the exponential upper bound in Theorem 3.3.1.

7.2 Lower bounds

t-corridors and t-rings. *t-corridors* and *t-rings* are particular types of subdiagrams that one gets in van Kampen diagrams over presentations $\langle \mathcal{A}, t \mid \mathcal{R}\rangle$ where the group presented retracts onto $\langle t\rangle$. We refer to [21] for a careful treatment, but point out that although this is where *t-corridors* were named and systematized, they were in use much earlier, e.g. in Rips's geometric proof of the unsolvability of the word problem (see the inside cover of [94]).

Consider the presentation $\Gamma = \langle \mathcal{A}, t_1, \dots, t_n \mid \mathcal{R}\rangle$, where the symbols t_j are not elements of $\mathcal{A}$ and the only relators involving any t_j are of the form $t_j u_i t_j^{-1} v_i \in \mathcal{R}$, where $u_i, v_i \in F(\mathcal{A})$. Consider a van Kampen diagram D over such a presentation and focus on an edge ε in the boundary labelled $t \in \{t_1, \dots, t_n\}$.

If this edge lies in the boundary of a 2-cell, then the boundary cycle of this 2-cell (read with suitable orientation from ε) has the form $tut^{-1}v$ with $u, v \in F(\mathcal{A})$. In particular, there is a unique edge other than ε in the boundary of the 2-cell that is labelled t; crossing this edge we enter another 2-cell with a similar boundary label; by iterating the argument we get a chain of 2-cells running across the diagram; this chain terminates at an edge of ∂D which (following the orientation of ∂D in the direction of our original edge ε) is labelled t^{-1}. This chain of 2-cells is called a *t-corridor.*

Topologically, a t-corridor is a map $[0,1] \times [0,1] \to D$ that is injective on $[0,1] \times (0,1)$. We make this map a morphism of labelled combinatorial 2-complexes by pulling back the cell structure and labelling from D. The labels on the 1-cells in $[0,1] \times \{0,1\}$ (the top and bottom of the corridor) are letters from $\mathcal{A}^{\pm 1}$; the remaining 1-cells are of the form $\{s\} \times [0,1]$, and these are labelled t.

A *t-ring* is defined similarly: it consists of a chain of 2-cells giving a combinatorial map $\phi : \mathbb{S}^1 \times [0,1] \to D$ that is injective on $\mathbb{S}^1 \times (0,1)$; in $\mathbb{S}^1 \times [0,1]$ the 1-cells of the form $\{\theta\} \times [0,1]$ are labelled t; the remaining 1-cells are contained in $\mathbb{S}^1 \times \{0,1\}$ and are labelled by letters from $\mathcal{A}^{\pm 1}$; the map ϕ is label-preserving.

Much of the utility of t-corridors and t-rings rests on the following observations:

Exercise 7.2.1 Let t_i, u_i, v_i, Γ and D by as in the preceding discussion. Prove:

(i) Distinct t-corridors and t-rings have disjoint interiors.

(ii) If P is the edge-path in D running along the top or bottom of a t-corridor, then P is labelled by a word in the letters $\mathcal{A}^{\pm 1}$ that is equal in Γ to the words labelling the subarcs of ∂D which share the endpoints of P (given appropriate orientations),

(iii) and if $k = \min\{\max_i |u_i|, \max_i |v_i|\}$, then the number of 2-cells in the t-corridor is at least $1/k$ times the length of P.

(iv) The words labelling the inner and outer boundary cycles of a t-ring are null-homotopic.

(v) If D contains a 2-cell that has an edge labelled t in its boundary, then D contains either a t-corridor or a t-ring.

Instead of indulging in a general discussion, let me give one proposition to illustrate the utility of t-corridors and one to illustrate the utility of t-rings.

Proposition 7.2.2 *Let ϕ be an automorphism of the finitely presented group $B = \langle \mathcal{A} \mid \mathcal{S} \rangle$. For each $a \in \mathcal{A}$, choose a word $v_a \in F(\mathcal{A})$ representing $\phi(a) \in B$. Let $\mathcal{R} = \mathcal{S} \cup \{t_j^{-1} a t_j = v_a \mid a \in \mathcal{A}, j = 1,2\}$ and define $\Gamma := \langle \mathcal{A}, t_1, t_2 \mid \mathcal{R} \rangle$. Then the Dehn function of Γ is $\simeq$ bounded below by*

$$n \mapsto n \max_b \{ d_{\mathcal{A}}(1, \phi^n(b)) \mid d_{\mathcal{A}}(1, b) \leqslant n \}.$$

Proof For each positive integer n, we choose a word β of length at most n in the generators $\mathcal{A}^{\pm 1}$ so as to maximize $d_{\mathcal{A}}(1, \phi^n(b))$, where b is the image of β in Γ. Let $u_n := t_1^{-n}\beta t_1^n$ and let $w_n := u_n(t_2t_1^{-1})^n u_n^{-1}(t_2t_1^{-1})^{-n}$, a word of length at most $10n$. Note that $w_n = 1$ in Γ. Note also that no proper subword of w_n is equal to $1 \in \Gamma$ (one sees this easily using the natural retraction $\Gamma \to F(t_1, t_2)$ and the fact that $t_j^{-i}\beta t_j^i \neq 1$ for all i). It follows that any van Kampen diagram for w_n is a disc, in particular every edge of ∂D lies in the closure of some 2-cell, and therefore a t_j-corridor emanates from each edge of ∂D labelled t_j, for $j = 1, 2$.

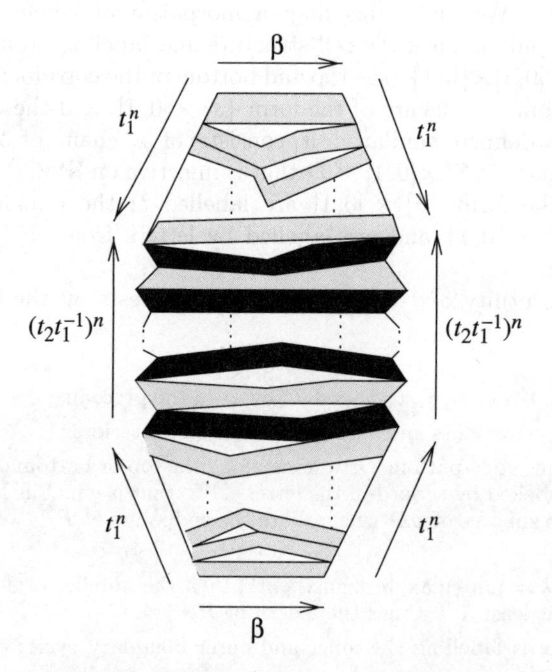

Fig. 7.2.3 The pattern of t_j-corridors

The simple fact that distinct t_j-corridors cannot cross (fact 7.2.1(i)) implies that the pattern of t_2-corridors in any van Kampen diagram for w_n must be as shown in Fig. 7.2.3. The words in the letters $\mathcal{A}^{\pm 1}$ labelling the bottom of each of each t_2-corridor is equal in Γ to u_n. Hence (fact 7.2.1(iii)) each of these corridors contains at least $(1/k)\, d_{\mathcal{A}}(1, \phi^n(b))$ 2-cells, where k is the length of the longest of the words v_a. And there are n such corridors. □

Exercises 7.2.4 (i) Let $\phi \in \mathrm{GL}(n, \mathbb{Z})$ be a unipotent matrix and let $\Gamma = \mathbb{Z}^m \rtimes F(t_1, t_2)$, where the generators t_1 and t_2 of the free group $F(t_1, t_2)$ both act on $\mathbb{Z}^m$ as ϕ. Deduce from the above proposition and your proof of (3.1.5) that the Dehn function of Γ is bounded below by a polynomial of degree $c + 1$, where c is the size of the largest elementary block in the Jordan form of ϕ. Adapt 7.1.4(i) to deduce that in fact $\delta_\Gamma(n) \simeq n^{c+1}$. (If you get stuck, refer to [18].)

(ii) Britton's lemma states that, given an HNN extension $G*_\phi = (G, t \mid t^{-1}st = \phi(s), \forall s \in S)$, and a generating set $\mathcal{A}$ for G, every null-homotopic word in the letters $(\mathcal{A} \cup \{t\})^{\pm 1}$ either contains no occurrences of $t^{\pm 1}$, or else contains a subword $t^\varepsilon u t^{-\varepsilon}$, where $\varepsilon = \pm 1$ and u is a word in the letters $\mathcal{A}^{\pm 1}$ that lies in $\langle S \rangle$ if $\varepsilon = -1$ and lies in $\langle \phi(S) \rangle$ if $\varepsilon = 1$.

Use t-corridors to prove Britton's lemma.

Proposition 7.2.5 *Let G be a finitely presented group, let $L, L' \subset G$ be finitely generated subgroups that are free, let $\phi : L \to L'$ be an isomorphism, and let $\Gamma = G*_\phi$ be the associated HNN extension. Then $\delta_\Gamma \succeq \delta_G$.*

Exercises 7.2.6 (i) Let D be a van Kampen diagram for a null-homotopic word w over a presentation $\langle \mathcal{A} \mid \mathcal{R} \rangle$, and let u be the label on a simple closed loop c in the 1-skeleton of D. Prove that if $\text{Area}_c D = \text{Area}_a w$, then the number of 2-cells in the sub-diagram enclosed by c is $\text{Area}_a u$. Extend this result to non-crossing loops.[28]

(ii) Prove Proposition 7.2.5. (Hint: $\Gamma = \langle \mathcal{A}, t \mid \mathcal{R}, t^{-1}lt = \phi(l), l \in S \rangle$ where $G = \langle \mathcal{A} \mid \mathcal{R} \rangle$ and $S \subset \mathcal{A}$ is a basis for L. Given a word $w \in F(\mathcal{A})$ with $w = 1$ in Γ, take a van Kampen diagram D with $\text{Area}_c D = \text{Area}_a w$. Use (i) and the fact that L is free to argue that D contains no t-rings and hence is a diagram over $\langle \mathcal{A} \mid \mathcal{R} \rangle$.)

Cohomological methods. Both Gersten and Gromov have developed cohomological methods for obtaining lower bounds on Dehn functions. In particular, Gersten [48] developed an ℓ_∞-cohomology theory which, among other things, allows one to recover results obtained using t-corridors in a more elegant and systematic manner. It would take too long to explain these ideas here, so we refer the reader to [48]. We content ourselves with a more simple-minded result that uses de Rham cohomology. (We give this result in part because it resonates with ideas in Section 5).

The statement of the following lemma is phrased in the vocabulary introduced in (4.3). The constant A_ω is defined to be the maximum of the integrals $\int_E \omega$ where E is a 2-cell mapped into $\tilde{M}$ by $\tilde{K}(\mathcal{A} : \mathcal{R}) \to \tilde{M}$.

Lemma 7.2.7 *Let M be a smooth, closed Riemannian manifold with fundamental group $\Gamma = \langle \mathcal{A} \mid \mathcal{R} \rangle$ and let ω be a Γ-invariant closed 2-form on $\tilde{M}$. If D is a van Kampen diagram with boundary label w, and h_D is the map described*

[28] A *non-crossing loop* is the restriction to $S^1 \times \{1\}$ of a map $S^1 \times [0,1] \to \mathbb{R}^2$ that is injective on $S^1 \times [0,1)$.

in (4.3), *then*

$$\int_D h_D^* \omega \leqslant A_\omega \ \mathrm{Area}_a(w).$$

Proof The integral $\int_D h_D^* \omega$ is well-defined because h_D is differentiable except on a set of measure zero. If D' is a second van Kampen diagram for w, then one can regard $-h_D \cup h_{D'}$ as a 2-cycle in $\tilde{M}$, and hence $\int_D h_D^* \omega = \int_{D'} h_{D'}^* \omega$. And when $\mathrm{Area}_c\, D = \mathrm{Area}_a(w)$ the inequality is clear. □

The utility of this lemma stems from the fact that one does not need to understand the nature of least-area van Kampen diagrams in order to get a lower bound on their area: if one can locate *any* van Kampen diagram D for a given word w, then one gets a lower bound on $\mathrm{Area}_a(w)$ by integrating $h_D^* \omega$ over D. Moreover, by Stokes theorem, if the 2-form ω is exact, say $\omega = d\alpha$, then one can simply calculate $\int_{\hat{w}} \alpha$, thus avoiding the construction of diagrams altogether.

Example 7.2.8 In the case where $\phi \in \mathrm{Sp}(m, \mathbb{Z})$, Bridson and Pittet [23] established the lower bound in Theorem 3.1.4 by applying Lemma 7.2.7 to the standard symplectic form on $\mathbb{R}^m$.

Exploiting asphericity. A group presentation $\langle \mathcal{A} \mid \mathcal{R} \rangle$ is called *aspherical* if the associated 2-complex $K(\mathcal{A}; \mathcal{R})$ is aspherical (i.e. its universal covering is contractible). One of the great joys of working with aspherical presentations is that when one finds an *embedded* van Kampen diagram one knows that it is of minimal area:

Lemma 7.2.9 *Suppose that* $X = K(\mathcal{A}; \mathcal{R})$ *is aspherical. Let* D *be a van Kampen diagram for* w. *If the associated map* $D \to \tilde{X}$ *is injective on the complement of the 1-skeleton* $D^{(1)}$, *then the number of 2-cells in* D *is* $\mathrm{Area}_a(w)$.

Proof Let D' be a second van Kampen diagram for w. One can regard $-D \cup D'$ as a 2-cycle in the cellular chain complex of $\tilde{X}$. Since there are no 3-cells and $H_2 \tilde{X}$ is trivial (by Hurewicz), this 2-cycle must be zero. And since the 2-cells in the image of D are all distinct, each must cancel with some 2-cell in D'. Hence $\mathrm{Area}_c D \leqslant \mathrm{Area}_c D'$. And since D' was arbitrary, $\mathrm{Area}_a(w) = \mathrm{Area}_c\, D$. □

Examples 7.2.10 (i) $\mathbb{Z}^2 = \langle a, b \mid [a, b] \rangle$ is aspherical. Hence the area of the obvious (square) diagram for $w_n = a^{-n} b^{-n} a^n b^n$ equals $\mathrm{Area}_a(w_n)$ (cf. 3.1.2).

(ii) The presentation of B_m described in (3.2.1) is aspherical. Explicit disc diagrams show that $\delta_{B_m}(n) \succeq \varepsilon_m(n)$ — see Exercise 7.2.11.

(iii) A celebrated theorem of Roger Lyndon shows that 1-relator presentations are aspherical if the relation is not a proper power [66].

(iv) The natural presentations of free-by-free groups are aspherical and provide interesting examples of Dehn functions [20].

Exercises 7.2.11 (i) Let X denote the universal covering of the standard 2-complex of the presentation of B_m described in (3.2.1). The 1-skeleton $X^{(1)}$ is identified with the Cayley graph of B_m. Show that the loop in $X^{(1)}$ labelled $x_1^{-n}x_0x_1^nx_0^{-2^n}$ bounds an embedded disc $\Delta_n(x_0, x_1)$ that has $(2^n - 1)$ faces (2-cells). By juxtaposing two copies of $\Delta_n(x_0, x_1)$, construct a disc D_1 showing that $x_1^{-n}x_0x_1^nx_0x_1^{-n}x_0^{-1}x_1^nx_0^{-1}$ is a null-homotopic word of area $2(2^n - 1)$.

(ii) Now suppose that $n = 2^r$. By attaching four copies of a disc diagram $\Delta_r(x_1, x_2)$ to the segments of ∂D_1 labelled x_1^n, construct a disc diagram for $(x_2^{-r}x_1^{-1}x_2^r)x_0(x_2^{-r}x_1x_2^r)x_0(x_2^{-r}x_1^{-1}x_2^r)x_0^{-1}(x_2^{-r}x_1x_2^r)x_0^{-1}$ that has more than 2^{2^r} faces (2-cells).

Iterate this construction and use Lemma 7.2.9 to deduce that $\delta_{B_m}(n) \succeq \varepsilon_m(n)$.

Reprove this inequality using t-corridors instead of asphericity.

The following exercises lead the reader through the proof that the Dehn function of the group $G_{p,q}$ described in 3.1.10 is $\succeq n^{2\log_2 2p/q}$. If you get stuck during these exercises, refer to [15].

Exercises 7.2.12 (i) Let $f, g : [0, \infty) \to [0, \infty)$ be non-decreasing functions and let (n_i) be an increasing sequence of positive integers with $n_0 = 0$ and $n_{i+1} \leqslant Cn_i$ for all i, where $C > 0$ is constant. Show that if $f(n_i) \leqslant g(n_i)$ for all i, then $f \preccurlyeq g$. (Thus we see that to establish lower bounds on Dehn functions $\delta(n)$, it is only necessary to look at fairly sparse sequences of integers (n_i).)

(ii) Consider the presentation of $G_{p,q}$ given in (3.1.10). Prove that this presentation is aspherical. (Hint: One can build the 2-complex of the presentation as follows. Start with a torus corresponding to the subgroup $\text{gp}\{a, b\} \subseteq G_{p,q}$ and fix a basepoint on it. Attach two cylinders (annuli) to the torus along simple curves through the basepoint – one end of each cylinder traces out a curve in the homotopy class a^q and the other ends trace out $a^p b^{\pm 1}$. The Seifert–van Kampen theorem shows that this complex has fundamental group $G_{p,q}$. The universal cover $\tilde{X}$ of this 2-complex is a contractible complex obtained by gluing planes indexed by the cosets of $\text{gp}\{a, b\} \subset G_{p,q}$ along strips (copies of the line cross an interval) covering the annuli in the quotient.)

(iii) Complete the following outline to a proof that the Dehn function of $G_{p,q}$ is $\succeq n^\alpha$ where $\alpha = 2\log_2 2p/q$.

Let $w_0 = a^q$ and let $w_1 = sa^qs^{-1}ta^qt^{-1}$. Define words $w_k = sw_{k-1}a^{\epsilon_{k-1}} \cdot s^{-1}tw_{k-1}a^{\epsilon_{k-1}}t^{-1}$ with $0 \leq \epsilon_{k-1} \leq q - 1$ so that $w_{k-1}a^{\epsilon_{k-1}}$ represents a power of a that is divisible by q. Show that $4(2^k) \leq |w_k| \leq (4q)2^k$ and that $w_k = a^{m_k}$ in $G_{p,q}$, where $m_k \geqslant q(2p/q)^k$.

Show that one can find embedded in $\tilde{X}$ a van Kampen diagram portraying the equality $w_k = a^{m_k}$. (See Fig. 7.2.13 – the large faces in this figure are diagrams over the subpresentation $\langle a, b \mid [a, b]\rangle$.)

Let $W_k = [sw_{k-1}a^{\epsilon_{k-1}}s^{-1}, tw_{k-1}a^{\epsilon_{k-1}}t^{-1}]$. Show that W_k represents the identity in $G_{p,q}$ and describe a van Kampen diagram for W_k that embeds in $\tilde{X}$. Deduce that there is a constant $C > 0$ such that

$$\text{Area}_a(W_k) \geq Cm_k^2 \geq Cq^2(2p/q)^{2k}.$$

Use (i) to conclude that the Dehn function of $G_{p,q}$ is bounded below by $n \mapsto n^{2\log_2 2p/q}$.

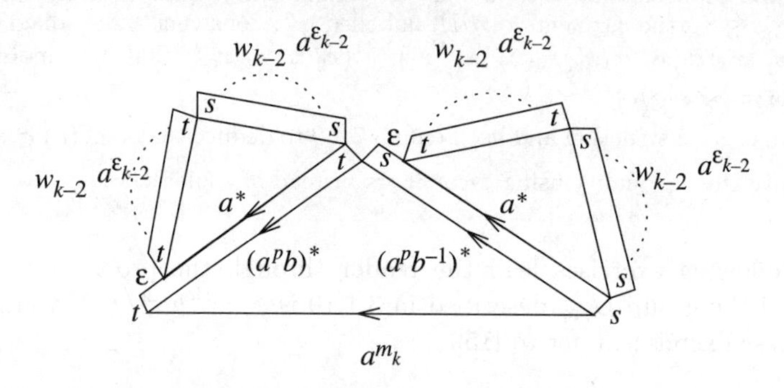

Fig. 7.2.13 The diagram portraying $w_k = a^{m_k}$

Calculating in abelian quotients. Let $\Gamma = \langle \mathcal{A} \mid \mathcal{R} \rangle$ and let K be the subgroup of $F = F(\mathcal{A})$ generated by the set of elements $\mathcal{C} = \{x^{-1}rx \mid x \in F(\mathcal{A}),\ r \in \mathcal{R}\}$. By definition, $\text{Area}_a\, w$ is the least number N for which there is an equality $w = c_1 \dots c_N$ with $c_i \in \mathcal{C}^{\pm 1}$. One anticipates that the task of estimating N would be easier if one were working with sums in abelian groups rather than products in free groups. With this in mind (and motivated by results of Gersten [45]) Baumslag, Miller, and Short [10] look at the projection of equalities such as the one above into the abelianization of K, i.e. the *relation module*[29] of the presentation $\langle \mathcal{A} \mid \mathcal{R} \rangle$. They also consider what happens when one projects further, onto $K/[K, F]$.

Thus they define the *abelianized isoperimetric function* Φ_Γ^{ab} by analogy with the Dehn function (1.2.2), replacing $\text{Area}_a\, w$ by $\text{Area}_a^{\text{ab}} w$, which is defined to be the least integer N for which there is an equality

$$w = \sum_{i=1}^{N} x_i^{-1} r_i x_i$$

in $K/[K, K]$, with $r_i \in \mathcal{R}^{\pm 1}$ and $x_i \in F$. And they define the *centralized isoperimetric function* $\Phi_\Gamma^{\text{cent}}$ by counting the minimum number of summands required to express w in $K/[K, F]$. Baumslag *et al.* prove that each of these functions is $\simeq$ independent of the chosen finite presentation of Γ.

[29] The conjugation action of F on K induces an action of Γ on $K/[K, K]$, hence the module structure.

Note the obvious inequalities

$$\Phi_\Gamma^{\text{cent}} \preceq \Phi_\Gamma^{\text{ab}} \preceq \delta_\Gamma.$$

From the Hopf formula ([25] page 41) one sees that $K/[K,F]$ is a direct sum of a free abelian group and $H_2(\Gamma,\mathbb{Z})$, and there is a well-developed technology for calculating in $H_2(\Gamma,\mathbb{Z})$ – in particular one has Fox's free differential calculus. By using this calculus Baumslag *et al.* obtain bounds on $\Phi_\Gamma^{\text{cent}}$ for various groups. In certain cases they are also able to show that $\Phi_\Gamma^{\text{cent}} \simeq \delta_\Gamma$. In this way they were able to calculate the Dehn functions of free nilpotent groups, thus exemplifying the merits of the aphorism that the homological approach works best for groups that contain a lot of commutivity.

Exercise 7.2.14 Observe that the argument given in (7.2.9) actually shows that if $D = \sum_{i=1}^N c_i$ in the cellular chain complex of $\tilde{X}$ then $N \geqslant \text{Area}_a(w)$ (where the c_i are 2-cells). Deduce from (7.2.12) that $\Phi_{G_{p,q}}^{\text{ab}}(n)$ is $\succeq n^{2\log_2(2p/q)}$.

8 Other decision problems and measures of complexity

8.1 Alternative analyses of the word problem

We saw in Section 1 that the Dehn function measures one's likelihood of success when one mounts a direct attack on the word problem for a finitely presented group. But there are other interesting ways to measure the complexity of the word problem. For example, instead of focusing on the area of van Kampen diagrams one might focus on some other aspect of their geometry, such as their diameter or the radius of the largest ball in the interior of the diagram. One might also bound the length of the intermediate words that arise during the process of applying relations to reduce a null-homotopic word to the empty word – 'filling length'. In Chapters 4 and 5 of [56] Gromov discusses many measures of complexity such as these, and there has been some interesting work on their interdependency (e.g. [46], [18], and [50]). Let me describe the most widely studied of these alternatives, which relates to the diameter of filling-discs in Riemannian manifolds.

Definition 8.1.1 *Let $\langle \mathcal{A} \mid \mathcal{R} \rangle$ be a finite presentation for the group Γ. Let w be a word that equals 1 in Γ and let D be a van Kampen diagram for w. Let p be the basepoint of D. Endow the 1-skeleton of D with a path metric ρ that gives each edge length 1. The diameter of w is defined by*

$$\text{diam}(w) := \min_D \max_q \{\rho(p,q) \mid q \textit{ a vertex of } D\}.$$

The (unreduced) isodiametric function of $\langle \mathcal{A} \mid \mathcal{R} \rangle$ is

$$\Psi(n) := \max_{|w| \leqslant n} \text{diam}(w).$$

The $\simeq$ equivalence class of Ψ depends only on Γ (see [46]) and is denoted Ψ_Γ.

Isodiametric functions turn out to be as unconstrained in nature as Dehn functions (3.1.11), see [95]. They can be interpreted in the following purely algebraic manner.

Proposition 8.1.2 $\mathrm{diam}(w) = \min_\Pi \max |x_i|$, *where the minimum is taken over all free equalities of the form*

$$w = \prod_{i=1}^{N} x_i^{-1} r x_i.$$

Exercises 8.1.3 (i) Deduce this proposition from the constructions in Section 4.

(ii) Use the diagrams constructed in (7.1.3) to show that if a group Γ admits a combing with the asynchronous fellow-traveller property, then $\Psi_\Gamma(n) \simeq n$.

(iii) Prove that $\Psi_\Gamma \preceq \delta_\Gamma$ for all finitely presented groups.

Steve Gersten and Daniel Cohen (independently) proved that for any group one can find constants $A, B > 0$ such that $\delta_\Gamma(n) \leqslant A^{B^{\Psi(n)}}$, and it is conjectured that in reality there is a single exponential bound. The relationship between Ψ_Γ and δ_Γ is complicated by the fact that in general the minima in the definitions of these functions will not be attained on the same family of diagrams: if one proceeds as in Definition 8.1.1 but quantifies only over least-area diagrams, then one obtains a function $\Psi_\Gamma^{\mathrm{ma}}$ that in general is $\prec \Psi_\Gamma$.

8.1.4 Extrinsic solutions to the word problem. In general, invariants based entirely on the geometry of van Kampen diagrams cannot give a full and accurate measure of the complexity of the word problem in a group because there might exist algorithms that require *extrinsic* structure that cannot be seen in a presentation. For example, one can solve the word problem for $B_1 = \langle x_0, x_1 \mid x_1^{-1} x_0 x_1 = x_0^2 \rangle$ in polynomial time by looking at the orbit of $\frac{1}{3} \in \mathbb{R}$ under the action $B_1 \to \mathrm{Aff}(\mathbb{R})$ described following (3.2.1), and yet $\delta_{B_1}(n) \simeq 2^n$.

If there is an embedding $\Gamma \hookrightarrow \hat{\Gamma}$ into a group whose Dehn function is smaller than that of Γ then one can apply the solution to the word problem in $\hat{\Gamma}$ to solve the word problem in Γ. Examples of this phenomenon are described in [8], [47] and [22] page 487. Remarkably, in [13] Birget, Ol'shanskii, Rips, and Sapir prove that such embeddings take full account of the complexity of the word problem in a precise sense that includes the following statement: the word problem of a finitely generated group G is an NP problem if and only if G is a subgroup of a finitely presented group that has a polynomial Dehn function.

8.2 Other decision problems

In this chapter we are concentrating on the word problem, but I should say a few words about the complexity of the other basic decision problems in group theory.

We fix a group Γ with a finite generating set $\mathcal{A}$. In order to solve the word problem one must decide which words in the letters $\mathcal{A}^{\pm 1}$ equal $1 \in \Gamma$. Two natural generalizations of this problem are:

(1) *The membership problem (relative word problem).* Instead of determining which words represent elements of the trivial subgroup, one is asked for an algorithm that decides which words represent elements of the subgroup $H \subset \Gamma$ generated by a specified finite subset of Γ.

(2) *The conjugacy problem.* Instead of determining which words represent elements conjugate to the identity, one is asked for an algorithm that decides which pairs of words represent conjugate elements of Γ.

Just as solving the word problem in Γ amounts to finding discs with a specified boundary loop in a closed manifold M with $\pi_1 M = \Gamma$, so the conjugacy problem amounts to finding annuli whose boundary is a specified pair of loops; minimizing the thickness of the annulus corresponds to bounding the length of the conjugating element. In the same vein, the membership problem is analogous to determining which paths can be homotoped (rel endpoints) into a given subspace of M.

There are various constructions connecting the word, conjugacy, and membership problems — see [73] and [9]. The following fibre product construction provides a particularly nice example as it can be modelled readily in geometric settings.

Exercise 8.2.1 Let $\Gamma = \langle \mathcal{A} \mid \mathcal{R} \rangle$ be a finitely presented group and let $D \subset F(\mathcal{A}) \times F(\mathcal{A})$ be the subgroup $\{(w, w') \mid w = w' \text{ in } \Gamma\}$. Show that D is finitely generated. Explain why solving the word problem for Γ is equivalent to solving the membership problem for D. Show that if one cannot solve the word problem in Γ then one cannot solve the conjugacy problem in D. (Hint: Fix $r \in \mathcal{R}$. Given a word w in the generators of $F(\mathcal{A}) \times \{1\}$, express the element $w^{-1}(r, r)w$ as a word in your chosen generators of D. When is the word you have created conjugate to (r, r) in D?)

If Γ is infinite then the group D in the above exercise is not finitely presentable (see [57]). For finitely presented examples and variations of a more geometric nature, see [9].

Remark 8.2.2 The conjugacy problem is considerably more delicate than the word problem in general. For example, in contrast to the fact that the complexity of the word problem for a group remains essentially unchanged when one passes to a subgroup or overgroup of finite index (1.3.5), Collins and Miller [32] constructed pairs of finitely presented groups $H \subset G$ such that $|G/H| = 2$

but H has a solvable conjugacy problem while G does not. They also show that one can arrange for G to have a solvable conjugacy problem when H does not.

The isomorphism problem. Roughly speaking, the isomorphism problem asks for an algorithm that will decide which finite presentations drawn from a specified list define isomorphic groups. The difficulty of this problem depends very much on the nature of the groups being presented. For example, Zlil Sela [96] proved that if one is given the knowledge that all of the groups being presented are the fundamental groups of closed negatively curved manifolds, then there is an algorithm that one can run to decide which of the groups are isomorphic. In contrast, it is unknown if there exists such an algorithm when one weakens the curvature condition to allow non-positively curved manifolds. Indeed there are very few natural contexts in which the isomorphism problem has been solved. (Note that in order to solve the isomorphism problem in a given class of groups it is not enough to have an algorithm that determines which presentations give the trivial group; for example, there is an algorithm to decide whether presentations of automatic groups determine the trivial group (chapter 5 of [42]) but this does not lead to a solution of the isomorphism problem in this class of groups.)

The following construction illustrates how HNN extensions can be used to translate word problems into other sorts of decision problems.

Exercise 8.2.3 Let $\Gamma = \langle \mathcal{A} \mid \mathcal{R} \rangle$ be a finitely presented group that is not free. Suppose that $\mathcal{A} = \{a_1, \dots, a_n\}$ where each a_i has infinite order in Γ (this can be arranged by replacing Γ with $\Gamma * \mathbb{Z}$ if necessary). Consider the following sequence of finite presentations indexed by words $w \in F(\mathcal{A})$:

$$G_w = \langle a_1, t_1, \dots, a_n, t_n \mid \mathcal{R}, t_i^{-1} a_i t_i = w \text{ for } i = 1, \dots, n \rangle.$$

Show that G_w is a free group if and only if $w = 1$ in Γ.

Assuming that there exists a group with an unsolvable word problem, use this construction (or a variation on it) to show that there exist (recursive) classes of finite presentations such that there are no algorithms to decide which of the groups presented are free, are torsion-free, contain $\mathbb{Z}^2$ (or any other specified subgroup), can be generated by three elements, or admit a faithful representation into $\mathrm{SL}(n, \mathbb{Z})$.

8.3 Subgroup distortion

Following Gromov [56], we define the distortion of a pair of finitely generated groups $H \subset \Gamma$ to be the function $\rho : \mathbb{N} \to \mathbb{N}$, where $\rho(n)$ is the radius of the set of vertices in the Cayley graph of H that are a distance at most n from the identity in Γ. (One shows that, up to $\simeq$ equivalence, this function does not depend on the choice of generating sets.)

If Γ has a solvable word problem, then the membership problem for $H \subset \Gamma$ is solvable if and only if the *distortion* of H in Γ is a recursive function.

Examples 8.3.1 (i) If $\phi \in \mathrm{GL}(r, \mathbb{Z})$ has an eigenvalue of absolute value greater than 1, then $\mathbb{Z}^r$ is exponentially distorted in $\mathbb{Z}^r \rtimes_\phi \mathbb{Z}$.

(ii) Let G_c be as in (3.1.7). In [19] I proved that for all positive integers $a > b$ the distortion of G_b in $G_a *_{\langle z \rangle} G_b$, the group formed by amalgamating G_a and G_b along their centres, is $\simeq n^{\frac{a}{b}}$. In [85] Osin proves that one can also obtain arbitrary positive rational exponents a/b by considering subgroups of finitely generated nilpotent groups.

(iii) Let $G_{p,q}$ be as in (3.1.10). In [15] Brady and I proved that the distortion function of the torus subgroup $\langle a, b \rangle$ in $G_{p,q}$ is equivalent to n^α, where $\alpha = \log_2(2p/q)$.

Ol'shanskii and Sapir have established comprehensive results, analogous to Theorem 3.1.11, concerning the possible distortion functions of finitely presented subgroups — see [82], [84].

See [22] page 507 for an interpretation of subgroup distortion in terms of Riemannian geometry as well as a connection between subgroup distortion and Dehn functions. See [43] for a discussion of relative Dehn functions.

A Geometric realizations of finitely presented groups

This appendix contains a brief description of some of the basic constructions of geometric group theory. There are two main (inter-related) strands in geometric group theory: one seeks to understand groups by studying their actions on appropriate spaces, and one seeks understanding from the intrinsic geometry of (discrete, finitely generated) groups endowed with word metrics. We begin by introducing the latter approach.

A.1 Finitely generated groups and quasi-isometries

The following constructions allow one to regard finitely generated groups as geometric objects.

A.1.1 Word metrics and Cayley graphs. Given a group Γ with generating set $\mathcal{A}$, the first step towards realizing the intrinsic geometry of the group is to give Γ the *word metric* associated to $\mathcal{A}$: this is the metric obtained by defining $d_{\mathcal{A}}(\gamma_1, \gamma_2)$ to be the shortest word in the letters $\mathcal{A}^{\pm 1}$ that equals $\gamma_1^{-1}\gamma_2$ in Γ. The action of Γ on itself by left multiplication gives an embedding $\Gamma \to \mathrm{Isom}(\Gamma, d_{\mathcal{A}})$. (The action of $\gamma_0 \in G$ by right multiplication $\gamma \mapsto \gamma\gamma_0$ is an isometry only if γ_0 lies in the centre of Γ.)

The *Cayley graph*[30] of Γ with respect to $\mathcal{A}$, denoted $\mathcal{C}_{\mathcal{A}}(\Gamma)$, has vertex set Γ and has an edge connecting γ to γa for every $\gamma \in \Gamma$ and $a \in \mathcal{A}$. The edges of $\mathcal{C}_{\mathcal{A}}(\Gamma)$ are endowed with local metrics in which they have unit length, and $\mathcal{C}_{\mathcal{A}}(\Gamma)$ is turned into a geodesic space by defining the distance between each pair of points to be equal to the length of the shortest path joining them.

The word metrics associated to different finite generating sets $\mathcal{A}$ and $\mathcal{A}'$ of Γ are Lipschitz equivalent, i.e. there exists $\ell \geqslant 1$ such that $(1/\ell)d_{\mathcal{A}}(\gamma_1, \gamma_2) \leqslant d_{\mathcal{A}'}(\gamma_1, \gamma_2) \leqslant \ell\, d_{\mathcal{A}}(\gamma_1, \gamma_2)$ for all $\gamma_1, \gamma_2 \in \Gamma$. One sees this by expressing the elements of $\mathcal{A}$ as words in the generators $\mathcal{A}'$ and vice versa – the constant ℓ is the length of the longest word in the dictionary of translation.

The Cayley graphs associated to different finite generating sets are not homeomorphic in general, but they are quasi-isometric in the following sense.

Definition A.1.2 *A (not necessarily continuous) map $f : X \to X'$ between metric spaces is called a* quasi-isometry *if there exist constants $\lambda \geqslant 1, \epsilon \geqslant 0$, $C \geqslant 0$ such that every point of X' lies in the C-neighbourhood of $f(X)$ and*

$$\frac{1}{\lambda}d(x,y) - \epsilon \leqslant d(f(x), f(y)) \leqslant \lambda d(x,y) + \epsilon$$

for all $x, y \in X$.

Exercises A.1.3 (i) If there exists a quasi-isometry $X \to X'$ then X and X' are said to be quasi-isometric. Prove that being quasi-isometric is an equivalence relation on any set of metric spaces.

(ii) Show that if $\mathcal{A}$ and $\mathcal{A}'$ are finite generating sets for Γ, then $(\Gamma, d_{\mathcal{A}})$, $\mathcal{C}_{\mathcal{A}}(\Gamma)$, and $\mathcal{C}_{\mathcal{A}'}(\Gamma)$ are quasi-isometric.

(iii) When is a homomorphism between finitely generated groups a quasi-isometry?

Since the quasi-isometry type of a finitely generated group does not depend on a specific choice of generators, statements such as 'the finitely generated group Γ is quasi-isometric to the metric space Y' or 'the finitely generated groups Γ_1 and Γ_2 are quasi-isometric' are unambiguous.

One may view the inclusion $\Gamma \hookrightarrow \mathcal{C}_{\mathcal{A}}(\Gamma)$ in the following light: Γ acts by isometries on $\mathcal{C}_{\mathcal{A}}(\Gamma)$, the action of $\gamma_0 \in \Gamma$ sending the edge with label $a \in \mathcal{A}$ emanating from the vertex γ to the edge labelled a emanating from the vertex $\gamma_0\gamma$, and $\Gamma \hookrightarrow \mathcal{C}_{\mathcal{A}}(\Gamma)$ is the map $\gamma \mapsto \gamma \cdot 1$. This is a simple instance of the important observation that quasi-isometries arise naturally from group actions (see [22] page 140).

[30] This graph was introduced by Arthur Cayley in 1878 to study 'the quasi-geometrical' nature of (in his case, finite) groups. It played an important role in the seminal work of Max Dehn (1910) who gave it the name Gruppenbild.

Proposition A.1.4 [The Švarc–Milnor lemma] *If a group* Γ *acts properly and cocompactly by isometries on a length space* X, *then for every choice of basepoint* $x_0 \in X$ *the map* $\gamma \mapsto \gamma \cdot x_0$ *is a quasi-isometry.*

The fundamental group of any (locally simply-connected) space acts by deck transformations on the universal covering. If the space is a compact geodesic space and the universal covering is endowed with the induced length metric ([22] page 42), then this action is proper, cocompact, and by isometries. Thus we have:

Corollary A.1.5 *The fundamental group of any closed Riemannian manifold* M *is quasi-isometric to the universal covering* $\widetilde{M}$.

We note one other corollary of the Švarc–Milnor lemma:

Corollary A.1.6 *If* X_1 *and* X_2 *are length spaces and there is a finitely generated group* Γ *that acts properly and cocompactly by isometries on both* X_1 *and* X_2, *then* X_1 *and* X_2 *are quasi-isometric.*

Dehn functions behave well with respect to quasi-isometries (see [4] and compare with Proposition 1.3.3 above and pages 143 and 415 of [22]).

Proposition A.1.7 *If* Γ *is a finitely presented group and* Γ' *is a finitely generated group quasi-isometric to* Γ, *then* Γ' *is also finitely presented and the Dehn functions of* Γ *and* Γ' *are* $\simeq$ *equivalent.*

By combining this proposition with the preceding corollaries and the Filling theorem we obtain:

Theorem A.1.8 *If the universal coverings of two closed, smooth, Riemannian manifolds* M_1 *and* M_2 *are quasi-isometric, then the isoperimetric functions* $\mathrm{Fill}_0^{M_1}$ *and* $\mathrm{Fill}_0^{M_2}$ *are* $\simeq$ *equivalent.*

One can prove this result more directly by following Alonso's proof of (A.1.7) using the combinatorial approximation techniques developed in Section 5.

A.2 Realizing the geometry of finite presentations

We now focus on finitely presented groups. The following category of complexes and maps is more rigid than the CW category and lends itself well to arguments such as those that we saw in the section on van Kampen's lemma. The discussion here follows that of Appendix I.8.A in [22].

A.2.1 Combinatorial complexes. These complexes are topological objects with a specified combinatorial structure. They are defined by a recursion on dimension; the definition of an open cell is defined by a simultaneous recursion. If K_1 and K_2 are combinatorial complexes, then a continuous map $K_1 \to K_2$ is said to be *combinatorial* if its restriction to each open cell of K_1 is a homeomorphism onto an open cell of K_2.

A combinatorial complex of dimension 0 is simply a set with the discrete topology; each point is an open cell. Having defined $(n-1)$-dimensional combinatorial complexes and their open cells, one constructs n-dimensional combinatorial complexes as follows.

Take the disjoint union of an $(n-1)$-dimensional combinatorial complex $K^{(n-1)}$ and a family $(e_\lambda \mid \lambda \in \Lambda)$ of copies of closed n-dimensional discs. Suppose that for each $\lambda \in \Lambda$ a homeomorphism is given from ∂e_λ (a sphere) to an $(n-1)$-dimensional combinatorial complex S_λ, and that a combinatorial map $S_\lambda \to K^{(n-1)}$ is also given; let $\phi_\lambda : \partial e_\lambda \to K^{(n-1)}$ be the composition of these maps. Define K to be the quotient of $K^{(n-1)} \cup \coprod_\Lambda e_\lambda$ by the equivalence relation generated by $t \sim \phi_\lambda(t)$ for all $\lambda \in \Lambda$ and all $t \in \partial e_\lambda$. Then K, with the quotient topology, is an n-dimensional combinatorial complex whose open cells are the (images of) open cells in $K^{(n-1)}$ and the interiors of the e_λ.

In the case $n = 2$, if the circle S_λ has k 1-cells then e_λ is called a k-gon.

A.2.2 The standard 2-complex $K(\mathcal{A} : \mathcal{R})$. Associated to any group presentation $\langle \mathcal{A} \mid \mathcal{R} \rangle$ one has a 2-complex $K = K(\mathcal{A} : \mathcal{R})$ that is compact if and only if the presentation is finite. K has one vertex and it has one edge ε_a (oriented and labelled a) for each generator $a \in \mathcal{A}$; thus edge loops in the 1-skeleton of K are in 1–1 correspondence with words in the alphabet $\mathcal{A}^{\pm 1}$: the letter a^{-1} corresponds to traversing the edge ε_a in the direction opposite to its orientation, and the word $w = a_1 \dots a_n$ corresponds to the loop that is the concatenation of the directed edges $\varepsilon_{a_1}, \dots, \varepsilon_{a_n}$; one says that w *labels* this loop. The 2-cells e_r of K are indexed by the relations $r \in \mathcal{R}$; if $r = a_1 \dots a_n$ (as a reduced word) then e_r is attached along the loop labelled $a_1 \dots a_n$. The map that sends the homotopy class of ε_a to $a \in \Gamma$ gives an isomorphism $\pi_1 K(\mathcal{A} : \mathcal{R}) \cong \Gamma$ (by the Seifert–van Kampen theorem).

Γ acts on the universal covering $\tilde{K}$ of $K(\mathcal{A} : \mathcal{R})$ by deck transformations and there is a natural Γ-equivariant identification of the Cayley graph $\mathcal{C}_\mathcal{A}(\Gamma)$ with the 1-skeleton of $\tilde{K}$: fix a base vertex $v_0 \in \tilde{K}(\mathcal{A} : \mathcal{R})$, identify $\gamma \cdot v_0$ with γ, and identify the edge of $\mathcal{C}_\mathcal{A}(\Gamma)$ labelled a issuing from γ with the (directed) edge at $\gamma \cdot v_0$ in the pre-image of ε_a. This identification is label-preserving: for all words w and all $\gamma \in \Gamma$, there is a unique edge-path labelled w beginning at $\gamma \in \mathcal{C}_\mathcal{A}(\Gamma)$ and the image of this path in $\tilde{K}$ is the lift at $\gamma \cdot v_0$ of the loop in $K(\mathcal{A} : \mathcal{R})$ labelled w.

Exercise A.2.3 Prove that if $\mathcal{A}$ is finite and w is a reduced word in which a and a^{-1} both occur exactly once, for every $a \in \mathcal{A}$, then $K = K(\mathcal{A} : w)$ is a closed surface.

A.3 4-manifolds associated to finite presentations

Proposition A.3.1 *Every finitely presented group is the fundamental group of a closed 4-dimensional manifold.*

We indicate two proofs of this proposition, leaving the details to the reader.

Exercises A.3.2 (i) Given a presentation $\langle a_1, \ldots, a_n \mid r_1, \ldots, r_m \rangle$, consider the compact 4-manifold obtained by taking the connected sum W of n copies of $\mathbb{S}^1 \times \mathbb{S}^3$ and identify $\pi_1 W$ with the free group on $\{a_1, \ldots, a_n\}$. Remove open tubular neighbourhoods about m disjoint embedded loops in W whose homotopy classes correspond to the relators $r_i \in \pi_1 W$. Let W' be the resulting manifold with boundary. Use the Seifert–van Kampen theorem to show that by attaching m copies of $\mathbb{S}^2 \times \mathbb{D}^2$ to W' along $\partial W'$ one obtains a closed manifold whose fundamental group is $\langle a_1, \ldots, a_n \mid r_1, \ldots, r_m \rangle$.

(ii) Show that if $n \geqslant 4$ then one can embed any compact combinatorial 2-complex in $\mathbb{R}^n$ by a piecewise linear map. Apply this construction to $K(\mathcal{A}\colon \mathcal{R})$ and consider the boundary M of a regular neighbourhood. Argue that the natural map $\pi_1 M \to \langle \mathcal{A} \mid \mathcal{R} \rangle$ is an isomorphism if $n \geqslant 5$.

By performing constructions of the above type more carefully one can force the manifold to have additional structure. For example, in [52] Bob Gompf proves:

Theorem A.3.3 *Every finitely presented group is the fundamental group of a closed symplectic 4-manifold.*

A.4 Obtaining presentations from group actions

Whenever one realizes a group as the fundamental group of a (semi-locally simply-connected) space one has the action of the group by deck transformations on the universal covering of the space. Thus the constructions of $K(\mathcal{A} : \mathcal{R})$ and the manifolds considered above may be viewed as means of constructing group actions out of presentations. The following theorem shows that, conversely, group actions give rise to presentations.

Theorem A.4.1 *Let X be a topological space, let Γ be a group acting on X by homeomorphisms, and let $U \subset X$ be an open subset such that $X = \Gamma \cdot U$.*

(1) *If X is connected, then the set $S = \{\gamma \in \Gamma \mid \gamma \cdot U \cap U \neq \emptyset\}$ generates Γ.*
(2) *Let $\mathcal{A}_S$ be a set of symbols a_s indexed by S. If X and U are both path-connected and X is simply connected, then $\Gamma = \langle \mathcal{A}_S \mid \mathcal{R} \rangle$, where*

$$\mathcal{R} = \{a_{s_1} a_{s_2} a_{s_3}^{-1} \mid s_i \in S;\ U \cap s_1.U \cap s_3.U \neq \emptyset;\ s_1 s_2 = s_3 \text{ in } \Gamma\}.$$

Corollary A.4.2 *If a group Γ acts by isometries on a complete Riemannian manifold M, and if every point of M is a distance less than r from a certain orbit $\Gamma \cdot p$, then Γ can be presented as $\Gamma = \langle \mathcal{A} \mid \mathcal{R} \rangle$ where $\mathcal{A}$ is the set of elements $a \in \Gamma$ such that $d(p, a \cdot p) < 2r$ and $\mathcal{R}$ is the set of words in the letters $\mathcal{A}^{\pm 1}$ that have length at most 3 and are equal to the identity in Γ.*

Proof Apply the theorem with U the open ball of radius r about p. □

The above theorem has a long history. In this form it is due to Murray Macbeath [68]. See [22] page 136 for a proof and further information.

Exercises A.4.3 Establish the following geometric characterization of finitely presented groups: a group is finitely presented if and only if it acts properly and cocompactly by isometries on a simply-connected geodesic space.

Give an example to show that part (2) of the above theorem can fail if X is not simply connected.

B Length spaces

For the benefit of the reader unfamiliar with non-Riemannian length spaces we list some of the basic vocabulary of the subject.

Length metrics

Definition B.0.1 *Let X be a metric space. The* length $l(c)$ *of a curve $c\colon [a,b] \to X$ is*

$$l(c) = \sup_{a=t_0 \leqslant t_1 \leqslant \cdots \leqslant t_n = b} \sum_{i=0}^{n-1} d(c(t_i), c(t_{i+1})),$$

where the supremum is taken over all possible partitions (no bound on n) with $a = t_0 \leqslant t_1 \leqslant \cdots \leqslant t_n = b$.

$l(c)$ is either a non-negative number or it is infinite. The curve c is said to be *rectifiable* if its length is finite, and it is called a *geodesic*[31] if its length is equal to the distance between its endpoints.

A (connected) *length space* is a metric space X in which every pair of points $x, y \in X$ can be joined by a rectifiable curve and $d(x,y)$ is equal to the infimum of the length of rectifiable curves joining them; X is called a *geodesic space* if this infimum is always attained, i.e. each pair of points $x, y \in X$ can be joined by a geodesic. A general form of the Hopf–Rinow theorem (see [6] or [22]) states that if a length space is complete, connected, and locally compact, then it is a geodesic space (and all closed balls in it are compact).

Upper curvature bounds. Let M_k^2 denote the complete simply-connected 2-manifold of constant sectional curvature $k \in \mathbb{R}$. If $k = 0$ then M_k^2 is the Euclidean plane; if $k < 0$ then M_k^2 is the hyperbolic plane with the metric

[31] This differs from the standard usage in differential geometry, where being geodesic is a local concept. For this reason, some authors use the term 'length-minimizing geodesic' in the context of length spaces.

scaled by a factor of $1/\sqrt{-k}$; and if $k > 0$ then M_k^2 is $\mathbb{S}^2$ with the metric scaled by $1/\sqrt{k}$.

A *triangle* Δ in a metric space consists of three points x_1, x_2, x_3 (the vertices) and a choice of geodesic connecting each pair of these points.

A geodesic space X is said to have *curvature* $\leqslant k$ if every point $x \in X$ has a neighbourhood in which all triangles Δ satisfy the following property: the distance from each vertex of Δ to the midpoint of the opposite side is no greater than the corresponding distance in a triangle $\overline{\Delta} \subset M_k^2$ that has the same edge lengths as Δ. This definition is due to A.D. Alexandrov.

We refer the reader to [22] for a comprehensive introduction to (singular) spaces with upper curvature bounds.

Pull-back length metrics. Let D be a topological space. Associated to any continuous map $f : D \to X$ to a metric space one has the length pseudo-metric on D: the length of each curve in D is defined to be the length of its image under f, and the distance between two points of D is defined to be the infimum of the lengths of paths connecting them. We write (D, d_f) to denote the length space obtained by forming the quotient of this pseudo-metric space by the relation that identifies points that are a distance 0 apart. In general one can say little about the underlying space of (D, d_f); it certainly need not be homeomorphic to D.

If X is a smooth Riemannian manifold and $f : D \to X$ is a least-area disc with piecewise geodesic boundary, then (D, d_f) will be a singular disc and its curvature will be bounded above by the sectional curvature of X; if $f|_{\partial D}$ is injective, then (D, d_f) will actually be a disc. It can also be that (D, d_f) is a disc when f is not injective, for example if f is the map $z \mapsto z^2$ from the unit disc to the complex plane, then (D, d_f) is the metric completion of the connected 2-fold covering of the punctured unit disc.

C A proof of the Cellulation lemma

This appendix contains a proof of the following technical result that was needed in Section 5. Recall that a *singular disc* is a space homeomorphic to the underlying space of a singular disc diagram, as defined in (4.1).

C.0.1 Cellulation lemma *Let D be a length space homeomorphic to a (perhaps singular) 2-disc, and suppose that D is ε-filled by a set Σ of cardinality N. Then there exists a combinatorial 2-complex Φ, homeomorphic to the standard 2-disc, and a continuous map $\phi : \Phi \to D$ such that:*

(1) *Φ has less than $8N$ faces (2-cells) and each is a k-gon with $k \leqslant 12$;*
(2) *the restriction of ϕ to each 1-cell in Φ is a path of length at most 2ε;*
(3) *$\phi|_{\partial\Phi}$ is a monotone parametrization of ∂D and $\Sigma \cap \partial D$ lies in the image of the 0-skeleton of $\partial\Phi$.*

For convenience we rescale the metric on D and assume that $\varepsilon = 1$. To avoid complicating the terminology, we also assume that D is a non-singular disc (the concerned reader will have little difficulty in making the adjustments needed in the general case). We fix a set Σ of cardinality N that 1-fills D and define $\Sigma_0 = \Sigma \cap \partial D$ and $\Sigma_1 = \Sigma \smallsetminus \Sigma_0$.

C.1 Reducing to the case of thin discs

Our aim in the first stage of the proof is to reduce to the case where $\Sigma = \Sigma_0$. We shall do this by cutting D open along a certain graph whose vertex set has cardinality less than $2N$ and includes Σ. To this end, we view ∂D as a graph $\mathcal{G}_0$ with vertex set Σ_0 and 1-cells the closures of the connected components of $\partial D \smallsetminus \Sigma_0$.

Since every point of the connected space D lies in the 1-neighbourhood of Σ, the open neighbourhoods of radius 1 about Σ_0 and Σ_1 cannot be disjoint. Hence there exists $s \in \Sigma_1$ and $s' \in \Sigma_0$ with $d(s, s') < 2$. Choose a geodesic $[s, s']$ and consider a minimal subarc $[s, v]$ with $v \in \mathcal{G}_0$. We augment $\mathcal{G}_0$ (which is ∂D subdivided) by adding s and v as vertices and adding $[s, v]$ as a new edge (if v is not a vertex of $\mathcal{G}_0$ then its introduction will also subdivide one of the existing edges). Call the new graph $\mathcal{G}_0'$ and define $\Sigma_0' = \Sigma_0 \cup \{s\}$.

By repeating the above argument with Σ_0' in place of Σ_0, and $\mathcal{G}_0'$ in place of $\mathcal{G}_0$, we obtain a connected graph with at most $|\Sigma_0| + 4$ vertices including Σ_0 and two elements of Σ_1. We iterate this argument a further $|\Sigma_1| - 2$ times to obtain a connected graph $\mathcal{G}$ whose vertex set consists of Σ and at most $2|\Sigma_1|$ other vertices; the important point is that this graph has less than $2N$ vertices in total, and less than $2N$ edges. Note that the edges of $\mathcal{G}$ all have length at most 2, that $E := D \smallsetminus \mathcal{G}$ is homeomorphic to an open 2-disc, and that $T := \mathcal{G} \smallsetminus \partial D$ is a forest (i.e. it is simply-connected, but not necessarily connected).

We now focus our attention on E, which we endow with the induced path metric from D. Let Δ be the space obtained by completing this metric. Δ is homeomorphic to a 2-disc; intuitively speaking, it is obtained by cutting D open along the branches of T (cutting along each edge of T forms two edges in the boundary of Δ). The inclusion $E \hookrightarrow D$ extends continuously to a map $\pi : \Delta \to D$ that preserves the lengths of all curves and sends (a monotone parametrization of) $\partial\Delta$ onto the boundary cycle of E in $\mathcal{G}$; we endow $\partial\Delta$ with the combinatorial structure induced from this identification. Thus Δ is a topological 2-disc endowed with a length metric such that $\partial\Delta$ is the concatenation of less than $4N$ geodesic segments, each of length at most 2. Moreover, every point of Δ is a distance at most 1 from $\partial\Delta$. This completes the first stage of the proof.

Definition C.1.1 *A singular disc of weight* n *consists of a singular disc* Δ *and* n *distinguished points* (vertices) $x_1 = f(t_1), \ldots, x_n = f(t_n)$ *in cyclic order on the boundary cycle* $f : \mathbb{S}^1 \to \partial\Delta$; *the restriction of* f *to the arc joining* t_i *to* t_{i+1} *(indices mod* n*) is required to be a geodesic of length at most* 2; *the images* *of these arcs are called* facets. Δ *is said to be* thin *if every point is a distance less than 1 from* $\partial\Delta$.

A partition *of* Δ *is a continuous map* $\phi : \Phi \to \Delta$, *where* Φ *is a combinatorial 2-complex that is homeomorphic to the standard disc and* $\phi|_{\partial\Phi}$ *is a monotone parametrization of* f *sending vertices to vertices and edges to facets.*

Φ *is called a* k-partition *if each of its 2-cells is an* m*-gon with* $m \leqslant k$. *And* Φ *is said to be* admissible *if the restriction of* ϕ *to each* 1*-cell in* Φ *is a path of length at most* 2. *The* area *of* Φ *is the number of* 2*-cells in* Φ.

The final stage in the proof of the cellulation lemma is:

Proposition C.1.2 *If* $k \geqslant 12$, *then every thin singular disc of weight* n *admits a* k-partition *of area at most* $2n - 8$.

Before turning to the proof of this proposition, let us see how it implies the Cellulation lemma.

End of the proof of the Cellulation lemma. In the first stage of the proof we showed that if a disc can be ε-filled with a set of cardinality N then one can construct in D a graph $\mathcal{G}$ with at most $2N$ vertices so that the edges of the graph have length less than 2ε and the space obtained by cutting D open along the forest $T = \mathcal{G} \smallsetminus \partial D$ is a thin disc X of weight less than $4N$. The natural map $\pi : \Delta \to D$ is length-preserving.

The above proposition furnishes a 12-partition $\phi_0 : \Phi_0 \to \Delta$ of area at most $8N - 8$. Define Φ to be the combinatorial complex obtained by taking the quotient of Φ_0 by the equivalence relation that identifies the pair of edges in the pre-image of each edge of T in the obvious manner. Φ is a disc whose area (number of 2-cells) is the same that of Φ_0. The map $\phi : \Phi \to D$ induced by $\pi \circ \phi_0 : \Phi_0 \to D$ satisfies the requirements of the cellulation lemma. □

C.2 Surgery on thin discs

We shall prove Proposition C.1.2 by induction on n, the weight of the singular disc being filled. In this induction we shall need the following surgery operation.

Let Δ be a singular disc of weight n with boundary cycle $f : \mathbb{S}^1 \to \partial\Delta$. Given two vertices $x, y \in \partial\Delta$ one can cut Δ along a geodesic $[x, y]$ to form two new singular discs. To do this, first note that one can choose $[x, y]$ so that its intersection with each facet of $\partial\Delta$ is a single arc, because given the first and last points of intersection of an arbitrary geodesic $[x, y]'$ with a facet, one can replace the corresponding subarc of $[x, y]'$ with a subarc of the facet. Having chosen $[x, y]$ in this way, express y as $f(t)$ and proceed in the positive direction around $\mathbb{S}^1$ from t to the first value t' such that $f(t') = x$; let α denote this arc from t to t' and call the complementary arc β'.

The first of the two singular subdiscs into which we cut Δ is that whose boundary cycle is the concatenation of $f|_\alpha$ and $[x, y]$. The boundary cycle of the second subdisc is the concatenation of $f|_\beta$ and $[y, x]$. We subdivide $[x, y]$ into the minimal possible number of subarcs of length less than 2 and define these subarcs to be facets of our two new singular discs.

The reader should have no difficulty in verifying:

Lemma C.2.1 *In the notation of the preceding paragraph: if Δ is thin then the singular discs obtained by surgery are thin; and if $d(x, y) < 4$, then the sum of the weights of the new singular discs is at most $n + 4$.*

In the course of the proof of Proposition C.1.2 we shall require the following fact.

> *Exercise C.2.2* Let $X = U_1 \cup U_2 \cup U_3 \cup U_4$ be a metric space. Assume that each of the sets U_i is path-connected, that $d(U_i, U_j) > 0$ when $|i - j| = 2$, and that $U_i \cap U_j \neq \emptyset$ otherwise. Construct a surjective homomorphism $\pi_1 X \to \mathbb{Z}$. (Hint: Consider the map to $\mathbb{R}/\mathbb{Z}$ that is constant on $X \smallsetminus U_2$ and is given on U_2 by $x \mapsto d(x, U_1)/(d(x, U_1) + d(x, U_3))$.)

The proof of Proposition C.1.2. Let Δ be a singular disc of weight n that is thin. We proceed by induction on n. If $n \leqslant k$ there is nothing to prove.

Assuming $n \geqslant 12$, we express the boundary cycle $f : \mathbb{S}^1 \to \Delta$ as the concatenation of four subpaths, namely the first three facets taken together, the next three facets, then the next three, and then the remaining $n - 9$ facets. Define U_1, U_2, U_3, U_4 to be the closed neighbourhoods of radius 1 about the images of these four arcs. The union of these neighbourhoods is the whole of Δ (because it is assumed to be thin). The U_i cannot satisfy the hypotheses of the preceding exercise because Δ is simply connected. Therefore $U_i \cap U_j \neq \emptyset$ for some $i - j = 2$. (Here we need the fact that the metric on Δ is a path metric in order to know that the U_i are path-connected.)

Since U_i and U_j intersect, one of the vertices along our i-th arc, say x, is a distance at most 4 from one of the vertices along our j-th arc, say y. We separate Δ by surgery along $[x, y]$. Because all four of our subarcs contained at least 3 facets, and because we need only divide $[x, y]$ into two facets, the weights n' and n'' of the new singular discs Δ' and Δ'' obtained by surgery are both strictly less than n. Also (see the lemma) $n' + n'' \leqslant n + 4$.

By induction, there exist admissible k-partitions $\Phi' \to \Delta'$ and $\Phi'' \to \Delta''$ whose areas are at most $2n' - 8$ and $2n'' - 8$ respectively. Let Φ be the combinatorial disc obtained by gluing Φ' and Φ'' along the pre-images of $[x, y]$ in the obvious manner. The given maps $\Phi' \to \Delta'$ and $\Phi'' \to \Delta''$ define an admissible k-partition $\Phi \to \Delta$ whose area is the sum of the areas of Φ' and Φ''. In particular the area of Φ is at most $2(n' + n'') - 16 \leqslant 2(n + 4) - 16 = 2n - 8$, so the induction is complete. □

The bound $k \geqslant 12$ in Proposition C.1.2 can be improved at the expense of complicating the proof.

Acknowledgement. I first drafted a proof of the Filling theorem (Section 4) in 1994 with the intention of saving it for a panorama such as this. I was on the faculty at Princeton then and Fred Almgren was my neighbour in Fine Hall. He answered my questions on geometric measure theory with patience and took a stimulating interest in the connections with group theory. Although he is no longer with us, I wish to record my gratitude to Fred for these conversations.

References

1. A.D. Alexandrov: A theorem on triangles in a metric space and some of its applications, *Trudy Mat. Inst. Steklov* **38** (1951), 5–23.
2. D.J. Allcock: An isoperimetric inequality for the Heisenberg groups, *Geom. Funct. Anal.* **8** (1998), 219–233.
3. F.J. Almgren Jr: 'Plateau's problem: An invitation to varifold geometry', W. A. Benjamin, Inc. (New York, 1966).
4. J.M. Alonso: Inégalités isopérimétriques et quasi-isométries, *C.R.A.S. Paris Série 1* **311** (1990), 761–764.
5. J. Alonso and M.R. Bridson: Semihyperbolic groups, *Proc. London Math. Soc.* **70** (1995), 56–114.
6. W. Ballmann: 'Lectures on spaces of nonpositive curvature', DMV Seminar 25, Birkhäuser (Basel, 1995).
7. G. Baumslag, S. Gersten, M. Shapiro and H. Short: Automatic groups and amalgams, *J. Pure and Appl. Alg.* **76** (1991), 229–316.
8. G. Baumslag, M.R. Bridson, C.F. Miller III, H. Short: Finitely presented subgroups of automatic groups and their isoperimetric functions, *J. London Math. Soc.* (2) **56** (1997), 292–304.
9. G. Baumslag, M.R. Bridson, C.F. Miller III, H. Short: Fibre products, non-positive curvature and decision problems, *Comment. Math. Helv.* **75** (2000), 457–477.
10. G. Baumslag, C.F. Miller III and H. Short: Isoperimetric inequalities and the homology of groups, *Invent. Math.* **113** (1993), 531–560.
11. V.N. Berestovskii and I.G. Nikolaev: Multidimensional generalized Riemannian spaces, in *Geometry IV*, Encyclopedia of Mathematical Sciences, (Yu. G. Reshetnyak, ed), Springer **70** (1993), 165–243.
12. M. Bestvina and M. Feighn: A combination theorem for negatively curved groups, *J. Diff. Geom.* **35** (1992), 85–101.
13. J-C. Birget, A.Yu. Ol'shanskii, E. Rips, M. Sapir: Isoperimetric functions of groups and computational complexity of the word problem, *Ann. of Math.*, to appear.
14. B.H. Bowditch: A short proof that a sub-quadratic isoperimetric inequality implies a linear one, *Mich. J. Math.* **42** (1995), 103–107.
15. N. Brady and M.R. Bridson: There is only one gap in the isoperimetric spectrum, *GAFA (Geom. Funct. Anal.)* **10** (2000), 1053–1070.
16. M.R. Bridson: On the geometry of normal forms in discrete groups, *Proc. London Math. Soc.* (3) **67** (1993), 596–616.
17. M.R. Bridson: Combings of semidirect products and 3-manifold groups, *Geom. Funct. Anal.* **3** (1993), 263–278.
18. M.R. Bridson: Asymptotic cones and polynomial isoperimetric inequalities, *Topology* **38** (1999), 543–554.

19. M.R. Bridson: Fractional isoperimetric inequalities and subgroup distortion, *J. Amer. Math. Soc.* **12** (1999), 1103–1118.
20. M.R. Bridson: Polynomial Dehn functions and the length of asynchronous automatic structures, *Proc. London Math. Soc.*, to appear.
21. M.R. Bridson and S.M. Gersten: The optimal isoperimetric inequality for torus bundles over the circle, *Quart. J. Math. Oxford Ser.* (2) **47** (1996), 1–23.
22. M.R. Bridson and A. Haefliger: 'Metric spaces of non-positive curvature', Grundlehren der math Wiss., 319, Springer (Berlin, 1999).
23. M.R. Bridson and Ch. Pittet: Isoperimetric inequalities for the fundamental groups of torus bundles over the circle, *Geom. Dedicata* **49** (1994), 203–219.
24. M.R. Bridson and K. Vogtmann: On the geometry of the automorphism group of a free group, *Bull. London Math. Soc.* **27** (1995), 544–552.
25. K.S. Brown: 'Cohomology of groups', GTM 87, Springer (New York, 1982).
26. J. Burillo and J. Taback: The equivalence of geometric and combinatorial Dehn functions, *Proc. Amer. Math. Soc.*, to appear.
27. J.W. Cannon: The combinatorial structure of cocompact discrete hyperbolic groups, *Geom. Dedicata.* **16** (1984), 123–148.
28. J.W. Cannon: The theory of negatively curved spaces and groups, in 'Ergodic theory, symbolic dynamics and hyperbolic spaces' (T. Bedford, M. Keane, C. Series eds.), Oxford Univ. Press (New York, 1991), 315–369.
29. A. Cayley: On the theory of groups, *Proc. London Math. Soc.* **9** (1878), 126–133.
30. B. Chandler and W. Magnus: 'The history of combinatorial group theory', Springer (New York, 1982).
31. R. Charney: Artin groups of finite type are biautomatic, *Math. Ann.* **292** (1992), 671–683.
32. D.J. Collins and C.F. Miller III: The conjugacy problem and subgroups of finite index, *Proc. London Math. Soc.* **34** (1977), 535–556.
33. M. Coornaert, T. Delzant and A. Papadopoulos: 'Notes sur les groupes hyperboliques de Gromov', Springer Lect. Notes in Math. **1441** (1990)
34. M. Dehn: 'Papers on Group Theory and Topology', translated and introduced by John Stillwell, Springer (Berlin, 1987).
35. M. Dehn: Transformationen der Kurven auf zweiseitigen Flächen, *Math. Ann.* **2** (1912), 413–421.
36. M. Dehn: Über unendliche diskontinuierliche Gruppen, *Math. Ann.* **71** (1912), 116–144.
37. J. Douglas: Solution of the problem of Plateau, *Trans. Amer. Math. Soc.* **33** (1931), 263–321.
38. C. Druţu: Filling in solvable groups and in lattices in semisimple groups, Preprint, Université de Lille, 2000.
39. C. Druţu: Cônes asymptotiques et invariants de quasi-isométrie pour des espaces métriques hyperboliques, *Ann. Inst. Fourier* **51** (2001), 81–97.
40. C. Druţu: Remplissage dans des réseaux de Q-rang 1 et dans des groupes résolubles, *Pacific J. Math.* **185** (1998), 269–305.
41. C. Druţu: Quasi-isometry invariants and asymptotic cones, *Internat. J. Algebra Comput.*, to appear.
42. D.B.A. Epstein, J.W. Cannon, D.F. Holt, S.V.F. Levy, M.S. Paterson, W.P. Thurston: 'Word Processing in Groups', Jones and Bartlett (Boston, 1992).
43. B. Farb: The extrinsic geometry of subgroups and the generalized word problem, *Proc. London Math. Soc.* **68** (1994), 577–593.

44. H. Federer: 'Geometric Measure Theory', Springer (New York, 1969).
45. S.M. Gersten: Dehn functions and l_1-norms of finite presentations, in 'Algorithms and classification in combinatorial group theory' MSRI Publ. 23, Berkeley, CA, 1989, 195–224.
46. S.M. Gersten: Isoperimetric and isodiametric functions of finite presentations, in 'Geometric group theory', vol. 1, LMS lecture notes 181, (G. Niblo and M. Roller, eds.), Camb. Univ. Press (1993).
47. S.M. Gersten: Preservation and distortion of area in finitely presented groups, *GAFA (Geom. Funct. Anal.)* **6** (1996), 301–345.
48. S.M. Gersten: Cohomological lower bounds for isoperimetric functions on groups, *Topology* **37** (1998), 1031–1072.
49. S.M. Gersten: The double exponential theorem for isodiametric and isoperimetric functions, *Internat. J. Algebra Comput.* **1** (1991), 321–327.
50. S.M. Gersten and T. Riley: Filling length in finitely presentable groups, *Geom. Dedicata.*, to appear.
51. E. Ghys and P. de la Harpe (ed): 'Sur les groupes hyperboliques d'après Mikhael Gromov', Prog. in Math. **83**, Birkhäuser (Boston, 1990).
52. R.E. Gompf: A new construction of symplectic manifolds, *Ann. of Math.* **142** (1995), 527–595.
53. M. Gromov: Infinite groups as geometric objects, Proc. ICM, Warsaw 1978, PWN (Warsaw, 1984), 385–392.
54. M. Gromov: 'Partial Differential Relations', Springer (Berlin, 1986).
55. M. Gromov: Hyperbolic groups, in 'Essays on group theory', MSRI Publ. No. 8 (S.M. Gersten, ed.), Springer (New York, 1987).
56. M. Gromov: 'Asymptotic invariants of infinite groups', Geometric group theory, vol. 2, LMS lecture notes 182, (G. Niblo and M. Roller, eds.), Camb. Univ. Press (1993).
57. F. Grunewald: On some groups which cannot be finitely presented, *J. London Math. Soc.* **17** (1978), 427–436.
58. A. Hatcher and K. Vogtmann: Isoperimetric inequalities for automorphism groups of free groups, *Pacific J. Math.* **173** (1996), 425–441.
59. C. Hidber: Isoperimetric functions of finitely generated nilpotent groups and their amalgams, PhD Thesis 12166, ETH Zürich, 1997.
60. S. Hildebrandt: Boundary value problems for minimal surfaces, in 'Geometry V', Encyclopaedia Math. Sci. vol. 90, Springer (Berlin, 1997), 153–237.
61. E.R. van Kampen: On some lemmas in the theory of groups, *Amer. J. Math.* **55** (1933), 268–273.
62. E. Leuzinger and Ch. Pittet: Isoperimetric inequalities for lattices in semisimple Lie groups of rank 2, *GAFA* (*Geom. Funct. Anal.*) **6** (1996), 489–511.
63. E. Leuzinger and Ch. Pittet: Quadratic Dehn functions for solvable groups, Preprint, Toulouse, 2000.
64. R.C. Lyndon: Cohomology theory of groups with a single defining relation, *Annals of Math.* **52** (1950), 650–665.
65. R.C. Lyndon: On Dehn's algorithm, *Math. Ann.*, **166** (1966), 208–228.
66. R.C. Lyndon and P.E. Schupp: 'Combinatorial Group Theory', Springer (New York, 1977).
67. I. G. Lysenok: On some algorithmic properties of hyperbolic groups, *Math. USSR Izv.* **35** (1990), 145–163. (English translation of Izv. **53** (1989))
68. A.M. Macbeath: Groups of homeomorphisms of a simply connected space, *Ann. of Math.* **79** (1964), 473–488.

69. N. Macura: Quadratic isoperimetric inequality for mapping tori of polynomially growing automorphisms of free groups, GAFA (*Geom. Funct. Anal.*) **10** (2000), 874–901.
70. A.A. Markov: Insolubility of the problem of homeomorphy, Proc. ICM Cambridge 1958, pp. 300–306, Cambridge Univ. Press., 1960.
71. W.H. Meeks III and S-T Yau: The classical Plateau problem and the topology of three-dimensional manifolds, *Topology* **21** (1982), 409–442.
72. C.F. Miller III: 'On group-theoretic decision problems and their classification', Annals of Math. Studies, No. 68, Princeton University Press (1971).
73. C.F. Miller III: Decision problems for groups: survey and reflections, in 'Algorithms and Classification in Combinatorial Group Theory' (eds. G. Baumslag and C.F. Miller III), MSRI Publications No. 23, Springer (New York, 1992), 1–59.
74. J. Milnor: A note on the fundamental group, *J. Diff. Geom.* **2** (1968), 1–7.
75. C.B. Morrey: The problem of Plateau in a Riemann manifold, *Ann. of Math.* **49** (1948), 807–851.
76. L. Mosher: Mapping class groups are automatic, *Ann. of Math.* **142** (1995), 303–384.
77. W. Neumann and L. Reeves: Central extensions of word hyperbolic groups, *Ann. of Math.* **145** (1997), 183–192.
78. I.G. Nikolaev: Solution of the Plateau problem in spaces of curvature at most K (Russian), *Sibirsk. Mat. Zh.* **20** (1979), 345–353.
79. J.C.C. Nitsche: 'Lectures on minimal surfaces, Vol 1.', Camb. Univ. Press, 1989.
80. I. Niven: 'Irrational Numbers', The Carus Mathematical Monographs **11**, MAA (1967).
81. A. Yu Ol'shanskii: Hyperbolicity of groups with subquadratic isoperimetric inequalities, *Intl. J. Alg. Comput.* **1** (1991), 281–290.
82. A. Yu Ol'shanskii: Distortion functions for subgroups, in 'Geometric Group Theory Down Under' (Canberra, 1996), 281–291, de Gruyter (Berlin, 1999).
83. A. Yu Ol'shanskii and M. Sapir: Quadratic isoperimetric functions for Heisenberg groups: a combinatorial proof, *J. Math. Sci.* (New York) **93** (1999), 921–927.
84. A. Yu Ol'shanskii and M. Sapir: Length and area functions on groups and quasi-isometric Higman embeddings, *Internat. J. Algebra Comput.*, to appear.
85. D.V. Osin: Subgroup distortions in nilpotent groups, *Comm. Algebra*, to appear.
86. R. Osserman: A proof of the regularity everywhere of the classical solution to Plateau's problem, *Ann. of Math.* **91** (1970), 550–569.
87. P. Papasoglu: On the sub-quadratic isoperimetric inequality, in 'Geometric group theory' (R. Charney, M. Davis, M. Shapiro, eds.), de Gruyter (Berlin, 1995).
88. P. Papasoglou: Asymptotic cones and the quadratic isoperimetric inequality, *J. Diff. Geom.* **44** (1996), 789–806.
89. Ch. Pittet: Isoperimetric inequalities in nilpotent groups, *J. London Math. Soc.* **55** (1997), 588–600.
90. J.A.F. Plateau: 'Statique Experimentale et Théorique des Liquides Soumis aux Seules Forces Moleculaires', Gauthier-Villars (Paris, 1873).
91. T. Rado: On Plateau's problem, *Ann. of Math.* **31** (1930), (457–469).
92. T. Rado: 'On the Problem of Plateau', Ergeb. der Math. Wiss. 2, Springer (Berlin, 1953).
93. Yu. G. Reshetnyak: Non-expansive maps in a space of curvature no greater than K, *Sib. Mat. J.* **9** (1968), 683–689.
94. J.J. Rotman: 'An introduction to the theory of groups', Fourth edition, GTM 148, Springer (New York, 1995).

95. M. Sapir, J-C. Birget, E. Rips: Dehn functions of finitely presented groups, *Ann. of Math.*, to appear.
96. Z. Sela: The isomorphism problem for hyperbolic groups I, *Ann. of Math.* **141** (1995), 217–283.
97. J-P. Serre: 'Trees', Springer (Berlin, 1977).
98. H. Short (ed.): Notes on word hyperbolic groups, in 'Group theory from a geometrical viewpoint' (E. Ghys, A. Haefliger, A. Verjovsky, eds.), pp. 3–64, World Scientific (Singapore, 1991).
99. W.P. Thurston: 'The geometry and topology of 3-manifolds', Lecture notes, Princeton University, 1979.
100. W.P. Thurston: Three dimensional manifolds, Kleinian groups and hyperbolic geometry, *Bull. Amer. Math. Soc.* **6** (1982), 357–381.

3
The Fuller index

M. C. CRABB AND A. J. B. POTTER

1 Introduction

In 1967 F. B. Fuller published a paper [33] which introduced a $\mathbb{Q}$-valued fixed-point index to 'count' the periodic orbits of differentiable flows. This work was in many respects ahead of its time. In the last decade or so many authors have re-examined Fuller's definition in the light of the developments in the Lefschetz fixed-point theory for maps and the Poincaré–Hopf index theory for vector fields which have taken place in the intervening years. (See, in alphabetical order: [5, 6, 8, 11, 12, 13, 14, 15, 16, 23, 24, 25, 26, 27, 31, 35, 36, 37, 41, 43, 44].) In this article we explain the construction of the Fuller index, as it has emerged from this large corpus of work, in the setting of contemporary equivariant fibrewise stable homotopy theory.

The methods will be entirely topological, and the homotopy-invariance of the index will be implicit in its definition. The classical fixed-point index of a map or degree of a vector field is defined as an integer which counts, with appropriate sign and multiplicity, the fixed-points of the map or the zeros of the vector field. There are two natural generalizations of the classical theory. First, one may study a family of maps or vector fields parametrized by a base space B. The work of Dold and others [19, 20, 22] shows that the index should then be defined as an element of the stable cohomotopy ring $\omega^0(B)$ of B, under conditions which we shall make precise later. Second, there are situations in which a compact Lie group G acts as a group of symmetries. In that case, the index will be an element of the Burnside ring $A(G)$ of G or, more generally, for a family parametrized by B, an element of the equivariant stable cohomotopy ring $\omega^0_G(B)$ (as in [45]).

We shall be concerned with semi-flows of the following type. Let X be a Euclidean Neighbourhood Retract (ENR). In practice, X is often a submanifold of some Euclidean space and is a retract of an open tubular neighbourhood. Let Ω be an open subset of $[0,\infty)\times X$, such that whenever $(t,x)\in\Omega$ then $(s,x)\in\Omega$ for all s with $0\leqslant s\leqslant t$, and let

$$\theta:\Omega\to X \tag{1.1}$$

be a continuous map satisfying:

(1.1)(i) *for all* $x\in X$, *we have* $(0,x)\in\Omega$ *and* $\theta(0,x)=x$;

(1.1)(ii) *if* $(t,x)\in\Omega$ *and* $(s,\theta(t,x))\in\Omega$, *then* $(s+t,x)\in\Omega$ *and* $\theta(s+t,x)=\theta(s,\theta(t,x))$.

Intuitively speaking, $\theta(t, x)$ is the location at time t of a point that began its journey along the flow at location x.

Let us assume, for the moment, that Ω is the whole of $[0, \infty) \times X$. In order to investigate periodic orbits we look at the set

$$\Pi = \{(T, x) \in (0, \infty) \times X \mid \theta(T, x) = x\}.$$

This will consist of the union of contributions:

$$(0, \infty) \times \{x_0\}$$

from each stationary point x_0 of the flow, and

$$\bigcup_{k \geqslant 1} \{kT_0\} \times C$$

from each periodic orbit, $C \subseteq X$, of minimal period T_0.

Typically, semi-flows arise as solutions of ordinary differential equations.

Example 1.2 Let X be the Möbius band $(\mathbb{R} \times \mathbb{R})/\sim$, where the equivalence relation $\sim$ identifies (x, y) with $(x + 1, -y)$. Writing (x, y) for the coordinates in $\mathbb{R} \times \mathbb{R}$, we define a flow, parametrized by $\lambda \in \mathbb{R}$, by the ordinary differential equation:

$$\dot{x} = 1, \qquad \dot{y} = -y(y^2 - \lambda^2).$$

(So $t \mapsto \theta(t, a)$ is the solution with initial value a.) For $\lambda = 0$, there is a single periodic orbit, with (minimal) period 1, at $y = 0$. For $\lambda \neq 0$, there are two orbits: (i) at $y = 0$, with period 1; (ii) at $y = \pm\lambda$, with period 2. Here is a picture of the set Π for $\lambda \neq 0$.

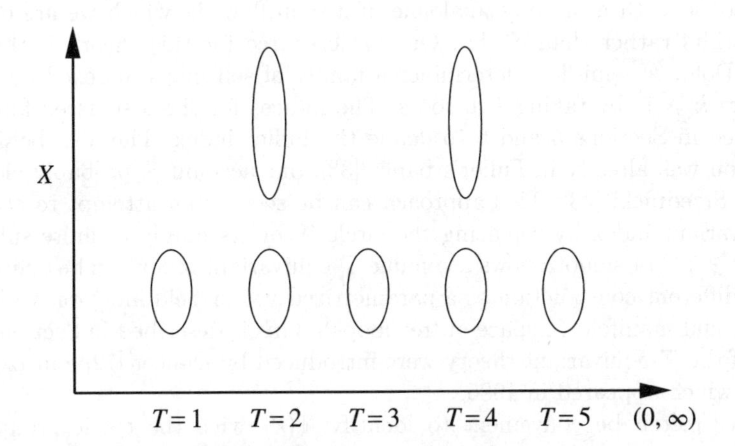

In an analytic context one might be able to count all the orbits by some sort of zeta-function (as in [32], for example). But since we are working topologically, we need essentially finite sums and suppose, therefore, that we are given an open subset $U \subseteq (0,\infty) \times X$ such that

$$F := \{(T,x) \in U \mid \theta(T,x) = x\} \tag{1.3}$$

is compact. This compactness condition has several important consequences:

Lemma 1.4 *If $(T,x) \in F$, then $(T,\theta(t,x)) \in F$ for all t. A (continuous) action of the circle $\mathbb{T} = \mathbb{R}/\mathbb{Z}$ on F can be defined by*

$$t \cdot (T,x) := (T, \theta(tT,x)), \quad t \in \mathbb{R} \ (\mathrm{mod}\ \mathbb{Z}).$$

If $(T,x) \in F$, then x is not a stationary point of the flow. In other words, the action of $\mathbb{T}$ on F is fixed-point-free: the fixed subspace $F^{\mathbb{T}}$ is empty. More precisely, $F^{\mathbb{T}(k)} = \emptyset$ for k large, where $\mathbb{T}(k) = \frac{1}{k}\mathbb{Z}/\mathbb{Z} \leqslant \mathbb{T}$ is the cyclic subgroup of order k.

The proof is an exercise in elementary topology. □

The *topological Fuller index* $\mathfrak{F}^*(\theta, U)$, which we shall define in Section 6 as an element of an inverse limit of equivariant stable homotopy groups:

$$\lim_{\substack{\longleftarrow \\ k \geqslant 1}} \omega^{-1}_{\mathbb{T}(k)}(*),$$

is, in a sense to be made precise, a measure of F regarded as a family of spaces parametrized by $(0,\infty)$ and equipped with a (fixed-point-free) action of the circle $\mathbb{T}$.

The first sections (2–4) deal with the fixed-point theory for iterates of a self-map, so with a discrete analogue of a semi-flow, in which we are counting finite orbits rather than circles. Our basic source for this theory is the paper [21] of Dold. A semi-flow determines a family of self-maps, indexed by natural numbers $k \geqslant 1$, by taking kth roots. The indices for the associated family are combined in Sections 5 and 6 to define the Fuller index. The idea behind this definition was already in Fuller's paper [33]; our account is probably closest to that of Srzednicki [43]. This approach can be seen as an attempt to capture a $\mathbb{T}$-equivariant index by replacing the circle $\mathbb{T}$ by its family of finite subgroups $\mathbb{T}(k)$, $k \geqslant 1$. For smooth flows a genuine $\mathbb{T}$-equivariant index can be defined by a rather different construction as a parametrized vector field index on an infinite-dimensional manifold (a space of free loops). This is described in Section 7. The ideas of the $\mathbb{T}$-equivariant theory were introduced by Dancer [12] and Ize [35] in papers which appeared in 1985.

It will often be convenient to identify $\mathbb{T}(k)$ with the cyclic group $\mathbb{Z}/k\mathbb{Z}$ of integers modulo k, so that the generator $1/k (\mathrm{mod}\ 1)$ of $\mathbb{T}(k) = \frac{1}{k}\mathbb{Z}/\mathbb{Z}$ corresponds to the generator $1\ (\mathrm{mod}\ k)$ of $\mathbb{Z}/k\mathbb{Z}$.

Notes on the exercises can be found at `www.maths.abdn.ac.uk/~crabb`

2 Iterates of maps: the theory for finite sets

Some of the fundamental ideas can be illustrated in an elementary, purely algebraic, setting.

We begin with a review of the relevant Lefschetz theory. For a finite group G the Burnside ring $A(G)$, which we shall interpret later as the equivariant stable cohomotopy group $\omega_G^0(*)$ of a point, is just the Grothendieck group of finite G-sets (with the multiplication determined by the product of sets). Every transitive G-set is isomorphic to a coset space G/H for some subgroup $H \leqslant G$, and two such spaces G/H and G/H' are isomorphic if and only if the subgroups H and H' are conjugate. Since any finite G-set is a disjoint union of orbits, $A(G)$ is a free abelian group with one generator for each conjugacy class of subgroups of G.

In particular, $A(\mathbb{Z}/k\mathbb{Z})$ has a basis

$$\sigma_{k,m} := [(\mathbb{Z}/k\mathbb{Z})/(m\mathbb{Z}/k\mathbb{Z})], \qquad m \mid k, \tag{2.1}$$

with one generator corresponding to each divisor m of k.

For a subgroup H of G we write $N_G(H)$ for its normalizer in G and $W_G(H)$ for the quotient $N_G(H)/H$. If M is a finite G-set, then the subset M^H of points fixed by H is naturally a $W_G(H)$-set, and this construction $[M] \mapsto [M^H]$ defines the *fixed-point homomorphism*

$$\rho_H : A(G) \to A(W_G(H)). \tag{2.2}$$

So, for example, we have for each l dividing k a homomorphism

$$\rho_{l\mathbb{Z}/k\mathbb{Z}} : A(\mathbb{Z}/k\mathbb{Z}) \to A(\mathbb{Z}/l\mathbb{Z}), \tag{2.3}$$

which maps $\sigma_{k,m} \in A(\mathbb{Z}/k\mathbb{Z})$ to $\sigma_{l,m} \in A(\mathbb{Z}/l\mathbb{Z})$ if $m \mid l$, and to 0 otherwise.

The homomorphism

$$h : A(G) \to \mathbb{Z} \tag{2.4}$$

given by the cardinality $[M] \mapsto \#M$ will appear in the general case as the *Hurewicz homomorphism*: $\omega_G^0(*) \to H_G^0(*)$, (3.4), from stable cohomotopy to G-equivariant (Borel) cohomology with integral coefficients.

Given a G-equivariant self-map $f : M \to M$ of a finite G-set M, we can define its equivariant *Lefschetz index* $L(f, M) \in A(G)$ simply as the class of the fixed-point set $\mathrm{Fix}(f)$. The Hurewicz image $h(L(f, M)) \in \mathbb{Z}$ is then equal to the number of fixed-points.

We turn now to the study of the iterates of a self-map $\phi : X \to X$ of a finite set X, using a construction which may be found in Fuller's original paper. Consider, for $k \geqslant 1$, the map $\pi_k(\phi) : X^k \to X^k$ defined by

$$\pi_k(\phi)(x_1, x_2, \ldots, x_k) := (\phi(x_k), \phi(x_1), \ldots, \phi(x_{k-1})) \tag{2.5}$$

(with ith entry $\phi(x_{i-1})$, $i \pmod{k}$). There is a natural bijection

$$\mathrm{Fix}(\phi^k) \to \mathrm{Fix}(\pi_k(\phi)) : x \mapsto (x, \phi(x), \phi^2(x), \ldots, \phi^{k-1}(x)) \tag{2.6}$$

from the fixed subspace of the kth iterate ϕ^k of ϕ to the fixed subspace of $\pi_k(\phi)$.

One should think of X^k as the space of maps, $\mathrm{map}(\mathbb{Z}/k\mathbb{Z}, X)$, from $\mathbb{Z}/k\mathbb{Z}$ to X. On X^k there is then a natural action of $\mathbb{Z}/k\mathbb{Z}$ by cyclic permutation: $(j \cdot x)_i = x_{i+j}$ for $i, j \in \mathbb{Z}/k\mathbb{Z}$. The generator $1 \in \mathbb{Z}/k\mathbb{Z}$ thus acts as the cyclic shift

$$(x_1, x_2, \ldots, x_k) \mapsto (x_2, x_3, \ldots, x_k, x_1). \tag{2.7}$$

The map $\pi_k(\phi)$ is $\mathbb{Z}/k\mathbb{Z}$-equivariant, and under the correspondence (2.6) $1 \in \mathbb{Z}/k\mathbb{Z}$ acts on $\mathrm{Fix}(\pi_k(\phi))$ as ϕ acts on $\mathrm{Fix}(\phi^k)$.

The equivariant Lefschetz index $L(\pi_k(\phi), X^k)$ is now defined as an element of $A(\mathbb{Z}/k\mathbb{Z}) = \omega^0_{\mathbb{Z}/k\mathbb{Z}}(*)$ for each $k \geqslant 1$. The various indices for different k are related by the fixed-point maps as follows.

Lemma 2.8 *If l divides k, then $L(\pi_l(\phi), X^l)$ is the image of $L(\pi_k(\phi), X^k)$ under the fixed-point homomorphism* (2.3) *above.*

Indeed, $\mathbb{Z}/k\mathbb{Z}$ acts on $X^k = \mathrm{map}(\mathbb{Z}/k\mathbb{Z}, X)$ and the subspace fixed by $l\mathbb{Z}/k\mathbb{Z}$ is naturally identified with $X^l = \mathrm{map}(\mathbb{Z}/l\mathbb{Z}, X)$ by the quotient map $\mathbb{Z}/l\mathbb{Z} \to \mathbb{Z}/k\mathbb{Z}$. Under this passage to the fixed-subspace, $\pi_k(\phi)$ restricts to $\pi_l(\phi)$. $\square$

Now we can express $L(\pi_k(\phi), X^k)$ in terms of the basis (2.1) as:

$$L(\pi_k(\phi), X^k) = \sum_{m \mid k} \mathcal{D}^m(\phi, X)\, \sigma_{k,m}. \tag{2.9}$$

It follows from the observation (2.8) and the calculation (2.3) of the fixed-point homomorphism that the coefficient $\mathcal{D}^m(\phi, X) \in \mathbb{Z}$ depends only on m. (We have adopted the notation '$\mathcal{D}$' in recognition of Dold's contributions [21] to the theory.)

The geometric significance of the indices $\mathcal{D}^m(\phi, X)$ is easy to understand in this simple case. We can represent the self-map ϕ diagrammatically by a quiver

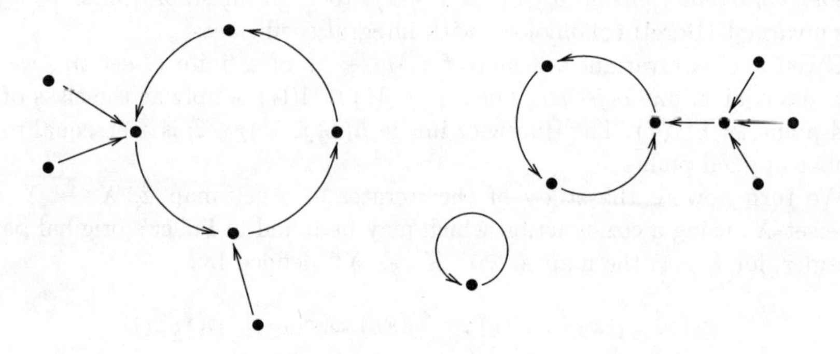

(that is, a directed graph) with vertices the elements x of the set X and arrows (or directed edges) from x to $\phi(x)$ for each $x \in X$. The diagram will show $\mathcal{D}^m(\phi, X)$ cycles of length m.

The equivariant Lefschetz index $L(\pi_k(\phi), X^k)$ determines the (non-equivariant) fixed-point index $L(\phi^k, X)$ of the iterated map by forgetting the group action.

Lemma 2.10 *The Lefschetz index $L(\phi^k, X) \in \mathbb{Z}$ is the image under the Hurewicz map:*

$$A(\mathbb{Z}/k\mathbb{Z}) \to \mathbb{Z}$$

of the class $L(\pi_k(\phi), X^k)$.

In this elementary case the result is immediate, because both classes count the fixed-points of ϕ^k. We shall see that it holds in greater generality. □

From (2.9) we obtain the equality:

$$L(\phi^k, X) = \sum_{m \mid k} m\mathcal{D}^m(\phi, X) \in \mathbb{Z}. \tag{2.11}$$

Now let us introduce the *zeta-function*:

$$Z(\phi, X; q) = \prod_{m \geqslant 1} (1 - q^m)^{\mathcal{D}^m(\phi, X)} \tag{2.12}$$

in the formal power series ring $\mathbb{Z}[[q]]$. For the moment, while we are working with finite sets, the zeta-function is a polynomial, but even here it is best regarded as a formal power series.

Exercises 2.13 (i) The equalities (2.11) may be combined in the identity:

$$Z(\phi, X; q) = \exp\left(-\sum_{k \geqslant 1} \frac{L(\phi^k, X)}{k} q^k \right) \in \mathbb{Q}[[q]].$$

(ii) Any formal power series $F(q) \in \mathbb{Z}[[q]]$ with $F(0) = 1$ can be factorized uniquely as an infinite product $\prod_{m \geqslant 1} (1 - q^m)^{d_m}$, with $d_m \in \mathbb{Z}$.

(iii) The system (2.11) can be inverted to give:

$$m\mathcal{D}^m(\phi, X) = \sum_{k \mid m} \mu\left(\frac{m}{k}\right) L(\phi^k, X),$$

where μ is the number-theoretic Möbius function. It follows, for example, that

$$L(\phi^{p^s}, X) \equiv L(\phi^{p^{s-1}}, X) \pmod{p^s},$$

for any prime $p > 1$ and integer $s \geqslant 1$.

The next so-called *contraction* property is also, at this naïve level, transparent.

Lemma 2.14 *Suppose that Y is a subset of X such that $\phi(X) \subseteq Y$. Then*

$$L(\pi_k(\phi), X^k) = L(\pi_k(\phi \,|\, Y), Y^k) \in A(\mathbb{Z}/k\mathbb{Z}),$$
$$\mathcal{D}^m(\phi, X) = \mathcal{D}^m(\phi \,|\, Y, Y) \in \mathbb{Z}.$$

□

Exercise 2.15 The contraction property (2.14) and direct computation for a cyclic permutation ϕ of a finite set X show that, in general,

$$Z(\phi, X; q) = \det(1 - qH_0(\phi)) \in \mathbb{Z}[[q]],$$

where $H_0(\phi)$ is the induced map on the rational homology group $H_0(X; \mathbb{Q})$ (that is, the vector space with basis the set X).

The inclusion $i : H \to G$ of a subgroup H of a finite group G determines an evident *restriction* homomorphism

$$i^* : A(G) \to A(H) \tag{2.16}$$

defined by restricting the action of G on a finite G-set to an action of H. To prepare the way for the definition of the Fuller index we now change our point of view and replace $\mathbb{Z}/k\mathbb{Z}$ by $\mathbb{T}(k) = \frac{1}{k}\mathbb{Z}/\mathbb{Z}$ so that we can look at restriction from $\mathbb{T}(k)$ to a subgroup $\mathbb{T}(l)$.

Lemma 2.17 *If l divides k, then the restriction homomorphism*

$$A(\mathbb{T}(k)) = \omega^0_{\mathbb{T}(k)}(*) \to A(\mathbb{T}(l)) = \omega^0_{\mathbb{T}(l)}(*)$$

maps $L(\pi_k(\phi), X^k)$ to $L(\pi_l(\phi^{k/l}), X^l)$.

For the generator $1/k$ of $\mathbb{T}(k)$ acts as ϕ on $\mathrm{Fix}(\phi^k)$ and the generator $1/l$ of $\mathbb{T}(l)$ acts as $\phi^{k/l}$ on the same fixed-point set $\mathrm{Fix}((\phi^{k/l})^l)$. □

Before leaving finite sets we discuss another important result, which involves the *induction* homomorphism

$$i_* : A(H) \to A(G)$$

associated to the inclusion $i : H \to G$ of a subgroup. Given a finite H-set N we may form the G-set $G \times_H N$: then $i_*[N] = [G \times_H N]$. (Recall that $G \times_H N$ is the quotient of $G \times N$ by the equivalence relation: $(gh, y) \sim (g, hy)$ for $h \in H$; the group G acts by multiplication on the left on the first component.) For example, for r dividing k the induction homomorphism

$$i_* : A(\mathbb{T}(k/r)) \to A(\mathbb{T}(k)) \tag{2.18}$$

maps $\sigma_{k/r,m} \in A(\mathbb{T}(k/r))$ to $\sigma_{k,rm} \in A(\mathbb{T}(k))$.

Suppose that the finite set X admits a decomposition as a disjoint union $X = \coprod_i X_i$, indexed by $i \in \mathbb{Z}/r\mathbb{Z}$, such that $\phi(X_i) \subseteq X_{i+1}$ for each i. Then the restriction of ϕ^r to a self-map of X_1, which might be called a *Poincaré map*, determines the index of ϕ. We think again in terms of $\mathbb{T}(k)$ rather than $\mathbb{Z}/k\mathbb{Z}$.

Lemma 2.19 *In the situation described in the text, if r divides k, the induction homomorphism*

$$A(\mathbb{T}(k/r)) = \omega^0_{\mathbb{T}(k/r)}(*) \to A(\mathbb{T}(k)) = \omega^0_{\mathbb{T}(k)}(*)$$

maps $L(\pi_{k/r}(\phi^r), X_1^{k/r})$ to $L(\pi_k(\phi), X^k)$.

Indeed, we can identify $\mathbb{Z} \times_{r\mathbb{Z}} \mathrm{Fix}(\phi^r|X_1)$ with $\mathrm{Fix}(\phi)$, where the generator r of $r\mathbb{Z}$ acts as ϕ^r and the generator 1 of $\mathbb{Z}$ acts as ϕ, by mapping $[i,x]$ to $\phi^i(x)$. Of course, ϕ^k has no fixed-points and $L(\pi_k(\phi), X^k) = 0$ if r does not divide k. □

It follows that

$$\mathcal{D}^m(\phi, X) = \begin{cases} \mathcal{D}^{m/r}(\phi^r, X_1) & \text{if } r \text{ divides } m \\ 0 & \text{otherwise,} \end{cases} \tag{2.20}$$

or in terms of zeta-functions: $Z(\phi, X; q) = Z(\phi^r, X_1; q^r)$.

To define the index $\mathcal{D}^m(\phi, X)$ we need not require the set X itself to be finite, only that the fixed-point set of ϕ^m be finite. For an infinite set X and a self-map ϕ such that $\mathrm{Fix}(\phi^m)$ is finite for all $m \geqslant 1$ the sequence $(\mathcal{D}^m(\phi, X))$ may contain infinitely many non-zero terms.

Exercises 2.21 (i) Let A be an endomorphism of a finite-dimensional vector space over a field K of characteristic zero. Then:

$$\exp\left(-\sum_{k\geqslant 1} \frac{\mathrm{tr}\, A^k}{k} q^k\right) = \det(1 - qA) \in K[[q]].$$

(ii) Consider a finite quiver, by which we mean a finite set V of vertices and a finite set E of arrows, together with a map $E \to V \times V$ assigning to each arrow its initial and terminal vertex. Let X be the set of infinite paths $(e_1, e_2, \ldots)$, with the terminal vertex of $e_i \in E$ equal to the initial vertex of e_{i+1}, for $i \geqslant 1$, and let $\phi : X \to X$ be the shift: $(e_1, e_2, \ldots) \mapsto (e_2, e_3, \ldots)$. Then the fixed-points of ϕ^m correspond to closed paths of length m, and $\mathcal{D}^m(\phi, X)$ is defined for each $m \geqslant 1$. (The group $\mathbb{Z}/m\mathbb{Z}$ acts by cyclic permutation on the set of closed paths of length m. Let us say that two closed paths are *equivalent* if they are in the same orbit and that a closed path is *non-repeating* if its isotropy group is trivial. Then $\mathcal{D}^m(\phi, X)$ is the number of equivalence classes of non-repeating closed paths of length m.) The indices are determined by the incidence matrix $A : V \times V \to \mathbb{Z}$, where $A(u, v)$ is the number of arrows from u to v:

$$Z(\phi, X; q) = \det(1 - qA) \in \mathbb{Z}[[q]].$$

Partially ordered finite sets

The theory just described extends rather easily from sets to partially ordered sets. (Any set is trivially partially-ordered by equality.)

Let G again be a finite group and let M be a finite G-set with an invariant partial order. Given an order-preserving G-map $f : M \to M$ (so $x \leqslant y \Rightarrow f(x) \leqslant f(y)$), we define its Lefschetz index to be the alternating sum

$$L(f, M) = \sum_{r \geqslant 0} (-1)^r [F_r] \in A(G), \tag{2.22}$$

where $F_r = \{(x_0, x_1, \ldots, x_r) \mid x_0 < x_1 < \cdots < x_r,\ x_i \in \mathrm{Fix}(f)\}$, with the evident action of G. Recall that the homology of the partially ordered set M is defined to be the homology of the complex $(C_r(M), \partial)$, where $C_r(M)$ is the free abelian group on the totally ordered $(r+1)$-element subsets written in increasing order as $(x_0, \ldots, x_r)$ with $x_0 < x_1 < \cdots < x_r$, and

$$\partial(x_0, \ldots, x_r) = \sum_{i=0}^{r} (-1)^i (x_0 \ldots, x_{i-1}, x_{i+1}, \ldots, x_r).$$

Exercise 2.23 The Lefschetz fixed-point theorem for partially ordered sets computes the Hurewicz image of $L(f, M)$ in $H^0_G(*) = \mathbb{Z}$, or the Euler characteristic of the fixed-subset, as the super-trace

$$\sum_{r \geqslant 0} (-1)^r \operatorname{tr} H_r(f) \in \mathbb{Z}$$

of the induced map $H_*(f)$ on the rational homology of M.

Now suppose that $\phi : X \to X$ is an order-preserving self-map of a finite partially ordered set. We order X^k pointwise: $(x_1, \ldots, x_k) \leqslant (y_1, \ldots, y_k)$ if and only if $x_i \leqslant y_i$ for all i. Then $\pi_k(\phi)$, as defined above, is order-preserving, and we form its Lefschetz index $L(\pi_k(\phi), X^k) \in A(\mathbb{Z}/k\mathbb{Z})$ as an order-preserving map. The indices $\mathcal{D}^m(\phi, X) \in \mathbb{Z}$ are defined as in (2.11), with the novelty that they may be negative.

Exercise 2.24 Using the Lefschetz theorem (2.23) and (2.21)(i), one can generalize the result (2.15) for sets to write the zeta-function, defined by (2.12), as the super-determinant of $1 - qH_*(\phi)$:

$$Z(\phi, X; q) = \prod_{r \geqslant 0} \det(1 - qH_r(\phi))^{(-1)^r} \in \mathbb{Z}[[q]].$$

Associated to a finite partially ordered set there is a finite polyhedron, its realization. The algebraic fixed-point theory extends to a topological theory for finite polyhedra, or more generally ENRs, to which we turn next.

3 Iterates of maps: the theory for ENRs

The indices with which we are now concerned arise in this section as equivariant stable maps and in the next as fibrewise equivariant stable maps. To fix notation we give the basic definitions. For details, the reader is referred to [17], [45], [9], [7].

Let G now be a compact Lie group. The one-point compactification of a finite-dimensional real G-module E is a sphere, which we denote by E^+, with basepoint at infinity. (The notation S^E, consistent with S^n for $E = \mathbb{R}^n$, is more usual.) For two G-modules E and F, $(E \oplus F)^+$ is naturally identified with the smash product $E^+ \wedge F^+$. A G-ENR is an equivariant retract of an open subspace of some finite-dimensional real G-module. Let P and Q be compact G-ENRs with basepoint. An equivariant stable map $P \to Q$ is represented by a pair (v, F), where F is a finite-dimensional real G-module and v is a basepoint-preserving G-map $v : F^+ \wedge P \to F^+ \wedge Q$. We introduce an equivalence relation $\sim$ on such pairs by requiring that (i) $(v, F) \sim (w, F)$ if v and w are homotopic (as pointed G-maps); (ii) for an isomorphism $a : F \to F'$ of G-modules, $(v, F) \sim (v', F')$, where $v' = (a^+ \wedge 1) \circ v \circ (a^+ \wedge 1)^{-1}$; (iii) for any G-module E, $(v, F) \sim (1 \wedge v, E \oplus F)$. An equivalence class is, by definition, an *equivariant stable map* $P \to Q$, and we denote the set (in fact, abelian group) of such stable maps by $\omega_G^0\{P;\, Q\}$. It is the set of morphisms $P \to Q$ in the *equivariant stable homotopy category*. There is, by construction, a suspension isomorphism

$$\omega_G^0\{P;\, Q\} \xrightarrow{\cong} \omega_G^0\{E^+ \wedge P;\, E^+ \wedge Q\}$$

for any G-module E. We use this suspension isomorphism to define, unambiguously, for any integer $i \in \mathbb{Z}$,

$$\omega_G^i\{P;\, Q\} = \omega_G^0\{(\mathbb{R}^{n+i})^+ \wedge P;\, (\mathbb{R}^n)^+ \wedge Q\},$$

where n is any natural number such that $n+i \geqslant 0$. The group G is understood to act trivially on $\mathbb{R}$. Stable homotopy and cohomotopy groups may then be defined as follows. Let B be a compact ENR and let $A \subseteq B$ be a closed sub-ENR. Using a subscript '+' for adjunction of a disjoint basepoint, we write

$$\omega_G^i(B) = \omega_G^i\{B_+;\, S^0\}, \qquad \omega_G^i(B, A) = \omega_G^i\{B/A;\, S^0\},$$
$$\omega_j^G(B) = \omega_G^{-j}\{S^0;\, B_+\}, \quad \omega_j^G(B, A) = \omega_G^{-j}\{S^0;\, B/A\}.$$

Similar notation, without the suffix 'G', is used for non-equivariant stable homotopy theory. In setting up the definition we have used the compactness of Q only in talking about the stable category; the definition makes sense for any pointed G-space Q, and we shall sometimes need this greater generality.

Consider a closed subgroup H of G, with normalizer $N_G(H)$, and as before write $W_G(H) = N_G(H)/H$. By taking fixed-points under the action of H we obtain a *fixed-point homomorphism*

$$\rho_H : \omega_G^0\{P;\, Q\} \to \omega_{W_G(H)}^0\{P^H;\, Q^H\}, \tag{3.1}$$

mapping the class of $v : F^+ \wedge P \to F^+ \wedge Q$ to the class of its restriction $v| : (F^H)^+ \wedge P^H \to (F^H)^+ \wedge Q^H$ to the fixed-subspaces.

The *restriction homomorphism*

$$i^* : \omega_G^*\{P;\, Q\} \to \omega_H^*\{P;\, Q\} \tag{3.2}$$

for the inclusion $i : H \to G$ is defined in the obvious way. The *induction homomorphism*

$$i_* : \omega_H^*\{\mathfrak{h}^+ \wedge P;\, Q\} \to \omega_G^*\{\mathfrak{g}^+ \wedge P;\, Q\}, \tag{3.3}$$

where $\mathfrak{g}$ and $\mathfrak{h}$ are the Lie algebras of G and H, is essentially given by the $G \times_H -$ construction, but the details are subtler when G/H is infinite. (See, for example, [10].)

Equivariant *Borel cohomology* $H_G^*(M)$, with integer coefficients say, of a G-ENR M is defined as the ordinary cohomology group $H^*(\mathbf{E}G \times_G M)$, where $\mathbf{E}G \to \mathbf{B}G$ is the universal principal G-bundle over the classifying space $\mathbf{B}G$. So $H_G^*(B)$, for a space B on which G acts trivially, is just $H^*(\mathbf{B}G \times B)$. In particular, $H_G^0(B) = \mathbb{Z}$ for a connected space B. More general (co)homology groups $H_G^*\{P;\, Q\}$ can also be defined (as, for example, in [7]) and there is a *Hurewicz transformation*

$$\omega_B^*\{P;\, Q\} \to H_G^*\{P;\, Q\} \tag{3.4}$$

from stable (co)homotopy to (co)homology.

We come now to the equivariant Lefschetz fixed-point theory for G-ENRs. Let M be a G-ENR. So there are G-maps $i : M \to N$ and $r : N \to M$ such that $r \circ i = 1 : M \to M$, for some open subset N of a finite-dimensional G-module E, which we may equip with a G-invariant Euclidean inner product. We consider an open G-subspace U of M and a G-map $f : U \to M$ with compact fixed-point set $\mathrm{Fix}(f) = \{x \in U \mid f(x) = x\}$. The fundamental fixed-point index of f is a stable G-map $S^0 \to U_+$ which we call the *Lefschetz–Hopf index*:

$$\tilde{L}(f, U) \in \omega_G^0\{S^0;\, U_+\} = \tilde{\omega}_0^G(U_+) = \omega_0^G(U). \tag{3.5}$$

The fixed-point index is defined, as a stable map, by an explicit geometric construction in Pontrjagin–Thom style (as in [17]). We require a map $c_\epsilon : E \to E^+$ which expands the open disc of radius $\epsilon > 0$, centre 0, to fill E and maps the complement to the point at infinity. Let $\kappa : [0, 1) \to [0, \infty)$ be a homeomorphism. (Any two are homotopic.) Then we define

$$c_\epsilon(x) = \begin{cases} \kappa(\|x\|/\epsilon)x & \text{if } \|x\| < \epsilon \\ * & \text{otherwise.} \end{cases}$$

Given the representation (i, r) above of M as a retract of an open subspace N of V, let us write $g = i \circ f \circ r : r^{-1}U \to V$. We choose an open neighbourhood W of the (compact) fixed subspace $\mathrm{Fix}(g)$ in E such that $\overline{W}$ is compact and

$\overline{W} \subseteq r^{-1}U$, and, then, a number $\epsilon > 0$ with $\|g(x) - x\| \geqslant \epsilon$ for all $x \in \overline{W} - W$. Let $l : E^+ \to E^+ \wedge U_+$ be the map defined on the one-point compactification E^+ of E by

$$l(x) = \begin{cases} [c_\epsilon(x - g(x)), r(x)] & \text{if } x \in \overline{W} \\ * \text{ (basepoint)} & \text{if } x \notin W. \end{cases} \tag{3.6}$$

It is straightforward to show that the stable map $S^0 \to U_+$ represented by l is independent of the various choices made. This is the definition of the Lefschetz–Hopf fixed-point index $\tilde{L}(f, U)$. The class of the stable map $S^0 \to S^0$ obtained by forgetting the second U_+ factor will be called the *Lefschetz index*

$$L(f, U) \in \omega_G^0(*).$$

For example, if G is a finite group, then any finite G-set M is a compact ENR and the identity map $f = 1$ has a compact fixed-point set. The Lefschetz index $[M] \mapsto L(1, M)$ gives a (ring) homomorphism

$$A(G) \to \omega_G^0(*)$$

from the algebraically defined Burnside ring to the equivariant stable homotopy ring. It is one of the first theorems of equivariant stable homotopy theory that this map is an isomorphism. (See [17].)

The Hurewicz homomorphism, (3.4), $\omega_G^0(*) \to H_G^0(*) = \mathbb{Z}$ to Borel cohomology maps $L(f, U)$ to the classical non-equivariant Lefschetz index defined via cohomology.

Various standard properties of the fixed-point index follow readily from the definition. It is localized at the fixed-point set in the sense that if we replace U by a smaller open neighbourhood we obtain essentially the same index. The index is additive in the sense that if U is a disjoint union of two pieces then the index is a sum of two terms. It is invariant under compactly-fixed homotopy: $\tilde{L}(f_t, U)$ is constant for a homotopy $f_t : U \to M$ such that the union of the fixed-subspaces of the maps f_t for $0 \leqslant t \leqslant 1$ is compact. (See [18].) The following is a specifically equivariant property.

Lemma 3.7 *Let $H \leqslant G$ be a closed subgroup, and write $W_G(H) := N_G(H)/H$. Taking H-fixed-points we get a $W_G(H)$-equivariant map $f^H : U^H \to M^H$. Then its index $\tilde{L}(f^H, U^H)$ is the image of $\tilde{L}(f, U)$ under the fixed-point map*

$$\rho_H : \omega_G^0\{S^0;\, U_+\} \to \omega_{W_G(H)}^0\{S^0;\, U_+^H\}.$$

Granted that the Lefschetz index is well-defined by the construction just described, the result is clear. For the construction is evidently compatible with taking fixed-points. □

Using this machinery, we can make an essentially routine extension of the fixed-point theory for iterates of a map from finite sets as in Section 2 to compact

ENRs. (The sources for this theory, listed again in alphabetical order, are [6, 28, 29, 30, 38, 39, 40, 42, 46].) Let X be an ENR and $\phi : U \to X$ a map defined on an open subset U. We define

$$\pi_k(\phi) : U^k \to X^k$$

by (2.5). In order to form the Lefschetz index $L(\pi_k(\phi), U^k) \in \omega^0_{\mathbb{Z}/k\mathbb{Z}}(*)$ we need to stipulate that $\mathrm{Fix}(\pi_k(\phi))$ be compact. Notice that we can define the kth iterate ϕ^k on the subset $(\phi^{-1})^{k-1}U$ of U, and that its fixed-point set is homeomorphic to that of $\pi_k(\phi)$. The generalization of (2.8) is clear:

Lemma 3.8 *Suppose that l divides k. Then $L(\pi_l(\phi), U^l)$ is the image of $L(\pi_k(\phi), U^k)$ under the fixed-point homomorphism*

$$\rho_{l\mathbb{Z}/k\mathbb{Z}} : \omega^0_{\mathbb{Z}/k\mathbb{Z}}(*) \to \omega^0_{\mathbb{Z}/l\mathbb{Z}}(*).$$

□

This allows us again to define integers $\mathcal{D}^m(\phi, U) \in \mathbb{Z}$, when $\mathrm{Fix}(\pi_m(\phi))$ is compact, such that

$$L(\pi_k(\phi), U^k) = \sum_{m|k} \mathcal{D}^m(\phi, U)\sigma_{k,m},$$

and we may combine them, when each $\pi_m(\phi)$ has a compact fixed-subspace, in a zeta-function:

$$Z(\phi, U; q) = \prod_{m \geqslant 1} (1 - q^m)^{\mathcal{D}^m(\phi,U)} \in \mathbb{Z}[[q]].$$

The generalization of (2.17), which is substantially in Fuller's paper [33], requires a little more effort.

Lemma 3.9 *Suppose l divides k. Then $L(\pi_k(\phi), U^k) \in \omega^0_{\mathbb{T}(k)}(*)$ restricts to $L(\pi_l(\phi^{k/l}), V^l) \in \omega^0_{\mathbb{T}(l)}(*)$, where $V = (\phi^{-1})^{k/l-1}U$.*

The proof is by a standard technique used in establishing the commutativity property of the Lefschetz index (as in [18]). From the way that the fixed-point index is defined, there is no loss of generality in assuming that X is a Euclidean space E. We consider $\pi_k(\phi)$ as a $\mathbb{T}(l)$-equivariant map

$$(x_1, \ldots, x_k) \mapsto (\phi(x_k), \phi(x_1), \ldots, \phi(x_{k-1})) : U^k \to E^k.$$

Its fixed-subspace lies in V^k, and we can localize to this open subset in order to compute the index. Writing $m = k/l$, we define two $\mathbb{T}(l)$-equivariant compactly-fixed homotopies F_t, $G_t : V^k \to E^k$, $0 \leqslant t \leqslant 1$, as follows.

$$F_t(x_1, \ldots, x_{lm}) = \Big((1-t)\phi(x_{lm}) + t\phi^m(x_{(l-1)m+1}), \phi(x_1), \ldots, \phi(x_{m-1}),$$
$$(1-t)\phi(x_m) + t\phi^m(x_1), \phi(x_{m+1}), \ldots\Big),$$
$$G_t(x_1, \ldots, x_{lm}) = \Big(\phi^m(x_{(l-1)m+1}), (1-t)\phi(x_1), \ldots, (1-t)\phi(x_{m-1}),$$
$$\phi^m(x_1), (1-t)\phi(x_{m+1}), \ldots\Big).$$

It is straightforward to check that the fixed-subspace of F_t is precisely the fixed-subspace of $F_0 = \pi_k(\phi)$ and that the fixed-subspace of G_t is homeomorphic, by a homeomorphism varying continuously with t, to the fixed-subspace of $G_0 = F_1$. By the homotopy-invariance of the Lefschetz index, we have $L(F_0, V^k) = L(G_1, V^k) \in \omega^0_{\mathbb{T}(l)}(*)$. But G_1 is the map:

$$(x_1, x_2, \ldots, x_{lm}) \mapsto (\phi^m(x_{(l-1)m+1}), 0, \ldots, 0, \phi^m(x_1), 0, \ldots, 0),$$

and its index is equal to $L(\pi_l(\phi^m, V^l))$, by multiplicativity of the Lefschetz index (or inspection of the definition). □

Exercise 3.10 There is an algebraic analogue of this result. Let E_i, $i \in \mathbb{Z}/k\mathbb{Z}$, be finite-dimensional vector spaces (over any field K) and let $a_i : E_i \to E_{i+1}$ be linear maps. Consider the linear map

$$\alpha : \bigoplus_{i=1}^{k} E_i \to \bigoplus_{i=1}^{k} E_i \quad : \quad (x_1, \ldots, x_k) \mapsto (a_k x_k, a_1 x_1, \ldots, a_{k-1} x_{k-1}).$$

Its characteristic polynomial is

$$\det(q - \alpha) = \det(q^k - \beta_1) \in K[q],$$

where β_1 is the (Poincaré) map $a_k a_{k-1} \cdots a_1 : E_1 \to E_1$.

It follows from (3.9) that we again have

$$L(\phi^k, (\phi^{-1})^{k-1}U) = \sum_{m|k} m\mathcal{D}^m(\phi, U) \in \mathbb{Z}, \tag{3.11}$$

and, when $\operatorname{Fix}(\pi_k(\phi))$ is compact for all $k \geqslant 1$,

$$\exp\left(-\sum_{k\geqslant 1} \frac{L(\phi^k, (\phi^{-1})^{k-1}U)}{k} q^k\right) = Z(\phi, U; q) \in \mathbb{Q}[[q]].$$

Exercise 3.12 If X is compact and $U = X$, we have, using the homological computation of the Lefschetz index as a super-trace,

$$Z(\phi, X; q) = \prod_{i \geqslant 0} \det(1 - qH_i(\phi))^{(-1)^i},$$

where $H_i(\phi) : H_i(X;\, \mathbb{Q}) \to H_i(X;\, \mathbb{Q})$ is the induced map in rational homology.

The extension of (2.19) depends on the construction of the group-theoretic induction map, (3.3).

Proposition 3.13 *Suppose that* $X = \coprod_{i \in \mathbb{Z}/r\mathbb{Z}} X_i$ *is a disjoint union of open subspaces* X_i, *indexed by* $\mathbb{Z}/r\mathbb{Z}$, *such that* $\phi(X_i \cap U) \subseteq X_{i+1}$ *for each* i. *Let* $U_i = X_i \cap (\phi^{-1})^{r-1}(U)$. *If* k *is a multiple of* r, *then* $L(\pi_k(\phi), U^k)$ *is the image of* $L(\pi_{k/r}(\phi^r), U_1^{k/r})$ *under the induction homomorphism*

$$i_* : \omega^0_{\mathbb{T}(k/r)}(*) \to \omega^0_{\mathbb{T}(k)}(*),$$

so that $\mathcal{D}^m(\phi, U) = \mathcal{D}^{m/r}(\phi^r, U_1)$ *if* $r \mid m$, *and is* 0 *otherwise.*

The fixed-subspace of $\pi_k(\phi)$ (which we are assuming to be compact) is contained in the disjoint union

$$\coprod_{i=1}^{r} (U_i \times U_{i+1} \times \cdots \times U_{i+k-1}),$$

which may be written as

$$\mathbb{T}(k) \times_{\mathbb{T}(k/r)} (U_1 \times U_2 \times \cdots \times U_k).$$

Essentially by the definition of induction i_* as the $\mathbb{T}(k) \times_{\mathbb{T}(k/r)}$-construction, it follows that $L(\pi_k(\phi), U^k) = i_* L(\pi_k(\phi)|, U_1 \times U_2 \times \cdots \times U_k)$. The argument used to establish (3.9) shows that

$$L(\pi_k(\phi)|, U_1 \times U_2 \times \cdots \times U_k) = L(\pi_{k/r}(\phi^r|), U_1^{k/r}) \in \omega^0_{\mathbb{T}(k/r)}(*).$$

□

Fixed-point theory in the differentiable setting has its own flavour and, as we shall see on several occasions, often leads to finiteness conditions.

Proposition 3.14 *Let* X *be a closed (smooth) manifold, and let* $\phi : X \to X$ *be a diffeomorphism of finite order* $k \geqslant 1$. *Then*

$$\mathcal{D}^m(X, \phi) = \begin{cases} \chi_c(X_m)/m & \text{if } m \mid k, \\ 0 & \text{otherwise,} \end{cases}$$

where χ_c *denotes the Euler characteristic with compact supports and* X_m *is the set of points whose orbit has order* m.

The set, $X_{(m)}$, of orbits of order m is the quotient of X_m by the free action of the cyclic group of order m generated by the restriction of ϕ and so is naturally a manifold: $\mathcal{D}^m(X, \phi) = \chi_c(X_{(m)})$ is its Euler characteristic.

To establish the result we choose a Riemannian metric on X. By averaging over the cyclic group of order k, we may arrange that ϕ is an isometry. We now use the fact that the Lefschetz index of an isometry is equal to the Euler characteristic of its fixed-submanifold. The reason is that the map can be deformed into the fixed-point set on a tubular neighbourhood without introducing new fixed-points. (See, for example, [9], Part II, (12.34).) So $L(\phi^m, X)$ is equal to the Euler characteristic of the fixed-point set of ϕ^m, which is $\coprod_{l \mid m} X_l$. By the additivity of the Euler characteristic,

$$L(\phi^m, X) = \sum_{l \mid m} \chi_c(X_l).$$

The result follows, by induction or using (2.13)(iii), from (2.11). □

Isolated fixed-points

Let us start again with a discussion of the general fixed-point theory. Suppose that a G-map $f : U \to M$, defined on an open G-subspace U of a G-ENR M, has precisely one fixed-point $a \in U$. (Of course, a must be fixed by G.) The Lefschetz index $L(f, U) \in \omega_G^0(*)$ is unchanged if we replace U by a smaller neighbourhood of a; it depends only on the germ of f at a, and we write it simply as

$$L(f, a) \in \omega_G^0(*).$$

When M is a G-manifold, the fixed-point index of an isolated fixed-point can be computed as a local degree. Without loss of generality, we may suppose that $M = E$ is a (Euclidean) G-module and that $a = 0 \in U \subseteq E$. Write $v : U \to E$ for the *vector field* $v(x) := x - f(x)$, which has an isolated zero at 0. Choose $\epsilon > 0$ such that the closed ϵ-disc in E, centre 0, is contained in U. Then $L(f, a) \in \omega_G^0(*)$ is equal to the (equivariant) *degree* of the self-map

$$a : x \mapsto \|v(\epsilon x)\|^{-1} v(\epsilon x) \ : \quad S(E) \to S(E) \tag{3.15}$$

of the unit sphere $S(E)$ in E. Radial extension of $a : S(E) \to S(E)$ ($tx \mapsto ta(x)$ for $t \in [0, \infty)$, $x \in S(E)$) produces a map $E^+ \to E^+$. This map, representing an element of $\omega_G^0\{S^0; S^0\} = \omega_G^0(*)$, defines the degree.

Suppose further that f is continuously differentiable with derivative $A = Df(0)$ at the fixed-point. If the fixed-point is non-degenerate, in the sense that $1 - A$ is non-singular, then the index is computed by K-theory. For $1 - A : E \to E$ defines an element of the real K-group $KO_G^{-1}(*)$. The map $E^+ \to E^+$ defined above as the radial extension of a is homotopic to the map $(1 - A)^+$ induced by $1 - A$. In other words, the local degree is the image of $[1 - A]$ under the equivariant *J-homomorphism*

$$J : KO_G^{-1}(*) \to \omega_G^0(*)^\times \ (\subseteq \omega_G^0(*)) \tag{3.16}$$

to the group of units in the stable cohomotopy ring. In the non-equivariant case, this is just the sign of the determinant:

$$\mathbb{Z}/2\mathbb{Z} \to \{\pm 1\} \subseteq \mathbb{Z},$$

and the local index is sign $\det(1 - A)$.

In the discussion below we shall meet the following situation, where f is again C^1 with an isolated fixed-point at 0. Suppose that $E = F_1 \oplus F_2$ splits as a direct sum (of G-modules) such that $f(U \cap F_1) \subseteq F_1$. Write $f_1 : U_1 \to F_1$ for the restriction of f to the first factor $U_1 = U \cap F_1$ and $A_2 : F_2 \to F_2$ for the $(2,2)$ component of $Df(0)$.

Lemma 3.17 *If, in the situation described in the text, $1 - A_2$ is non-singular, then*

$$L(f, 0) = L(f_1, 0) \cdot L(A_2, 0) \in \omega_G^0(*).$$

One shows that f is homotopic (on a neighbourhood of 0) through maps with an isolated fixed-point at 0 to the product $f_1 \times A_2$. The result follows from the multiplicativity of the index. □

Let us return once more to the study of the iterates of a map $\phi : U \to X$. Suppose that $a \in U$ is an isolated fixed-point of ϕ^k. If a is the only fixed-point in U, the indices $\mathcal{D}^m(\phi, U)$, for $m \mid k$, depend only on a, and we write them as

$$\mathcal{D}^m(\phi, a) \in \mathbb{Z} \tag{3.18}$$

(defined whenever a is an isolated fixed-point of ϕ^m). If a is an isolated fixed-point of ϕ^m for all $m \geqslant 1$, we have a corresponding zeta-function

$$Z(\phi, a; q) = \prod_{m \geqslant 1} (1 - q^m)^{\mathcal{D}^m(\phi, a)} = \exp\left(-\sum_{k \geqslant 1} \frac{L(\phi^k, a)}{k} q^k \right). \tag{3.19}$$

Exercises 3.20 (i) Let $d \in \mathbb{Z}$, and define $\phi : \mathbb{C} \to \mathbb{C}$ by $\phi(re^{i\theta}) = 2re^{id\theta}$. Then ϕ^k, for each $k \geqslant 1$, has just one fixed-point, at 0, and

$$Z(\phi, 0; q) = 1 - dq \in \mathbb{Z}[[q]].$$

(ii) Consider the polynomial mapping $\phi : \mathbb{C} \to \mathbb{C}$ given by $\phi(z) = \zeta z + z^{dk+1}$, where $k > 1$, $d \geqslant 1$ and ζ is a primitive kth root of unity. Then each power ϕ^m has an isolated fixed-point at 0 and

$$Z(\phi, 0; q) = (1 - q)(1 - q^k)^d.$$

(iii) Suppose that a is a fixed-point of a polynomial mapping $\phi : \mathbb{C} \to \mathbb{C}$ of degree > 1. If $\phi'(a)$ is a primitive kth root of unity, then $Z(\phi, a; q) = (1-q)\cdot(1-q^k)^d$ for some $d \geqslant 1$. If $\phi'(a)$ is not a root of unity, then $Z(\phi, a; q) = 1$.

We next look in more detail at differentiable maps. Suppose that $X = E$ is a Euclidean space and that $0 \in U$ is an isolated fixed-point of ϕ^k for all $k \geqslant 1$. In the non-singular case the calculation is straightforward.

Proposition 3.21 *Suppose that ϕ, as above, is C^1 and that no eigenvalue of $A := D\phi(0)$ is a root of unity. Set $d = \text{sign}\det(1-A) \in \{\pm 1\}$. Then*

$$Z(\phi, 0; q) = \begin{cases} (1-q)^d & \text{if sign}\det(1+A) = +1 \\ (1-q)^d(1-q^2)^{-d}(= (1+q)^{-d}) & \text{if sign}\det(1+A) = -1. \end{cases}$$

It is easy to determine $\text{sign}\det(1-A^k)$. The result then follows from (3.19). Alternatively, one can analyse $\pi_k(A)$ as in the proof of (3.22) below. □

In general, we have the following elegant result of Shub and Sullivan [42] showing that $\mathcal{D}^m(\phi, 0)$ vanishes for all but finitely many m. The example (3.20)(i) (also taken from [42]) shows that this need not be the case for a non-differentiable map. (See also [6].)

Proposition 3.22 *Under the hypotheses above, suppose that ϕ is C^1. Let $\Lambda \subseteq \mathbb{N}$ be the smallest set of natural numbers which is closed under formation of least common multiples and contains 1, 2 and the order of each eigenvalue of $A := D\phi(0)$ which is a root of unity. Then $\mathcal{D}^k(\phi, 0) = 0$ if $k \notin \Lambda$.*

Consider a natural number $k \notin \Lambda$. Then there is a proper divisor l of k which is even if k is even and such that no kth root of unity which is not an lth root of unity is an eigenvalue of A.

We have to examine the index of $\pi_k(\phi) : U^k \to E^k$. As a $\mathbb{Z}/k\mathbb{Z}$-module, $E^k = E \otimes V_k$, where V_k is the representation $\mathbb{R}^k$ on which $\mathbb{Z}/k\mathbb{Z}$ acts by cyclic permutation. The derivative of $\pi_k(\phi)$ at 0 is $\pi_k(A) = A \otimes S$, where $S : V_k \to V_k$ is given by the action of the generator $1 \in \mathbb{Z}/k\mathbb{Z}$ (the cyclic shift by one step).

Now

$$V_k = \begin{cases} \mathbb{R} \oplus (\bigoplus_{1 \leqslant r \leqslant (k-1)/2} L_r) & \text{if } k \text{ is odd,} \\ \mathbb{R} \oplus (\bigoplus_{1 \leqslant r < k/2} L_r) \oplus L & \text{if } k \text{ is even,} \end{cases}$$

where L_r is the 1-dimensional complex representation $\mathbb{C}$ on which S acts as multiplication by $e^{2\pi i r/k}$ and L, if k is even, is the 1-dimensional real representation $\mathbb{R}$ on which S acts as -1.

We split $E^k = E \otimes V_k$ as the direct sum $F_1 \oplus F_2$ of $F_1 = E \otimes V_l$ and $F_2 = E \otimes V_l^{\perp}$, where V_l is the submodule fixed by $l\mathbb{Z}/k\mathbb{Z}$ and $V_l^{\perp}$ is the direct sum of the complex lines L_r for r not divisible by l. Now apply (3.17) to $f = \pi_k(\phi) : U^k \to F_1 \oplus F_2$. In the notation introduced there, A_2 is $\mathbb{C}$-linear, and so the class of $1 - A_2$ in $KO^{-1}_{\mathbb{Z}/k\mathbb{Z}}(*)$ is trivial. This implies that $L(\pi_k(\phi), 0)$ is equal to the index of the restriction of $\pi_k(\phi)$ to the subspace $U^k \cap F_1$ fixed by $l\mathbb{Z}/k\mathbb{Z}$.

It follows that $L(\pi_k(\phi), 0)$ lies in the image of the homomorphism

$$p^* : \omega^0_{\mathbb{Z}/l\mathbb{Z}}(*) \to \omega^0_{\mathbb{Z}/k\mathbb{Z}}(*)$$

induced by the projection $p : \mathbb{Z}/k\mathbb{Z} \to \mathbb{Z}/l\mathbb{Z}$, and hence, by an elementary calculation, that $\mathcal{D}^k(\phi, 0) = 0$. (The homomorphism p^* splits the fixed-point homomorphism $\rho_{l\mathbb{Z}/k\mathbb{Z}}$ in (2.3).) □

Leaving the differentiable setting, consider a self-map $\phi : X \to X$ defined on the whole of X. We say that $a \in X$ is a *periodic point* of ϕ if $\phi^r(a) = a$ for some $r \geqslant 1$. If r is minimal, then the orbit of a is the finite set $[a] = \{\phi^i(a) \mid 0 \leqslant i < r\}$ of cardinality r. Suppose that a is an isolated fixed-point of ϕ^r. Then we may choose an open neighbourhood U of $[a]$ in which the only fixed-points of ϕ^r are the points of the orbit. In circumstances described by (3.13) the indices $\mathcal{D}^{m/r}(\phi^r, a)$ of the *monodromy* (or *Poincaré*) map ϕ^r determine the indices $\mathcal{D}^m(\phi, U)$ on U.

Exercise 3.23 Suppose that $\phi : X \to X$ is a self-map of a compact ENR such that each power ϕ^k has only isolated fixed-points. Then

$$Z(\phi, X; q) = \prod_{[a]} Z(\phi^{r(a)}, a; q^{r(a)}) \in \mathbb{Z}[[q]],$$

where the product runs over the (possibly infinite) set of orbits of periodic points, and $r(a)$ is the number of points in the orbit $[a]$ of $a \in X$. (See [29].)

4 Iterates of maps: the fibrewise version

A few preliminary remarks on fibrewise equivariant stable homotopy theory are necessary; more information may be found in [9], [45]. We work throughout this section over a compact ENR base space B. Let G, as in Section 3, be a compact Lie group, acting trivially on B, and let P and Q be pointed compact G-ENRs. A fibrewise equivariant stable map $B \times P \to B \times Q$ over B should be thought of as a family of equivariant stable maps $P \to Q$ parametrized by the base B. To be precise, such a fibrewise stable map is represented by a pair (v, F), where F is a G-module and $v : B \times (F^+ \wedge P) \to B \times (F^+ \wedge Q)$ is a basepoint-preserving G-map over B, that is, a continuous family of basepoint-preserving G-maps $v_b : F^+ \wedge P \to F^+ \wedge Q$, $(b \in B)$. A *fibrewise stable map* is defined as an appropriate equivalence class. (The situation we are considering, while sufficient for the present discussion, is special for two reasons. First, we are working with trivial bundles $B \times P$ and $B \times Q$ rather than with fibre bundles, or more general fibrewise spaces. Second, the group G is acting trivially on the base B.) We write

$$ {}_B\omega_G^0\{B \times P;\ B \times Q\} $$

for the group of fibrewise stable G-maps over B. Given a compact sub-ENR A of B, we define, also, a *relative group* of fibrewise stable G-maps which are zero over A:

$$ {}_{(B,A)}\omega_G^0\{B \times P;\ B \times Q\} $$

in terms of pairs (v, F) such that v_a is the null map for all $a \in A$. (Homotopies are similarly required to be null over A.) Groups ${}_B\omega_G^i\{-;-\}$ and ${}_{(B,A)}\omega_G^i\{-;-\}$ indexed by $i \in \mathbb{Z}$ are introduced as in Section 3.

It is easy to see, from the definition, that there is a natural identification

$$ {}_{(B,A)}\omega_G^*\{B \times S^0;\ B \times S^0\} = \omega_G^*(B, A), $$

specializing to

$$ {}_B\omega_G^*\{B \times S^0;\ B \times S^0\} = \omega_G^*(B). $$

(In fact, the group ${}_B\omega_G^*\{B \times P;\ B \times Q\}$ can be identified just as readily with $\omega_G^*\{B_+ \wedge P;\ Q\}$; but it would obscure the fibrewise theory to take this as a definition.)

The splitting of the Burnside ring $A(G)$ of a finite group G generalizes to a decomposition

$$ \omega_G^*(B, A) = \bigoplus_{(H)} \omega^*\{B/A;\ (\mathbf{B}W_G(H))^{\lambda(H)}\}, \tag{4.1} $$

indexed by the conjugacy classes of closed subgroups H of G, where, in the notation of (3.7), $(\mathbf{B}W_G(H))^{\lambda(H)}$ is the Thom space of the Lie algebra bundle $\lambda(H) = \mathbf{E}W_G(H) \times_{W_G(H)} \mathfrak{w}_G(H)$, associated to the adjoint representation $\mathfrak{w}_G(H)$ of $W_G(H)$, over the classifying space $\mathbf{B}W_G(H)$.

The development of a parametrized fixed-point theory is straightforward. Suppose that M is a G-ENR and that $f : U \to B \times M$ is a fibrewise G-map over B defined on an open G-subspace U of $B \times M$. Assume that $\mathrm{Fix}(f) = \{(b, x) \in U \mid f(b, x) = (b, x)\}$ is compact. Then the fibrewise Lefschetz–Hopf index is defined as a fibrewise stable G-map $B \times S^0 \to U_{+B}$ over B, where $U_{+B} = U \sqcup B$ is obtained by adjoining a basepoint to each fibre of U:

$$ \tilde{L}_B(f, U) \in {}_B\omega_G^0\{B \times S^0;\ U_{+B}\}. \tag{4.2} $$

Although we omitted the precise definition of a stable map $B \times S^0 \to U_{+B}$, both the meaning of the group and the construction of the index should be essentially clear if f is regarded as a family of maps $f_b : U_b \to M$, where U_b for $b \in B$ is the open subset $\{x \in M \mid (b, x) \in U\}$. In the same way, the fibrewise Lefschetz index is a class

$$ L_B(f, U) \in {}_B\omega_G^0\{B \times S^0;\ B \times S^0\} = \omega_G^0(B). \tag{4.3} $$

Suppose further that $A \subseteq B$ is a closed sub-ENR such that f has no fixed-points over A (so $f_a : U_a \to M$ has no fixed-points for $a \in A$). Then the Lefschetz index is naturally defined as a stable map which is zero over A:

$$L_{(B,A)}(f, U) \in \omega^0_G(B, A). \tag{4.4}$$

(We refer again to [9], Part II, Section 6, for details of the construction.)

Having completed the preparation, let us look again at the fixed-point theory for iterates of a map. Consider an ENR X and a fibrewise map $\phi : U \to B \times X$ over B defined on an open subspace U of $B \times X$. Let U^k_B, for $k \geqslant 1$, denote the k-fold fibre product over B with fibre at $b \in B$ the product U^k_b. The construction (2.5) can be performed in fibres to define a fibrewise $\mathbb{Z}/k\mathbb{Z}$-equivariant map

$$\pi_k(\phi) : U^k_B \to B \times X^k \tag{4.5}$$

over B. Supposing its fixed-point set to be compact, we have a fibrewise Lefschetz index

$$L_B(\pi_k(\phi), U^k_B) \in \omega^0_{\mathbb{Z}/k\mathbb{Z}}(B),$$

or, if the fixed-subspace is empty over A,

$$L_{(B,A)}(\pi_k(\phi), U^k_B) \in \omega^0_{\mathbb{Z}/k\mathbb{Z}}(B, A).$$

As in (3.8), the index of $\pi_l(\phi)$ for a divisor l of k is the image of the index of $\pi_k(\phi)$ under the fixed-point homomorphism and, using the decomposition (4.1):

$$\omega^0_{\mathbb{Z}/k\mathbb{Z}}(B, A) = \bigoplus_{m \mid k} \omega^0\{B/A;\ (\mathbf{B}(\mathbb{Z}/m\mathbb{Z}))_+\},$$

we may write

$$L_{(B,A)}(\pi_k(\phi), U^k_B) = \sum_{m \mid k} \mathcal{D}^m_{(B,A)}(\phi, U), \tag{4.6}$$

where

$$\mathcal{D}^m_{(B,A)}(\phi, U) \in \omega^0\{B/A;\ (\mathbf{B}(\mathbb{Z}/m\mathbb{Z}))_+\}. \tag{4.7}$$

To make this consistent with our earlier terminology we must identify the stable homotopy group $\omega^0\{S^0;\ (\mathbf{B}(\mathbb{Z}/m\mathbb{Z}))_+\}$ with $\mathbb{Z}$. Of course, when A is empty we write $\mathcal{D}^m_B(\phi, U) \in \omega^0\{B_+;\ (\mathbf{B}(\mathbb{Z}/m\mathbb{Z}))_+\}$.

The fibrewise version of (3.9) will be crucial for the definition of the Fuller index in the next section.

Proposition 4.8 *Suppose that l divides k. Then $L_{(B,A)}(\pi_l(\phi^{k/l}), V)$, where $V = (\phi^{-1})^{k/l-1}U$ is the image of $L_{(B,A)}(\pi_k(\phi), U)$ under the restriction map: $\omega^0_{\mathbb{T}(k)}(B, A) \to \omega^0_{\mathbb{T}(l)}(B, A)$.*

The proof of (3.9) generalizes by simply adding a parameter. □

5 Construction of the topological Fuller index

We make the definition of the Fuller index in some generality, only in the next section specializing to the basic case considered in Section 1. Let B be a compact ENR, and $A \subseteq B$ a closed sub-ENR. We consider a family of semi-flows, parametrized by B, on an ENR X. Suppose that $W \subseteq [0,\infty) \times B \times X$ is open and that

$$\Phi : W \to X \tag{5.1}$$

is a continuous map. For $b \in B$, we write $\Phi_b : W_b \to X$ for the map $\Phi_b(t,x) = \Phi(t,b,x)$ defined on $W_b = \{(t,x) \in [0,\infty) \times X \mid (t,b,x) \in W\}$. Each Φ_b is required to be a semi-flow, as in (1.1).

Our basic Fuller index will count orbits of period 1 (or, more accurately, fixed-points and orbits of period $1/r$ for some positive integer r). Let $U \subseteq B \times X$ be an open subset such that $[0,1] \times U \subseteq W$. We assume that $F := \{(b,x) \in U \mid \Phi(1,b,x) = x\}$ is compact and contained in $(B - A) \times X$. As we have already observed in Section 1, there is a natural (continuous) action of the circle $\mathbb{T}$ on the fixed-point set F:

$$t \cdot (b,x) = (b, \Phi(t,b,x)), \quad \text{for } t \in \mathbb{R}/\mathbb{Z}. \tag{5.2}$$

We now apply the theory of Section 4 to the map

$$\phi : U \to B \times X$$

over B given by $\phi(b,x) := (b, \Phi(1,b,x))$. Its fixed-subspace is $\mathrm{Fix}(\phi) = F$, so compact by hypothesis, and we can form at once the fixed-point indices of ϕ over B:

$$\tilde{L}_{(B,A)}(\phi, U) \in {}_{(B,A)}\omega^0\{B \times S^0;\, U_{+B}\}$$

and

$$L_{(B,A)}(\phi, U) \in \omega^0(B,A)$$

(which have been exploited in bifurcation theory by Alexander and others, [1, 2, 3, 4]). But now we can take kth roots for each $k \geqslant 1$, defining

$$\phi^{1/k} : U \to B \times X \tag{5.3}$$

by: $\phi^{1/k}(b,x) := \Phi(1/k, b, x)$. We carry out the construction (4.5) over B to form

$$\pi_k(\phi^{1/k}) : U_B^k \to B \times X^k. \tag{5.4}$$

The fixed-point set $\mathrm{Fix}(\pi_k(\phi^{1/k}))$ is naturally identified with F, and under this identification the action of $\mathbb{T}(k)$ corresponds to restriction of the $\mathbb{T}$-action (5.2) to the subgroup.

So we obtain a family of indices

$$\mathcal{F}^{(k)}_{(B,A)}(\Phi, U) := L_{(B,A)}(\pi_k(\phi^{1/k}), U_B^k) \in \omega^0_{\mathbb{T}(k)}(B,A), \tag{5.5}$$

for $k \geqslant 1$. They are related by (4.8).

Lemma 5.6 *If $l \mid k$ then $\mathcal{F}^{(k)}_{(B,A)}(\Phi, U)$ maps to $\mathcal{F}^{(l)}_{(B,A)}(\Phi, U)$ under restriction to the subgroup $\mathbb{T}(l)$ of $\mathbb{T}(k)$:*

$$\omega^0_{\mathbb{T}(k)}(B, A) \to \omega^0_{\mathbb{T}(l)}(B, A).$$

□

We call the element

$$\mathcal{F}_{(B,A)}(\Phi, U) \in \varprojlim \omega^0_{\mathbb{T}(k)}(B, A) \tag{5.7}$$

determined by the family $\mathcal{F}^{(k)}_{(B,A)}(\Phi, U)$, $k \geqslant 1$, the *topological Fuller index*. The inverse system is, of course, that defined by the restriction homomorphisms associated to the lattice of finite subgroups $\mathbb{T}(k)$ of $\mathbb{T}$.

Remark 5.8 Let $\mathcal{L}X$ denote the free loop space of all continuous maps $\alpha : \mathbb{T} \to X$, with $\mathbb{T}$ acting by rotating loops, and write $\mathcal{L}_B U$ for the subspace of $B \times \mathcal{L}X$ consisting of the pairs (b, α) with $(b, \alpha(t)) \in U$ for all $t \in \mathbb{T}$. With a little more effort, the Fuller index can be lifted to a class

$$\tilde{\mathcal{F}}_{(B,A)}(\Phi, U) \in \varprojlim {}_{(B,A)}\omega^0_{\mathbb{T}(k)}\{B \times S^0;\ (\mathcal{L}_B U)_{+B}\}.$$

For consider the map $p : \mathcal{L}_B U \to U^k_B$ given by restriction from $\mathbb{T}$ to $\mathbb{T}(k)$. If k is sufficiently large, we may embed X as a retract of some Euclidean space and use piecewise linear approximation to define, on some open neighbourhood N of F, a $\mathbb{T}(k)$-equivariant map $s : N \to \mathcal{L}_B U$ such that composition $p \circ s$ is the inclusion $N \to U^k_B$. Now form

$$\tilde{L}_{(B,A)}(\pi_k(\phi^{1/k}), N) \in {}_{(B,A)}\omega^0_{\mathbb{T}(k)}\{B \times S^0;\ N_{+B}\},$$

and compose with s. This defines $\tilde{\mathcal{F}}_{(B,A)}(\Phi, U)$. (Compare the definition of the Nielsen–Reidemeister index in [9], Part II, (6.12).) Composing with the inclusion of U in $B \times X$, we get an index in $\varprojlim \omega^0_{\mathbb{T}(k)}\{B/A;\ (\mathcal{L}X)_+\}$.

The inverse limit appearing in (5.7) is in itself rather intractable, but restriction from the circle $\mathbb{T}$ to its finite subgroups gives a monomorphism

$$\omega^0_{\mathbb{T}}(B, A) \to \varprojlim \omega^0_{\mathbb{T}(k)}(B, A). \tag{5.9}$$

(See [10] for the proof of injectivity.) In all our examples the index will lie in this subgroup.

As a special case of (4.1) we find that $\omega^0_{\mathbb{T}}(*) = \mathbb{Z}$.

Exercise 5.10 When B is a point (and $A = \emptyset$), the Fuller index is simply the Lefschetz index

$$L(\phi, U) \in \mathbb{Z} = \omega^0_{\mathbb{T}}(*) = \varprojlim\ \omega^0_{\mathbb{T}(k)}(*) = \varprojlim\ A(\mathbb{T}(k)).$$

The 1-dimensional groups are more interesting. The group $\omega_{\mathbb{T}}^{-1}(*)$ can be written as $\mathbb{Z}/2\mathbb{Z} \oplus \omega_1^{\mathbb{T}}(\mathbf{E}\mathfrak{F})$, where $\mathfrak{F}$ (see [7, 17]) is the family of finite subgroups of $\mathbb{T}$, and

$$\omega_1^{\mathbb{T}}(\mathbf{E}\mathfrak{F}) = \bigoplus_{n \geqslant 1} \mathbb{Z}\sigma_n = \bigoplus_{n \geqslant 1} \omega_0(\mathbf{B}(\mathbb{T}/\mathbb{T}(n))). \tag{5.11}$$

The class σ_n can be represented geometrically by the framed $\mathbb{T}$-manifold $\mathbb{T}/\mathbb{T}(n)$, framed by the trivialization of the tangent bundle. The restriction map

$$\omega_{\mathbb{T}}^{-1}(*) \to \omega_{\mathbb{T}(k)}^{-1}(*) = \bigoplus_{m \mid k} \omega_1(\mathbf{B}(\mathbb{Z}/\tfrac{k}{m}\mathbb{Z})) \tag{5.12}$$

(by (4.1) again) can be described as follows. We have $\omega_1(\mathbf{B}(\mathbb{Z}/l\mathbb{Z})) = (\mathbb{Z}/2\mathbb{Z}) \oplus (\mathbb{Z}/l\mathbb{Z})$, where the first factor is $\omega_1(*)$ and the second factor is the integral homology group $H_1(\mathbf{B}(\mathbb{Z}/l\mathbb{Z}))$.

Lemma 5.13 *The restriction map* (5.12) *maps* σ_n *to* $(1, n/m)$ *in the* m*-summand, where* $m = (n,k)$ *is the highest common factor.* (*Note that* n/m *is prime to* k/m.)

For consider the lth power homomorphism: $\mathbb{T} \to \mathbb{T}$. It maps σ_m to σ_{lm}. The lth power on $\mathbb{T}(k)$ induces multiplication by l on the homology group. The first component is non-trivial, because the circle framed by the trivialization of its tangent bundle represents the Hopf element in $\omega_1(*)$. □

We finish this section with an explicit computation (due, in effect, to Ize, [35]). Consider a continuous map $a : B \to \mathrm{End}(E)$ from B to the space of endomorphisms of a finite-dimensional real vector space E. We suppose that, for each $b \in B$, no eigenvalue of $a(b)$ is a multiple of $2\pi i$. Then the flow defined on $X = E$ by

$$\Phi(t, b, x) = e^{ta(b)}x$$

has a single stationary point, namely 0. We take the open set U, as in the definition, to be the whole of $B \times E$, and then $F = B \times \{0\}$. To calculate $\mathfrak{F}_B(\Phi, B \times E)$, we examine $\pi_k(\phi^{1/k})$, using ideas and notation from the proof of (3.22). It is the linear map

$$e^{a/k} \otimes S : E \otimes V_k \to E \otimes V_k.$$

Since B is compact, the eigenvalues of $a(b)$ are bounded, say by $2\pi N$. We may assume that k is even and $k \geqslant 4\pi N$. It is the index of an isolated fixed-point that is to be computed, and we evaluate it as a fibrewise local degree. On the summand L_r, S acts as $e^{2\pi i r/k}$. So the rth component of the index, for $0 < r < k/2$, is given by

$$1 - e^{(a + 2\pi i r)/k} : E \otimes \mathbb{C} \to E \otimes \mathbb{C},$$

as a map $f_r : B \to GL(E \otimes \mathbb{C})$ defining an element of the complex K-group $K^{-1}(B)$. Clearly f_r is homotopic to 1 if $|r| > N$, since the a/k term can be deformed to 0, and does not contribute to the index. (For the same reason the $E \otimes L$ factor is also trivial.) Using the inequality $|(e^z - 1) - z| \leqslant (e-2)|z|$ for $|z| \leqslant 1$, it is easy to see that, if $|r| \leqslant N$, f_r is homotopic to $-(a + 2\pi i r)/k$, and so to $-a - 2\pi i r$. On the trivial summand E we have $1 - e^{a/k}$ which is homotopic to $-a$. Now L_r is the restriction to $\mathbb{T}(k)$ of the rth power $Z^{\otimes r}$ of the standard 1-dimensional complex representation Z of $\mathbb{T}$, and the maps $-(a + 2\pi i r) \otimes 1 : E \otimes Z^{\otimes r} \to E \otimes Z^{\otimes r}$ are manifestly $\mathbb{T}$-equivariant. We see, therefore, that the index for large k is naturally the restriction of a $\mathbb{T}$-equivariant class which is independent of k. In the statement of the result thus proved we have taken the complex conjugate to replace $-2\pi i r$ by $2\pi i r$.

Proposition 5.14 *In the setting described above in the text, the Fuller index $\mathfrak{F}_B(\Phi, B \times E)$ is the image under the J-homomorphism*

$$J : KO_{\mathbb{T}}^{-1}(B) \to \omega_{\mathbb{T}}^0(B)^{\times} \subseteq \omega_{\mathbb{T}}^0(B)$$

of the class

$$([-a], ([2\pi i r - a])) \in KO^{-1}(B) \oplus \left(\bigoplus_{r \geqslant 1} K^{-1}(B)[Z^{\otimes r}] \right).$$

When the base B is a suspension, the product structure of $\omega_{\mathbb{T}}^0(B)$ is easy to describe. Let us specialize to the case that B is the circle S^1. Then $\omega_{\mathbb{T}}^0(S^1) = \omega_{\mathbb{T}}^0(*) \oplus \omega_{\mathbb{T}}^{-1}(*)$ is $\mathbb{Z} \oplus \mathbb{Z}/2\mathbb{Z} \oplus (\bigoplus_{r \geqslant 1} \mathbb{Z}\sigma_r)$. The index is $d(1, e, \sum_r a_r \sigma_r)$, where $d = \text{sign}\det(-a) \in \{\pm 1\}$, e is the element of $\pi_1(SO) = \mathbb{Z}/2\mathbb{Z}$ given by ab^{-1} where b is the constant map taking the value of a at the basepoint of S^1, and $a_r \in \mathbb{Z}$ is the degree of the map $P(a; 2\pi i r) : S^1 \to \mathbb{C}^{\times} (\to S^1)$, which evaluates the characteristic polynomial $P(a; \lambda) = \det(\lambda - a)$.

Exercise 5.15 Suppose that $a, b \in GL(E)$ are automorphisms of a finite-dimensional $\mathbb{C}$-vector space E, such that $p(x, y) := \det(xa + yb)$ is non-zero for all $(x, y) \in \mathbb{R}^2 - \{(0, 0)\}$ (which means that no eigenvalue of $a^{-1}b$ is real). Then the degree of the map $S^1 \to \mathbb{C}^{\times}(\to S^1) : (x, y) \mapsto p(x, y)$ is equal to $N_+ - N_-$, where N_+ and N_- are the numbers of eigenvalues (counted with multiplicity) of $a^{-1}b$ in the upper and lower half-planes respectively. More symmetrically, N_+ and N_- count points $[x, y]$ where $p(x, y) = 0$ in the hemispheres of the Riemann sphere $P(\mathbb{C}^2)$. (See [15] for related results.)

6 The index for a semi-flow

In this section we consider the index for a single semi-flow as specified in (1.1). We suppose that we are given an open subset U of $(0, \infty) \times X$ with $U \subseteq \Omega$ such that $F := \{(T, x) \in U \mid \theta(T, x) = x\}$ is compact.

Then there is a closed interval $[\alpha, \beta] \subseteq (0, \infty)$, such that $F \subseteq (\alpha, \beta) \times X$. There is no loss of generality in assuming that U is a subset of $(\alpha, \beta) \times X$.

We next fit this into the framework of Section 5. Put $B := [\alpha, \beta]$, $A := \{\alpha, \beta\} = \partial B$, $W := \{(t, b, x) \in [0, \infty) \times B \times X \mid (tb, x) \in \Omega\}$ and define

$$\Phi : W \to X \tag{6.1}$$

by $\Phi(t, b, x) := \theta(tb, x)$. Then $U \subseteq B \times X$ and $[0, 1] \times U \subseteq W$, and we can identify the subspace F above with the set $\{(b, x) \in U \mid \Phi(1, b, x) = x\}$.

The homeomorphism $t \mapsto (1 - t)\alpha + t\beta : [0, 1] \to [\alpha, \beta]$ determines an isomorphism of $\omega^{-1}_{\mathbb{T}(k)}(*)$ with $\omega^{0}_{\mathbb{T}(k)}(B, A)$. We define the *Fuller index*

$$\mathfrak{F}^*(\theta, U) \in \lim_{\leftarrow} \omega^{-1}_{\mathbb{T}(k)}(*), \tag{6.2}$$

to be the class corresponding to $\mathfrak{F}_{(B,A)}(\Phi, U)$. It is clearly independent of the choice of the interval $[\alpha, \beta]$, depending only on θ and the original set U. In the examples that we shall consider the index will lie in the subgroup $\omega^{\mathbb{T}}_1(\mathbf{E}\mathfrak{F}) \subseteq \omega^{-1}_{\mathbb{T}}(*)$, and, in fact, we know of no example in which it does not.

By its construction as a Lefschetz index, the Fuller index has the usual properties of a fixed-point index. It is localized at the set F of periodic orbits: if $F \subseteq U' \subseteq U$, then $\mathfrak{F}^*(\theta, U') = \mathfrak{F}^*(\theta, U)$. It is additive: if the subspace $U = U_1 \sqcup U_2$ is a disjoint union of two open sets, then $\mathfrak{F}^*(\theta, U) = \mathfrak{F}^*(\theta, U_1) + \mathfrak{F}^*(\theta, U_2)$. It is invariant under homotopy: if θ_λ, $0 \leqslant \lambda \leqslant 1$, is a family of semi-flows defined on Ω such that the union of the fixed-subspaces F_λ in U is compact, then $\mathfrak{F}^*(\theta_0, U) = \mathfrak{F}^*(\theta_1, U)$.

We shall give two situations in which the index can be computed, reasonably explicitly, as an element of $\omega^{\mathbb{T}}_1(\mathbf{E}\mathfrak{F})$, and, in particular, show that our definition coincides with Fuller's for a C^1-flow when F is a finite union of isolated periodic orbits. (The results have close parallels in the theory of the vector field index discussed in [8].)

We look first at an analogue of (3.13). Suppose that $\psi : Y \to Y$ is a diffeomorphism of a closed manifold Y. Let X be the mapping torus $\mathbb{R} \times_{\mathbb{Z}} Y$ of ψ (where $1 \in \mathbb{Z}$ acts as ψ) with the (globally defined) flow $\theta(t, -)$ induced from $u \mapsto t + u$ on $\mathbb{R}$. The fixed-point set of the associated map $(t, x) \mapsto (t, \theta(t, x)) : (0, \infty) \times X \to (0, \infty) \times X$ is

$$\bigcup_{k \geqslant 1} \{k\} \times (\mathbb{R} \times_{\mathbb{Z}} \mathrm{Fix}(\psi^k)) \subseteq (0, \infty) \times X. \tag{6.3}$$

For each $k \geqslant 1$ we have an index $L(\pi_k(\psi), Y^k)$ in $\omega^{0}_{\mathbb{T}(k)}(*)$, as described in Section 3. Let $k \geqslant 1$, $0 < \epsilon < 1$, and take $U = (k - \epsilon, k + \epsilon) \times X$.

Proposition 6.4 *The Fuller index $\mathfrak{F}^*(\theta, U)$ of the flow described above lies in $\omega^{\mathbb{T}}_1(\mathbf{E}\mathfrak{F})$ and is equal to the image of the index $L(\pi_k(\psi), Y^k)$ under the induction homomorphism,* (3.3),

$$\omega^{0}_{\mathbb{T}(k)}(*) \to \omega^{-1}_{\mathbb{T}}(*)$$

associated to the inclusion $\mathbb{T}(k) \to \mathbb{T}$. *In terms of the indices* $\mathcal{D}^m(\psi, Y)$, *we have*

$$\mathcal{F}^*(\theta, (k-\epsilon, k+\epsilon) \times X) = \sum_{n|k} \mathcal{D}^{k/n}(\psi, Y) \cdot \sigma_n \in \omega_1^{\mathbb{T}}(\mathbf{E}\mathfrak{F}).$$

The proof of this and of (6.5) below can be found in [10]. □

Next we look at the closely related case of an isolated periodic orbit. We suppose that X is a smooth manifold and θ a smooth flow. Let $C \subseteq X$ be an isolated periodic orbit of minimal period T_0. Choose an embedding $N \to X$ of the normal bundle N of C in X as an open tubular neighbourhood of C meeting no other periodic orbit and containing no stationary point of the flow. Fix a point $x_0 \in C$. By following the flow around close to C we obtain the *Poincaré map* $\psi : V \to N_{x_0}$ defined on an open neighbourhood V of 0 in N_{x_0}. Thus, $\psi(x)$, for $x \in V \subseteq X$, will be $\theta(T_0 + \tau(x), x)$, where $\tau : V \to \mathbb{R}$ is smooth and $\tau(0) = 0$. For $k \geqslant 1$, ψ^k has an isolated fixed-point at 0. Replacing N if necessary by a smaller tubular neighbourhood, we may assume that θ is defined on $U = (kT_0 - \epsilon, kT_0 + \epsilon) \times N$, where $0 < \epsilon < T_0$. Then we have $F = \{kT_0\} \times C$.

Proposition 6.5 *For an isolated periodic orbit as described above, the Fuller index is given by*

$$\mathcal{F}^*(\theta, (kT_0 - \epsilon, kT_0 + \epsilon) \times N) = \sum_{n|k} \mathcal{D}^{k/n}(\psi, 0) \cdot \sigma_n \in \omega_1^{\mathbb{T}}(\mathbf{E}\mathfrak{F}).$$

□

Example 6.6 This allows us to determine the indices in our introductory example (1.2). For $\lambda = 0$, the single periodic orbit, with (minimal) period 1, has the monodromy index $\mathcal{D}^1(\psi, 0) = 1$, $\mathcal{D}^m(\psi, 0) = 0$ for $m > 1$. At period $k \geqslant 1$, as in (6.5), the Fuller index is σ_k.

For $\lambda \neq 0$, there are two orbits: (i) at $y = 0$, with period 1 and index: $\mathcal{D}^1(\psi, 0) = 1$, $\mathcal{D}^2(\psi, 0) = -1$, $\mathcal{D}^m(\psi, 0) = 0$ for $m > 2$; (ii) at $y = \pm\lambda$, with period 2 and index: $\mathcal{D}^1(\psi, \pm\lambda) = 1$, $\mathcal{D}^m(\psi, \pm\lambda) = 0$ for $m > 1$. The contribution to the Fuller index at period $k \geqslant 1$ is: from $y = 0$, σ_k if k is odd, $\sigma_k - \sigma_{k/2}$ if k is even; and from $y = \pm\lambda$, 0 if k is odd, $\sigma_{k/2}$ if k is even.

Notice how as λ moves from 0 and the $y = 0$ orbit splits into two the index is preserved:

$$\sigma_k = \begin{cases} \sigma_k + 0 & \text{if } k \text{ is odd} \\ (\sigma_k - \sigma_{k/2}) + \sigma_{k/2} & \text{if } k \text{ is even.} \end{cases}$$

We can now explain the connection with Fuller's definition in [33] of a $\mathbb{Q}$-valued index. As explained in [7], the equivariant *integral* Borel (co)homology group $H_1^{\mathbb{T}}(\mathbf{E}\mathfrak{F})$, constructed as a direct limit, is equal $\mathbb{Q}$, and the Hurewicz map from equivariant stable homotopy to integral homology:

$$\omega_1^{\mathbb{T}}(\mathbf{E}\mathfrak{F}) \to H_1^{\mathbb{T}}(\mathbf{E}\mathfrak{F}) = \mathbb{Q} \tag{6.7}$$

maps $\sum a_n\sigma_n$ to $\sum a_n/n$. For the isolated periodic orbit this gives the index

$$\sum_{n|k} \frac{k}{n} \cdot \mathcal{D}^{k/n}(\psi, 0) \cdot \frac{1}{k} = \frac{L(\psi^k, 0)}{k} \in \mathbb{Q}, \tag{6.8}$$

by (3.19) (or (3.11)). Fuller took this as his starting point and defined the index for a general smooth flow by using the result that any smooth flow can be deformed into one with only isolated periodic orbits. He then established that his index was well-defined and homotopy-invariant essentially by using the ideas that we have reformulated as the definition, (6.2).

Corollary 6.9 *For a smooth flow θ on a manifold X, the Fuller index $\mathcal{F}^*(\theta, U)$ lies in $\omega_1^{\mathbb{T}}(\mathbf{E}\mathfrak{F})$.*

For, by homotopy-invariance and additivity, the index will be a sum of terms of the type (6.5). We shall describe a more general result, with a topological proof, in Section 7. □

As a final example, an analogue of (3.14), we suppose that X is a closed smooth $\mathbb{T}$-manifold and take the flow θ given by the circle action: $\theta(t,x) := t\cdot x$. Consider a rational number $j/k \in (0,\infty)$, written in its lowest terms, with $k \geqslant 1$. We take $U = (j/k - \epsilon, j/k + \epsilon) \times X$, where $\epsilon > 0$ is chosen so small that the fixed-point set F is precisely $\{j/k\} \times X^{\mathbb{T}(k)}$.

Proposition 6.10 *The Fuller index of the circle action given above is*

$$\mathcal{F}^*(\theta, (\tfrac{j}{k} - \epsilon, \tfrac{j}{k} + \epsilon) \times X) = \sum_{n\,:\,k|n} \chi_c(X_{(\mathbb{T}(n))}/\mathbb{T}) \cdot \sigma_{jn/k} \in \omega_1^{\mathbb{T}}(\mathbf{E}\mathfrak{F}),$$

where $X_{(\mathbb{T}(n))}$ is the subspace of points of X with isotropy group exactly $\mathbb{T}(n)$.

We recall that χ_c denotes the Euler characteristic with compact supports. See [10] again for a discussion of the proof. □

Bifurcation

The primary use of the Fuller index has been to establish the existence of bifurcation of periodic solutions of differential equations. Suppose that we have a family of semi-flows θ_λ, $0 \leqslant \lambda \leqslant 1$, defined on $\Omega \subseteq (0,\infty) \times X$. Let $V \subseteq X$ be an open subspace and let $[\alpha, \beta] \subseteq (0, \infty)$ be an interval such that $[\alpha,\beta] \times V \subseteq \Omega$ and (i) for $\lambda = 0$, 1, the set $\{(t,x) \in (\alpha,\beta) \times V \mid \theta_\lambda(t,x) = x\}$ is compact; (ii) for all $\lambda \in [0,1]$, $x \in V$, we have $\theta_\lambda(\alpha, x) \neq x$ and $\theta_\lambda(\beta, x) \neq x$. This means that the Fuller indices $\mathcal{F}^*(\theta_\lambda, (\alpha,\beta) \times V)$ are defined for $\lambda = 0$, 1. Their difference can be computed as follows. Take the base B to be the boundary of the rectangle $[\alpha, \beta] \times [0,1]$. We can define a family of flows Φ on X parametrized by B by:

$$\Phi(t, (T, \lambda), x) = \theta_\lambda(t/T, x).$$

Making the natural identification of B with the circle S^1, up to homotopy, and noting that the restriction of the index to a point of the base is 0, we can regard $\mathfrak{F}_B(\Phi, B \times V)$ as an element of

$$\underleftarrow{\lim}\ \omega^{-1}_{\mathbb{T}(k)}(*) \subseteq \underleftarrow{\lim}\ \omega^{0}_{\mathbb{T}(k)}(S^1) = \underleftarrow{\lim}\ \omega^{0}_{\mathbb{T}(k)}(B).$$

From the definition of addition in stable homotopy we obtain:

Lemma 6.11 *The change in the Fuller index is given by:*

$$\mathfrak{F}^*(\theta_0, (\alpha,\beta) \times V) - \mathfrak{F}^*(\theta_1, (\alpha,\beta) \times V) = \mathfrak{F}_B(\Phi, B \times V) \in \underleftarrow{\lim}\ \omega^{-1}_{\mathbb{T}(k)}(*).$$

□

If the difference is non-zero, we deduce that the set of points $(T, \lambda, x) \in (\alpha,\beta) \times [0,1] \times V$ with $\theta_\lambda(T,x) = x$ is not compact, which means that some sort of bifurcation of periodic orbits occurs as the parameter λ varies from 0 to 1. See [35] for an application of (5.14) to this situation.

The index of Dancer and Toland

In a series of papers [14, 15, 16], Dancer and Toland investigate the following situation. Suppose that X is a smooth manifold with a Riemannian metric and that w is a smooth vector field on X defining a flow $\theta : \Omega \to X$. We assume that there is a smooth function $\rho : X \to \mathbb{R}$ with no critical points, such that $d\rho(w) = 0$. Let $T > 0$ be a fixed period, and suppose that $V \subseteq X$ is an open subspace such that $(0,T] \times \overline{V} \subseteq \Omega$ and the set of points $x \in V$ with $\theta(T,x) = x$ is compact.

For $\lambda \in \mathbb{R}$, consider the vector field $w_\lambda = w + \lambda \operatorname{grad} \rho$. If λ is sufficiently small, say $|\lambda| \leqslant \epsilon$, the corresponding flow θ_λ will also be defined on $(0,T] \times V$. But for $\lambda \neq 0$ it will have no periodic orbits, because on a trajectory $x(t)$

$$\frac{d}{dt}\rho(x) = d\rho\left(\frac{dx}{dt}\right) = \lambda \|d\rho\|^2$$

is of constant sign.

We take $B = [-\epsilon, \epsilon]$, $A = \partial B$, $\Phi(t, \lambda, x) = \theta_\lambda(t/T, x)$. Then

$$\mathfrak{F}_{(B,A)}(\Phi, B \times V) \in \underleftarrow{\lim}\ \omega^{-1}_{\mathbb{T}(k)}(*), \tag{6.12}$$

depending on ρ but not on the choice of ϵ, is the index considered by Dancer and Toland.

7 The Fuller index of a smooth flow

In this final section we explain how to define the Fuller index of a smooth flow directly as a $\mathbb{T}$-equivariant vector field index, rather than as a fixed-point index. Let X be a smooth manifold.

For the preliminary discussion consider a smooth vector field w on X. The space M of smooth loops $\alpha : \mathbb{T} \to X$ is an infinite-dimensional manifold. Its tangent space $T_\alpha M$ at α is the space of smooth sections of $\alpha^* TX$ over $\mathbb{T}$. The group $\mathbb{T}$ acts on M by rotating loops: to be consistent with our usage in earlier sections we define $(s \cdot \alpha)(t) := \alpha(s+t)$, for $s, t \in \mathbb{T} = \mathbb{R}/\mathbb{Z}$. Now the vector field w on X determines a $\mathbb{T}$-equivariant vector field v on M: $v(\alpha) = \dot{\alpha} - w(\alpha)$, where the dot, as usual, denotes differentiation. A zero of this vector field is a periodic solution α of the differential equation $\dot{\alpha} = v(\alpha)$, of period 1. In other words, zeros of v correspond to periodic orbits of the flow Φ on X determined by w.

More generally, we may consider a family w_b, $(b \in B)$, of smooth vector fields on X parametrized by a compact ENR B. We obtain a corresponding family of vector fields v_b on M. Let $A \subseteq B$ be a closed sub-ENR, and suppose that $U \subseteq B \times M$ is an open $\mathbb{T}$-subspace such that the *zero-set*

$$Z := \{(b, \alpha) \in U \mid v_b(\alpha) = 0\} \tag{7.1}$$

is compact and also that there are are no points of Z over A. By analogy with the (fibrewise, equivariant) index theory for vector fields on finite-dimensional manifolds (as described in [9, 8], for example), one would hope to be able to define a Poincaré–Hopf index $I_B(v, U) \in \omega^0_{\mathbb{T}}(B, A)$. In order to define such an index one requires additional structure. We shall explain here how to define such an index for our particular vector field. It will be necessary to make some technical changes to the definition of both the manifold M and the vector field v.

For simplicity we shall suppose that $X = E$ is a Euclidean space. (See [10] for the general case.) It will be enough to assume that the vector field w on E is C^1 (rather than C^∞). We redefine the space $M = C^1(\mathbb{T};\, E)$ to be the space of C^1-loops in E; it is a Banach space.

Let $\lambda : [0, \infty) \to \mathbb{R}$ be a non-negative smooth function with compact support such that $\int_0^\infty \lambda = 1$. We define an operator $P_\lambda : C^0(\mathbb{T};\, E) \to C^1(\mathbb{T};\, E)$ on continuous loops x by:

$$(P_\lambda x)(t) := \int_0^\infty \lambda(s) \left\{ \int_{t-s}^{t} x(u)\, du \right\} ds, \quad t \in \mathbb{R}. \tag{7.2}$$

(All the functions are interpreted as periodic functions on $\mathbb{R}$.) Now P_λ is a $\mathbb{T}$-invariant parametrix (inverse modulo compact operators) for differentiation $D : C^1(\mathbb{T};\, E) \to C^0(\mathbb{T};\, E)$. We have $P_\lambda D = 1 - K_\lambda$ where $K_\lambda : C^0(\mathbb{T};\, E) \to C^1(\mathbb{T};\, E)$ is the compact convolution operator

$$(K_\lambda x)(t) := \int_0^\infty \lambda(s) x(t-s)\, ds;$$

and $DP_\lambda = 1 - iK_\lambda$, where $i : C^1(\mathbb{T};\, E) \to C^0(\mathbb{T};\, E)$ is the (compact) inclusion.

Lemma 7.3 *The parametrix P_λ is injective:* ker $P_\lambda = 0$.

For suppose that $P_\lambda x = 0$. Then $x = K_\lambda x$. Choose t_0 such that $\|x(t_0)\|$ is maximal, and let S be the closed set $\{t \in \mathbb{R} \mid x(t) = x(t_0)\}$. If $t \in S$, it follows that $u \in S$ whenever $\lambda(t-u) > 0$. In particular, there is an irrational a such that $a + S \subseteq S$. By periodicity, we find that $S = \mathbb{R}$. So x is constant. Since $\int_0^\infty s\lambda(s)\, ds > 0$, we must have $x = 0$. □

We define a (new) vector field v on $M = C^1(\mathbb{T}; E)$, depending on λ, by

$$v(\alpha) := P_\lambda(D\alpha - w(\alpha)). \tag{7.4}$$

By the injectivity of P_λ, the zero-set of v is the set of periodic solutions of $\dot\alpha = v(\alpha)$ of period 1. It is also well behaved from the point of view of fixed-point theory.

Lemma 7.5 *The (non-linear) map $f = 1 - v : C^1(\mathbb{T}; E) \to C^1(\mathbb{T}; E)$, is compact.*

For $f(\alpha) = K_\lambda(\alpha) - P_\lambda i(w(\alpha))$, and both K_λ and $P_\lambda i$ are compact maps. □

From a family w_b of C^1 vector fields on $X = E$ we obtain, by the same construction, a family of continuous $\mathbb{T}$-equivariant vector fields v_b on $M = C^1(\mathbb{T}; E)$ (depending still on our fixed choice of λ).

Now suppose that the vector field w defines a flow $\Phi : W \to X$, as in (5.1), on an open set $W \subseteq [0,\infty)\times B\times X$. Let $U \subseteq B\times X$ be an open subset such that $[0,1]\times U \subseteq W$. We require, as in Section 5, that $F := \{(b,x) \in U \mid \Phi(1,b,x) = x\}$ be compact and contained in $(B - A) \times X$.

Let U_B^∞ (corresponding to the subspaces U_B^k of Section 5) be the open subspace $\{(b,\alpha) \mid (b,\alpha(\mathbb{T})) \subseteq U\}$ of $B \times M$. (For continuous loops, in (5.8), we used the notation $\mathcal{L}_B U$.) The zero set $Z := \{(b,\alpha) \in U_B^\infty \mid v_b(\alpha) = 0\}$ is naturally homeomorphic to the fixed-subspace F.

We can now define the vector field index $I_{(B,A)}(v, U_B^\infty)$ as a Lefschetz index. For the map $f_b : M \to M$ given by $f_b(\alpha) = \alpha - v_b(\alpha)$ is compact. The family $f = (f_b)$ over B is defined on U_B^∞ and has a compact fixed-subspace Z, with no fixed-points over A. The Leray–Schauder theory, as formulated in [34], for compactly-fixed self-maps of (metrisable) ANRs extends in a routine way to a fibrewise equivariant theory. (See [9], Part II, Section 7.) The construction proceeds by approximating the compact map f by a map to a finite-dimensional Euclidean space. It provides us with a fixed-point index

$$L_{(B,A)}(f, U_B^\infty) \in \omega^0_{\mathbb{T}}(B, A);$$

we take this as our *ad hoc* definition of the vector field index $I_{(B,A)}(v, U_B^\infty)$.

The $\mathbb{T}$-equivariant fibrewise vector field index so defined is, indeed, the Fuller index.

Proposition 7.6 *The vector field index $I_{(B,A)}(v, U_B^\infty)$ maps to the Fuller index $\mathcal{F}_{(B,A)}(\Phi, U)$ under the inclusion:*

$$\omega^0_{\mathbb{T}}(B, A) \to \lim_{\substack{\longleftarrow \\ k}} \omega^0_{\mathbb{T}(k)}(B, A).$$

We refer to [10] for the details of the proof. □

Remark 7.7 In the absence of stationary points, the $\mathbb{T}$-action is fixed-point-free, and the vector field index lies in $\omega^0_{\mathbb{T}}\{B/A;\, (\mathbf{E}\mathfrak{F})_+\}$.

This result confirms the assertion (6.9) that the Fuller index $\mathcal{F}^*(\theta, U)$ for a smooth flow θ lies in $\omega_1^{\mathbb{T}}(\mathbf{E}\mathfrak{F})$ and so gives a rational index in $H_1^{\mathbb{T}}(\mathbf{E}\mathfrak{F}) = \mathbb{Q}$, by (6.7). The proof is topological and does not depend on perturbing the flow to one with only isolated orbits. (It is hard to see how such an argument might be extended to the parametrized theory.) We conclude our account with a sketch of the proof.

Working $\mathbb{T}(k)$-equivariantly, we have to show that a vector field index on a space of maps $\mathbb{T} \to E$ coincides with a fixed-point index defined on the space $\text{map}(\mathbb{T}(k), E) = E^k$. Every map $\mathbb{T}(k) \to E$, specified by $(x_1, \ldots, x_n) \in E^k$, extends to a piecewise linear loop $\mathbb{T} = \mathbb{R}/\mathbb{Z} \to E$:

$$(1-t)\frac{i}{k} + t\frac{(i+1)}{k} \mapsto (1-t)x_i + tx_{i+1}, \quad \text{for } 0 \leqslant t \leqslant 1.$$

By enlarging our loop-space to include both the C^1 and the piecewise linear loops, we can, when k is sufficiently large, take a finite-dimensional approximation on $\text{map}(\mathbb{T}(k), E)$ to define the infinite-dimensional ANR fixed-point index.

The discrete approximation which arises is not exactly of the form $\pi_k(\phi)$ which we used to define the Fuller index in Section 5 but of the following shape. For ease of notation, we drop the parametrization by B. Consider a continuous map $\phi : U \to E$ defined on an open subspace U of the Euclidean space E, such that $\text{Fix}(\pi_k(\phi))$ is compact. Let $(\lambda_i)_{i \geqslant 1}$ be a sequence of non-negative real numbers, only a finite number non-zero, with sum $\sum_i \lambda_i = 1$. Suppose also that $\lambda_1 > 0$. Regarding $E^k = \text{map}(\mathbb{Z}/k\mathbb{Z}, E)$ as the set of sequences $\mathbb{Z} \to E$ of period k, we define a map

$$\pi_k^\lambda(\phi) : U^k \to E^k, \quad (x_1, \ldots, x_k) \mapsto (y_1, \ldots, y_k), \tag{7.8}$$

where

$$y_i = x_i - \sum_{j \geqslant 1} \lambda_j \left(\sum_{m=i-j}^{i-1} ((x_{m+1} - x_m) - (\phi(x_m) - x_m) \right).$$

This is the discrete analogue of the map f: $f(\alpha) = \alpha - P_\lambda(D\alpha - w(\alpha))$.

If $\lambda = (1, 0, \ldots)$, $y_i = \phi(x_{i-1})$ and $\pi_k^\lambda(\phi) = \pi_k(\phi)$ from (2.5).

Exercise 7.9 For any λ, the fixed-subspace of $\pi_k^\lambda(\phi)$ coincides with $\text{Fix}(\pi_k(\phi))$.

By the convexity of the space of sequences (λ_i), the Lefschetz index of $\pi_k^\lambda(\phi)$ is independent of (λ_i) and equal to $L(\pi_k(\phi), U^k)$. So we have made the connection between the vector field index, via discrete approximation, and the definition of the Fuller index through the π_k-construction.

References

1. J.C. Alexander: Bifurcation of zeroes of parametrized functions, *J. Funct. Anal.* **29** (1978), 37–53.
2. J.C. Alexander and J.A. Yorke: Global bifurcation of periodic orbits, *Amer. J. Math.* **100** (1978), 263–292.
3. T. Bartsch: A global index for bifurcation of fixed points, *J. reine angew. Math.* **391** (1988), 181–197.
4. T. Bartsch: Bifurcation theory for metric parameter spaces, *Springer LNM* **1411** (1989), 1–8.
5. S-N. Chow and J. Mallet-Paret: The Fuller index and global Hopf bifurcation, *J. Diff. Equations* **29** (1978), 66–85.
6. S-N. Chow, J. Mallet-Paret and J.A. Yorke: A periodic orbit index which is a bifurcation invariant, *Springer LNM* **1007** (1983), 109–131.
7. M.C. Crabb: Invariants of fixed-point-free circle actions, *LMS Lecture note series* **139** (1989), 21–47.
8. M.C. Crabb: The Fuller index and $\mathbb{T}$-equivariant stable homotopy theory, *Astérisque* **191** (1990), 71–86.
9. M.C. Crabb and I.M. James: 'Fibrewise Homotopy Theory', Springer (Berlin 1998).
10. M.C. Crabb and A.J.B. Potter: 'The topological Fuller index', in preparation.
11. E.N. Dancer: Perturbation of zeros in the presence of symmetries, *J. Australian Math. Soc.* **A36** (1984), 106–125.
12. E.N. Dancer: A new degree for S^1-invariant gradient mappings, *Ann. Inst. Henri Poincaré* **2** (1985), 329–370.
13. E.N. Dancer: Remarks on S^1 Symmetries and a Special Degree for S^1-Invariant Gradient Mappings, *Proc. Symp. Pure Math.* **45** (1986), 353–358.
14. E.N. Dancer and J.F. Toland: Degree theory for orbits of prescribed period of flows with a first integral, *Proc. London Math. Soc.* **60** (1990), 549–580.
15. E.N. Dancer and J.F. Toland: Equilibrium states in the degree theory of periodic orbits with a first integral, *Proc. London Math. Soc.* **63** (1991), 569–594.
16. E.N. Dancer and J.F. Toland: The index change and global bifurcation for flows with a first integral, *Proc. London Math. Soc.* **66** (1993), 539–567.
17. T. tom Dieck: 'Transformation Groups', Walter de Gruyter (Berlin, 1987).
18. A. Dold: 'Lectures on Algebraic Topology', Springer (Berlin, 1972).
19. A. Dold: The fixed point index of fibre-preserving maps, *Invent. Math.* **25** (1974), 281–297.
20. A. Dold: The fixed point transfer of fibre-preserving maps, *Math. Zeit.* **148** (1976), 215–244.
21. A. Dold: Fixed point indices of iterated maps, *Invent. math.* **74** (1983), 419–435.
22. A. Dold and D. Puppe: 'Duality, trace and transfer', in Proc. of the Internat. Conf. on Geom. Topology, 81–102, Warszawa, 1980.
23. G. Dylawerski, K. Geba, J. Jodel and W. Marzantowicz: An S^1-equivariant degree and the Fuller index, *Annales Polonici Mathematici* **LII** (1991), 243–280.
24. L.H. Erbe, K. Geba and W. Krawcewicz: Equivariant fixed point index and the period-doubling cascades, *Canadian J. Math.* **43** (1991), 738–747.
25. C.C. Fenske: A simple-minded approach to the index of periodic orbits, *J. Math. Anal. Appl.* **129** (1988), 517–532.
26. C.C. Fenske: An index for periodic orbits of functional differential equations, *Math. Ann.* **285** (1989), 381–392.

27. C.C. Fenske: Fuller's index for periodic solutions of functional differential equations, *Springer LNM* **1411** (1989), 69–74.
28. P.M. Fitzpatrick and J. Pejsachowicz: Parity and generalized multiplicity, *Trans. Amer. Math. Soc.* **326** (1991), 281–305.
29. J. Franks: Period doubling and the Lefschetz formula, *Trans. Amer. Math. Soc.* **287** (1985), 275–283.
30. J. Franks and D. Fried: The Lefschetz function of a point, *Springer LNM* **1411** (1989), 83–87.
31. R.D. Franzosa: An homology index generalizing Fuller's index for periodic orbits, *J. Diff. Equations* **84** (1990), 1–14.
32. D. Fried: Lefschetz formulas for flows, *Contemp. Math.* **58 III** (1987), 19–69.
33. F.B. Fuller: An index of fixed point type for periodic orbits, *Amer. J. Math.* **89** (1967), 133–148.
34. A. Granas: The Leray–Schauder index and the fixed point theory for arbitrary ANRs, *Bull. Soc. Math. France* **100** (1972), 209–228.
35. J. Ize: Obstruction theory and multiparameter Hopf bifurcation, *Trans. Amer. Math. Soc.* **289** (1985), 757–792.
36. J. Ize, I. Massabó and A. Vignoli: Degree theory for equivariant maps. I, *Trans. Amer. Math. Soc.* **315** (1989), 433–510.
37. J. Ize, I. Massabó and A. Vignoli: 'Degree theory for equivariant maps', Memoirs of the AMS, Volume 100, no. 481 (1992).
38. K. Komiya: Fixed point indices of equivariant maps and Möbius inversion, *Invent. Math.* **91** (1988), 129–135.
39. K. Komiya: Congruences for fixed point indices of equivariant maps and iterated maps, *Springer LNM* **1411** (1989), 130–136.
40. T. Matsuoka: The number of periodic orbits of smooth maps, *Springer LNM* **1411** (1989), 148–154.
41. A.J.B. Potter: Approximation methods and the generalised Fuller index for semi-flows in Banach spaces, *Proc. Edinburgh Math. Soc.* **29** (1986), 299–308.
42. M. Shub and D. Sullivan: A remark on the Lefschetz fixed point formula for differentiable maps, *Topology* **13** (1974), 189–191.
43. R. Srzednicki: Periodic orbits indices, *Fundamenta Math.* **125** (1990), 147–173.
44. R. Srzednicki: Topological invariants and detection of periodic orbits, *J. Diff. Equations* **111** (1994), 283–298.
45. H. Ulrich: 'Fixed Point Theory of Parametrized Equivariant Maps', Springer LNM 1343 (Berlin, 1988).
46. P.P. Zabreiko and M.A. Krasnosel'skii: Iterations of operators, and fixed points, *Soviet Math. Dokl.* **12** (1971), 1006–1009.

4
The Borel–Weil theorem for complex projective space

MICHAEL EASTWOOD[1] AND JUSTIN SAWON

Introduction

Our aim, in this chapter, is to give a complete classification of the irreducible finite-dimensional holomorphic representations of the complex general linear group. We shall construct them as spaces of holomorphic sections of certain vector bundles over complex projective space. This construction is essentially due to Borel and Weil (with exposition due to Serre [18]).

Our account is complete in the sense that we shall assume that the reader is familiar with basic analysis and the theory of functions of several complex variables (for example, as in [10] or [19]), but we shall assume the reader knows practically nothing of representation theory. This combination is apparently not so uncommon. (For example, one of us (ME) fell into this category for many years.)

The Borel–Weil theorem is much more general than the version we shall explain in this chapter. Moreover, the literature already contains many fine proofs (see, for example, [13]). Everything in this chapter is well known. However, we hope that, by sticking to the most familiar of groups (the general linear group) and the most familiar of spaces (complex projective space), we can avoid some of the hurdles which present themselves in a more general treatment.

The existence of such hurdles is already lamented in Knapp's book [13, p. xv] where, also, the strategy is to explain via examples. A similar strategy is found in Fulton and Harris' book [8]. We highly recommend these sources for further reading in representation theory.

The text is punctuated with exercises. We hope that they will be found helpful but certainly they are not essential to the exposition and can be omitted on first reading. In an ideal world, paper would be engineered with exercises replaced by icons. Pressing on an icon would reveal the exercise. Some of the exercises use well-known notation or terminology which is not explained in the text. Towards the end of the article we discuss some variations, in particular Bott's cohomological variation and how to adapt the construction to other groups. These final sections have no details and are meant only as an introduction to further reading.

---

[1] ARC Senior Research Fellow.

One of us (ME) remembers his very first undergraduate tutorial at Hertford College back in 1970. 'Our tutor was Brian Steer and my fellow tutee was Paul Woodruff. Brian told us about groups acting on sets. By our second tutorial, two weeks later, we were expected to have understood Sylow's theorems from this point of view. Of course, knowing no better, we thought this was a perfectly reasonable request and so we just did it. Given this sort of start, we went on to learn an enormous amount from Brian during the next three years!' The Borel–Weil theorem concerns one of the finest examples of groups acting on sets. We dedicate this article to Brian Steer.

We thank Michael Murray for several useful conversations.

1 Preliminaries

In this section we shall describe background material and state some of the results that we shall need later. The following theorem is due to Cartan and Serre [7] (though, nowadays, it is more usually regarded as a special case of the general theory of elliptic complexes (see, for example, [9])):

Theorem 1.1 *If X is a compact complex manifold and $\mathcal{V}$ is a holomorphic vector bundle on X, then $\Gamma(X, \mathcal{V})$, the space of holomorphic sections of $\mathcal{V}$, is a finite-dimensional vector space.*

Complex projective space, $\mathbb{CP}_n$, is the space of complex lines through the origin in $\mathbb{C}^{n+1}$. It is a complex manifold and can be identified as $\mathrm{GL}(n+1, \mathbb{C})/P$, where P is the subgroup of $\mathrm{GL}(n+1, \mathbb{C})$ consisting of matrices of the form

$$\begin{pmatrix} \zeta & * & \cdots & * \\ 0 & & & \\ \vdots & & A & \\ 0 & & & \end{pmatrix} \quad \text{for } A \in \mathrm{GL}(n, \mathbb{C}). \tag{1.2}$$

We shall write $e_1, e_2, \ldots, e_n, e_{n+1}$ for the standard basis vectors in $\mathbb{C}^{n+1}$. Then P is the stabilizer of e_1 up to scale where $\mathrm{GL}(n+1, \mathbb{C})$ acts on $\mathbb{C}^{n+1}$ by matrix multiplication on column vectors in the usual way.

Exercises 1.3 Find a set of complex coordinate charts for $\mathbb{CP}_n$ and figure out the action of $\mathrm{GL}(n+1, \mathbb{C})$ in these charts. In particular, if $n = 1$ then show how $\mathbb{CP}_1$ may be regarded as the extended complex plane $\mathbb{C} \cup \{\infty\}$ and how the action of $\mathrm{GL}(2, \mathbb{C})$ is given by Möbius transformations. (It is necessary to relax the notion of action slightly.) The action of $\mathrm{GL}(n+1, \mathbb{C})$ on $\mathbb{CP}_n$ is not effective. Is there a subgroup which acts effectively and transitively? Does $\mathrm{GL}(n+1, \mathbb{R})$ act transitively? If not, what are its orbits and corresponding stabilizer subgroups? Answer the same questions for $\mathrm{SU}(n+1)$.

We shall write $\mathfrak{gl}(n+1,\mathbb{C})$ for the space of $(n+1)\times(n+1)$ complex matrices. Then the exponential mapping

$$\exp : \mathfrak{gl}(n+1,\mathbb{C}) \to \mathrm{GL}(n+1,\mathbb{C})$$

is defined by

$$\exp Z = 1 + Z + \tfrac{1}{2}Z^2 + \tfrac{1}{3!}Z^3 + \cdots .$$

This series always converges and the following is well known:

Proposition 1.4 *The mapping* $\exp : \mathfrak{gl}(n+1,\mathbb{C}) \to \mathrm{GL}(n+1,\mathbb{C})$ *is surjective.*

Exercise 1.5 Find a proof of Proposition 1.4 (perhaps by using the functional calculus). Is the corresponding statement with complex numbers replaced by real numbers true?

The subgroup $\mathrm{U}(n+1) \subset \mathrm{GL}(n+1,\mathbb{C})$ consisting of unitary matrices is a real compact submanifold. There is a non-zero integral defined over $\mathrm{U}(n+1)$ which is *invariant* in the sense that

$$\int_{g\in\mathrm{U}(n+1)} \phi(g) = \int_{g\in\mathrm{U}(n+1)} \phi(hg) \quad \text{for any } h \in \mathrm{U}(n+1).$$

If $X \in \mathfrak{gl}(n+1,\mathbb{C})$ is skew Hermitian (i.e. $X^{\mathrm{tr}} = -\overline{X}$), then $\exp X$ is unitary.

Exercises 1.6 Define this invariant integral (perhaps by regarding $\mathrm{U}(n+1)$ as a submanifold of Euclidean space $\mathbb{R}^N$ where $N = 4(n+1)^2$). As a warm-up exercise, the analogous case of $\mathrm{O}(n+1)$ might be slightly easier. Suppose $X \in \mathfrak{gl}(n+1,\mathbb{C})$ has zero trace. What does this imply about $\exp X$?

2 Representation theory

Let V be a finite-dimensional complex vector space. Denote by $\mathrm{GL}(V)$ the group of linear automorphisms of V and by $\mathfrak{gl}(V)$ the vector space of linear endomorphisms of V. We shall consider only finite-dimensional complex representations:

Definition 2.1 *A representation of a group G is a homomorphism $\rho : G \to \mathrm{GL}(V)$ for some finite-dimensional complex vector space V. If $g \in G$ and $v \in V$, then we shall often write gv instead of $\rho(g)v$ and regard G as acting on V by linear transformations. Thus, we shall say V is a representation of G and often omit any reference to ρ. If ρ is injective, this gives a way of 'representing' G as a group of matrices. A homomorphism of representations of G is a linear transformation which commutes with the action of G.*

If $G = \mathrm{GL}(n+1,\mathbb{C})$, then we shall also insist that ρ be holomorphic. The *standard* representation of $\mathrm{GL}(n+1,\mathbb{C})$ is its action on $\mathbb{C}^{n+1}$ by matrix multiplication on column vectors.

If V and W are representations of G, then $V \oplus W$ is also a representation where $g(v,w) = (gv, gw)$. Thinking of $\mathrm{GL}(V)$ and $\mathrm{GL}(W)$ as matrices, this direct sum just combines them as block-diagonal matrices in the obvious way. Also, $V \otimes W$ is a representation of G with action induced by $g(v \otimes w) = gv \otimes gw$ and the dual space V^* is a representation with action given by $(g\ell)(v) = \ell(g^{-1}v)$ for $\ell \in V^*$.

Given a representation $\rho : \mathrm{GL}(n+1,\mathbb{C}) \to \mathrm{GL}(V)$, define $\dot{\rho} : \mathfrak{gl}(n+1,\mathbb{C}) \to \mathfrak{gl}(V)$ by

$$\dot{\rho}(Z) = \frac{d}{dt}\rho(\exp(tZ))\bigg|_{t=0}.$$

Since ρ is holomorphic, $\dot{\rho}$ is *complex* linear. As with a representation, we shall often write Zv rather than $\dot{\rho}(Z)v$. Note that a tensor product representation of $\mathrm{GL}(n+1,\mathbb{C})$ satisfies the *Leibniz rule*:

$$Z(v \otimes w) = Zv \otimes w + v \otimes Zw \quad \text{for } Z \in \mathfrak{gl}(n+1,\mathbb{C}). \tag{2.2}$$

Exercise 2.3 Check that (2.2) holds.

Lemma 2.4 *A representation ρ of $\mathrm{GL}(n+1,\mathbb{C})$ is determined by its derivative $\dot{\rho}$.*

Proof Consider

$$f(s) \equiv \rho(\exp(sZ))(\exp(s\dot{\rho}(Z)))^{-1} = \rho(\exp(sZ))\exp(-s\dot{\rho}(Z)).$$

A short calculation yields $f'(s) = d/dt f(s+t)|_{t=0} = 0$. So $f(1) = f(0) = 1$. Hence,

$$\rho(\exp(Z)) = \exp(\dot{\rho}(Z)). \tag{2.5}$$

The result now follows from Proposition 1.4. □

Exercises 2.6 Perform the calculation which gives $f'(s) = 0$. Find a way around using Proposition 1.4 in the proof of Lemma 2.4. (There are several ways of doing this, one of which starts by using the Gram–Schmidt procedure to write an invertible complex matrix as a product of a unitary matrix, a diagonal matrix with real positive entries along the diagonal, and an upper triangular matrix with 1's along the diagonal. This step generalizes to become the *Iwasawa decomposition* [13].)

Definition 2.7 *A representation V of G is said to be irreducible if the only G-invariant linear subspaces of V are $\{0\}$ and V itself. A representation is said to be completely reducible if it can be written as a direct sum of irreducible subspaces.*

Theorem 2.8 *Every representation $\rho : \mathrm{GL}(n+1,\mathbb{C}) \to \mathrm{GL}(V)$ is completely reducible.*

Proof The following argument is usually known as 'Weyl's unitary trick.' In view of (2.5) and the observation that the exponential of a block-diagonal matrix is again block-diagonal, it suffices to show that if U is a $\mathrm{GL}(n+1,\mathbb{C})$-invariant subspace of V, then we can find a complementary subspace W such that $\dot{\rho}$ has block-diagonal form with respect to $V = U \oplus W$. Choose any Hermitian inner product $\langle \ , \ \rangle$ on V. Then

$$\langle\langle u, v \rangle\rangle \equiv \int_{g \in \mathrm{U}(n+1)} \langle \rho(g)u, \rho(g)v \rangle$$

is a $\mathrm{U}(n+1)$-invariant Hermitian inner product in the sense that

$$\langle\langle \rho(h)u, \rho(h)v \rangle\rangle = \langle\langle u, v \rangle\rangle$$

for all $h \in \mathrm{U}(n+1)$. Thus, if we take W to be the orthogonal complement of U with respect to this invariant inner product, then $V = U \oplus W$ as a representation of $\mathrm{U}(n+1)$. It follows that $\dot{\rho}(X)$ is block-diagonal when X is skew Hermitian. Now write a general $Z \in \mathfrak{gl}(n+1,\mathbb{C})$ as $Z = X + iY$ where X and Y are skew Hermitian. Then $\dot{\rho}(Z) = \dot{\rho}(X) + i\dot{\rho}(Y)$ is also block-diagonal. □

Remark 2.9 Further argument along these lines shows that the theory of holomorphic finite-dimensional representations of $\mathrm{GL}(n+1,\mathbb{C})$ is essentially *equivalent* to the finite-dimensional representation theory of $\mathrm{U}(n+1)$.

If $X \in \mathfrak{gl}(n+1,\mathbb{C})$ or, more generally, $X, Y \in \mathfrak{gl}(V)$, write

$$[X, Y] \equiv XY - YX.$$

Proposition 2.10 *If ρ is a representation of $\mathrm{GL}(n+1,\mathbb{C})$, then*

$$\dot{\rho}([X, Y]) = [\dot{\rho}(X), \dot{\rho}(Y)] \quad \textit{for all } X, Y \in \mathfrak{gl}(n+1,\mathbb{C}).$$

Proof Differentiate the identity

$$\rho(\exp(sX)\exp(tY)\exp(-sX)) = \rho(\exp(sX))\rho(\exp(tY))\rho(\exp(-sX))$$

with respect to s and t and set $s = t = 0$. □

Exercise 2.11 Perform the suggested differentiation.

It is easy to see (for example, by considering $\dot{\rho}$) that the irreducible representations of $\mathrm{GL}(1,\mathbb{C})$ are one-dimensional, of the form

$$\mathrm{GL}(1,\mathbb{C}) \ni \zeta \mapsto \zeta^a \quad \text{where } a \text{ is an arbitrary } \textit{integer}.$$

As usual, we are omitting the brackets of a 1×1 matrix. Theorem 2.8 now implies that every representation of $\mathrm{GL}(1,\mathbb{C})$ may be written as a direct sum of such one-dimensional representations. A similar argument applies to $D \subset \mathrm{GL}(n+1,\mathbb{C})$, the subgroup of diagonal matrices. Thus, if we start with a representation of $\mathrm{GL}(n+1,\mathbb{C})$, then V may be written as a direct sum

$$V = \bigoplus_{\lambda\in\mathbb{Z}^{n+1}} V_\lambda$$

where $v \in V_\lambda$ for $\lambda = a,b,\ldots,d \in \mathbb{Z}^{n+1}$ if and only if

$$\begin{pmatrix} \zeta_1 & & & \\ & \zeta_2 & 0 & \\ & & \ddots & \\ 0 & & & \zeta_{n+1} \end{pmatrix} v = \zeta_1^a\zeta_2^b\cdots\zeta_{n+1}^d v. \tag{2.12}$$

(We shall write elements $\lambda \in \mathbb{Z}^{n+1}$ as *strings* of integers, i.e. without enclosing brackets, so as to reserve the notation $(a,b,\ldots,d)$ for a separate purpose later.) The subspace V_λ is called a *weight space* of *weight* λ. An element $v \in V$ is called a *weight vector* if v is non-zero and lies in V_λ for some λ which is then called the *weight* of v.

Exercise 2.13 Let h_i denote the matrix with 1 as its ith diagonal entry and zeroes elsewhere. Crucial to the previous paragraph is the fact that all $\dot{\rho}(h_i)$ are diagonalizable. Prove this. Then prove that they are simultaneously diagonalizable.

Suppose V is a representation of $\mathrm{GL}(2,\mathbb{C})$. The following matrices

$$h_1 = \begin{pmatrix} 1 & 0 \\ 0 & 0 \end{pmatrix} \quad h_2 = \begin{pmatrix} 0 & 0 \\ 0 & 1 \end{pmatrix} \quad x = \begin{pmatrix} 0 & 1 \\ 0 & 0 \end{pmatrix} \quad y = \begin{pmatrix} 0 & 0 \\ 1 & 0 \end{pmatrix}$$

form a basis for $\mathfrak{gl}(2,\mathbb{C})$ and satisfy the 'commutation relations'

$$[h_1,h_2]=0 \quad [h_1,x]=x \quad [h_2,x]=-x$$
$$[h_1,y]=-y \quad [h_2,y]=y \quad [x,y]=h_1-h_2.$$

Decompose V into its weight spaces

$$V = \bigoplus_{a,b\in\mathbb{Z}^2} V_{a,b}.$$

If $v \in V_{a,b}$, then differentiating (2.12) implies $h_1 v = av$ and $h_2 v = bv$ in which case, using Proposition 2.10, we may conclude that

$$\begin{aligned} h_1 yv &= [h_1, y]v + yh_1 v = -yv + ayv = (a-1)yv \\ h_2 yv &= [h_2, y]v + yh_2 v = \ \ yv + byv \ \ = (b+1)yv \end{aligned}$$

so either $yv = 0$ or $yv \in V_{a-1,b+1}$. Similarly, $xv = 0$ or $xv \in V_{a+1,b-1}$. Thus, the quantity $a - b$ is either decreased or increased by 2. For this reason, y is called a *lowering* operator and x is called a *raising* operator. Since V is finite-dimensional, there is a weight vector $v_0 \in V$ with $yv_0 = 0$. This is called a *minimal* weight vector. For such a v_0 of weight a, b, it is easy to check (again using the commutation relations), that for $j = 1, 2, 3, \ldots$,

$$yx^j v_0 = j(b - a + 1 - j)x^{j-1} v_0. \tag{2.14}$$

We know that $x^j v_0 = 0$ if j is sufficiently large. The only way that this can be consistent with (2.14) is if $a \leqslant b$ and $x^j v_0 = 0$ for $j \geqslant b - a + 1$. Another conclusion of this typical 'raising and lowering' argument is that the subspace of V obtained by taking arbitrary linear combinations of raising operators (powers of x) applied to v_0 is invariant under the action of $\mathfrak{gl}(2, \mathbb{C})$ and hence under $\mathrm{GL}(2, \mathbb{C})$ (by (2.5)). In particular, if V is irreducible, then v_0 is unique up to scale in which case it is called a *lowest* weight vector. Its weight is called the *lowest* weight of V.

Exercises 2.15 Prove (2.14) by using the commutation relations, as suggested. The standard representation of $\mathrm{GL}(2, \mathbb{C})$ induces a representation on the symmetric tensor product $\bigodot^k \mathbb{C}^2$. Show that this representation is irreducible and find its lowest weight. Decompose the representation of $\mathrm{GL}(2, \mathbb{C})$ on $\bigotimes^2 \mathbb{C}^2$ into irreducibles and find their highest weights. Decompose $\bigotimes^k \mathbb{C}^2$ into irreducibles.

A similar analysis applies to a representation of $\mathrm{GL}(n+1, \mathbb{C})$. For raising operators in $\mathfrak{gl}(n+1, \mathbb{C})$ one can take the n elements $x_1, x_2, \ldots, x_n$ where x_j has a 1 in the jth position of the superdiagonal and zeroes elsewhere. For lowering operators one can take the corresponding elements $y_1, y_2, \ldots, y_n$ with respect to the subdiagonal. As before, raising and lowering operators will either kill a weight vector or yield from it another weight vector. If $v \in V$ is a weight vector of weight $a, b, \ldots, d$, then the quantity $na + (n-2)b + \cdots - nd$ increases (respectively decreases) by 2 under the action of any raising operator (respectively lowering operator). Hence, there is a minimal weight vector $v_0 \in V$, i.e. a weight vector annihilated by all the lowering operators. The subspace generated by raising v_0 is $\mathrm{GL}(n+1, \mathbb{C})$-invariant. Thus, if V is irreducible, then V contains a lowest weight vector unique up to scale. The weight $a, b, \ldots d$ of any minimal weight vector must be *dominant* in the sense that

$$a \leqslant b \leqslant \cdots \leqslant d. \tag{2.16}$$

This follows immediately from the $\mathrm{GL}(2,\mathbb{C})$ case by restricting the representation V to the subgroups of $\mathrm{GL}(n+1,\mathbb{C})$ consisting of matrices of the form

$$\begin{pmatrix} 1 & & & & & & \\ & \ddots & 0 & & 0 & & \\ & & \begin{matrix} 1 & 0 & 0 & 0 \\ 0 & * & * & 0 \\ 0 & * & * & 0 \\ 0 & 0 & 0 & 1 \end{matrix} & & & & \\ & 0 & & & 0 & & \\ & & & & \ddots & & \\ & & & & & \ddots & \\ & 0 & 0 & & & & 1 \end{pmatrix} \tag{2.17}$$

We can now state the classification theorem that we aim to prove:

Theorem 2.18 *Each irreducible (finite-dimensional holomorphic) representation of* $\mathrm{GL}(n+1,\mathbb{C})$ *is determined up to isomorphism by its lowest weight* $a, b, \ldots, d \in \mathbb{Z}^{n+1}$. *These weights are dominant* (2.16) *and every dominant weight arises in this way. In other words, there is a one-to-one correspondence between the irreducible representations of* $\mathrm{GL}(n+1,\mathbb{C})$ *up to isomorphism and the dominant elements of* $\mathbb{Z}^{n+1}$.

Exercise 2.19 Prove this theorem for $n = 1$. Hint: look in the previous exercise box.

Theorem 2.18 is more usually stated in terms of *highest* weights (and for more general groups: e.g. [13, Theorem 4.28] or [8, Theorem 14.18]). We need to prove existence and uniqueness. Existence will be proved as a consequence of the Borel–Weil theorem for complex projective space (Theorem 4.1). Uniqueness is straightforward as follows:

Theorem 2.20 *If* U *and* V *are irreducible representations of* $\mathrm{GL}(n+1,\mathbb{C})$ *with the same lowest weight, then* $U \cong V$.

Proof Choose lowest weight vectors $u_0 \in U$ and $v_0 \in V$. Then (u_0, v_0) is a lowest weight vector in $U \oplus V$ and the subspace generated by raising this element is an irreducible representation which we shall denote by W. Projection onto the first factor induces a homomorphism of $\mathrm{GL}(n+1,\mathbb{C})$-representations. As $(u_0, v_0) \mapsto u_0$, this homomorphism is non-zero. Its kernel is a $\mathrm{GL}(n+1,\mathbb{C})$-invariant subspace of W which must be $\{0\}$ by irreducibility. Similarly, the range must be all of U by irreducibility of U. Hence, $W \cong U$. Similarly, $W \cong V$. □

Lemma 2.21 *In a representation of* $\mathrm{GL}(n+1,\mathbb{C})$, *a weight vector is minimal, if and only if it is fixed by matrices of the form* $1 + L$ *with* L *strictly lower triangular.*

Proof Suppose v_0 is a weight vector fixed by matrices of the form $1 + L$ for L strictly lower triangular. In particular, for every lowering operator y_j,

$$(\exp(1 + t y_j)) v_0 = v_0 \quad \text{for all } t.$$

Differentiating with respect to t and setting $t = 0$ implies $y_j v_0 = 0$, as required.

Conversely, suppose v_0 is minimal. Then it is not only annihilated by the y_j's but also by their commutators—for example:

$$[y_2, y_1]v_0 = 0 \quad [y_3, [y_2, y_1]]v_0 = 0.$$

Taking linear combinations shows that $yv_0 = 0$ for *any* strictly lower triangular matrix y. Now, if L is strictly lower triangular, then $1 + L = \exp y$ for

$$y = L - \tfrac{1}{2}L^2 + \tfrac{1}{3}L^3 - \tfrac{1}{4}L^4 + \cdots,$$

a *finite* sum. Since

$$\frac{d}{ds}(\exp(sy))v_0 = \frac{d}{dt}(\exp((s+t)y))v_0\Big|_{t=0} = (\exp(sy))(yv_0) = 0,$$

it follows that $(\exp(sy))v_0$ does not depend on s. Evaluating at $s = 1$ and $s = 0$ gives $(1 + L)v_0 = v_0$, as required. □

Exercise 2.22 Use the identity

$$1 + (s+t)L = (1 + tL)(1 + s\underbrace{(L - tL^2 + t^2L^3 - t^3L^4 + \cdots)}_{\text{a finite sum}}$$

to construct an alternative proof of Lemma 2.21.

We shall write $(a, b, \ldots, d)$ for the representation of $\mathrm{GL}(n+1, \mathbb{C})$ with lowest weight $a, b, \ldots, d$ (if such a representation exists). For example, the standard representation is generated by raising the $(n+1)^{\text{st}}$ standard basis vector e_{n+1}. More specifically,

$$\begin{aligned} e_j &= x_j e_{j+1} = x_j x_{j+1} e_{j+2} = \cdots = x_j x_{j+1} \cdots x_n e_{n+1} \\ e_j &= y_{j-1} e_{j-1} = y_{j-1} y_{j-2} e_{j-2} = \cdots = y_{j-1} y_{j-2} \cdots y_1 e_1. \end{aligned} \tag{2.23}$$

This is the irreducible representation $(0, \ldots, 0, 1)$ (since $0, \ldots, 0, 1$ is the weight of e_{n+1}). The following examples are similarly verified:

- The skew symmetric tensors $\bigwedge^2 \mathbb{C}^{n+1}$ form the irreducible representation
$$(0, \ldots, 0, 1, 1).$$
- The symmetric tensors $\bigodot^2 \mathbb{C}^{n+1}$ form the irreducible representation
$$(0, \ldots, 0, 2).$$
- The decomposition of a 2-tensor into its skew and symmetric parts corresponds to the decomposition into irreducibles:
$$(0, \ldots, 0, 1) \otimes (0, \ldots, 0, 1) = (0, \ldots, 0, 1, 1) \oplus (0, \ldots, 0, 2).$$

- The endomorphisms of $\mathbb{C}^{n+1}$ split into trace-free and pure-trace parts. This corresponds to the decomposition into irreducibles:

$$(-1,0,\ldots,0) \otimes (0,\ldots,0,1) = (-1,0,\ldots,0,1) \oplus (0,\ldots,0).$$

These decompositions generalise according to (4.3).

Exercise 2.24 Verify these examples.

3 Holomorphic Induction

Recall that complex projective space $\mathbb{CP}_n$ may be written as G/P where G is the complex general linear group $\mathrm{GL}(n+1,\mathbb{C})$ and P consists of matrices of the form (1.2). Much of the following construction applies more generally (when G is a complex Lie group with complex co-compact subgroup P (a more detailed discussion can be found in [14])).

Definition 3.1 *A homogeneous holomorphic vector bundle on G/P is a holomorphic vector bundle whose total space is equipped with a holomorphic action of G compatible with its action on G/P and with the vector space structure on each fibre.*

The fibre of such a bundle over the identity coset yields a representation of P. Let us denote it by V. Conversely, this representation determines the homogeneous bundle—it may be identified with

$$G \times_P V \xrightarrow{\pi} G/P$$

where $G \times_P V$ is the product $G \times P$ modulo the equivalence relation $(g, pv) \sim (gp, v)$ for $p \in P$ and π is induced by projection onto the first factor. We shall write $\mathcal{O}(V)$ for the homogeneous vector bundle on G/P induced by the representation V of P. The action of G on the total space of $\mathcal{O}(V)$ induces a linear action of G on $\Gamma(G/P, \mathcal{O}(V))$. This is a representation of G (finite-dimensionality is guaranteed by Theorem 1.1). Thus, from a representation of the subgroup P, we obtain a representation of G itself. This process is called *holomorphic induction.*

It is often useful to have a more explicit description of $\Gamma(G/P, \mathcal{O}(V))$. It may be written as

$$\{\text{holomorphic } \phi : G \to V \text{ s.t. } \phi(gp) = p^{-1}\phi(g)\ \forall p \in P\} \tag{3.2}$$

and the action of $g \in G$ is given by $(g\phi)(h) = \phi(g^{-1}h)$.

Exercise 3.3 Check that this is so.

Lemma 3.4 *Suppose W is a representation of G and hence of P by restriction. Then*

$$\Gamma(G/P, \mathcal{O}(W)) = W$$

as representations of G. More generally, if V is a representation of P and W is a representation of G, then

$$\Gamma(G/P, \mathcal{O}(V \otimes W)) = \Gamma(G/P, \mathcal{O}(V)) \otimes W.$$

Proof The mapping

$$\begin{array}{ccc} G \times_P W & \longrightarrow & G/P \times W \\ \Cup & & \Cup \\ (g, w) & \longmapsto & (gP, gw) \end{array}$$

is well-defined and trivializes $\mathcal{O}(W)$ as a holomorphic vector bundle. □

Finally, it is easy to check that holomorphic induction is functorial: a homomorphism of P-representations $V \to W$ gives rises to a homomorphism of homogeneous bundles $\mathcal{O}(V) \to \mathcal{O}(W)$ (i.e. one which commutes with the action of G) and hence a homomorphism

$$\Gamma(G/P, \mathcal{O}(V)) \to \Gamma(G/P, \mathcal{O}(W))$$

of representations of G.

Exercise 3.5 Check that this is so.

4 The Borel–Weil theorem

The aim of this section is to realize the representations of $\mathrm{GL}(n+1, \mathbb{C})$ as sections of certain homogeneous vector bundles on $\mathbb{CP}_n$. The construction will use induction on n, the case $n = 0$ being trivial. Thus, suppose we have constructed all the representations of $\mathrm{GL}(n, \mathbb{C})$ (in other words, we have the classification of Theorem 2.18), and we wish to proceed to $\mathrm{GL}(n+1, \mathbb{C})$. For any dominant weight $b, \ldots, d \in \mathbb{Z}^n$, consider the representation $(b, \ldots, d)$ of $\mathrm{GL}(n, \mathbb{C})$. For any integer a, we may extend this to a representation of P by decreeing that ζ in the general form (1.2) act by multiplication by ζ^a. Let us write $(a|b, \ldots, d)$ for the resulting representation of P.

Theorem 4.1 [Borel–Weil] *For any dominant weight $a, b, \ldots, d \in \mathbb{Z}^{n+1}$,*

$$\Gamma(\mathbb{CP}_n, \mathcal{O}(a|b, \ldots, d)) = (a, b, \ldots, d),$$

i.e. this holomorphically induced representation is irreducible with $a, b, \ldots, d$ as its lowest weight. If $a > b$, then the induced representation vanishes.

Proof We maintain that $\Gamma(\mathbb{CP}_n, \mathcal{O}(a|b,\ldots,d))$ is either zero or irreducible with lowest weight $a, b, \ldots, d$. View $\Gamma(\mathbb{CP}_n, \mathcal{O}(a|b,\ldots,d))$ as in (3.2) and suppose that $\phi_0 : \mathrm{GL}(n+1,\mathbb{C}) \to (a|b,\ldots,d)$ is a minimal weight vector. By Theorem 2.8, it suffices to show that there is only one such vector up to scale (and that ϕ_0 has the expected weight). Let $v_0 = \phi_0(1)$. Then, by Lemma 2.21 and the formula for the action of $\mathrm{GL}(n+1,\mathbb{C})$,

$$\phi_0\left(\left(\begin{array}{cccc} 1 & 0 & \cdots & 0 \\ z_1 & & & \\ \vdots & & 1 & \\ z_n & & & \end{array}\right)\right) = v_0$$

for all $(z_1, \ldots, z_n) \in \mathbb{C}^n$. The corresponding P-cosets describe a standard affine chart on $\mathbb{CP}_n$. Thus, v_0 determines ϕ_0 uniquely and, in particular, $v_0 \neq 0$ else ϕ_0 would vanish. Also, from the explicit description (3.2), we obtain

$$pv_0 = p\phi_0(1) = \phi_0(p^{-1}1) = (p\phi_0)(1) \tag{4.2}$$

for $p \in P$. Thus, minimality of ϕ_0 under $\mathrm{GL}(n+1,\mathbb{C})$ implies that v_0 is a minimal weight vector under $\mathrm{GL}(n,\mathbb{C}) \subset P$. As such, v_0 is unique up to scale. Since v_0 determines ϕ_0, the same is true of ϕ_0. That ϕ_0 has weight $a, b, \ldots, d$ also follows from (4.2), applied with p diagonal. The last statement in the theorem is now immediate else (2.16) would be violated.

Thus, it remains to show that $\Gamma(\mathbb{CP}_n, \mathcal{O}(a|b,\ldots,d))$ is non-zero when $a, b, .., d$ is dominant. In order to do this, we shall need the following special case of the *Littlewood–Richardson* rules

$$\begin{aligned}(a, b, \ldots, c, d, e) &\otimes (0, \ldots, 0, 1) = \\ &(a, b, \ldots, c, d, e+1) \oplus \underbrace{(a, b, \ldots, c, d+1, e)}_{\text{if } d<e} \oplus \cdots \oplus \underbrace{(a+1, b, \ldots, c, d, e)}_{\text{if } a<b}\end{aligned} \tag{4.3}$$

(cf. [8]). In fact, we shall prove these rules by induction on n as we go along. Thus, we shall assume that they are true for representations of $\mathrm{GL}(n,\mathbb{C})$ and the proof of the Borel–Weil theorem for $\mathbb{CP}_n$ will also yield the Littlewood–Richardson rules for $\mathrm{GL}(n+1,\mathbb{C})$. (Of course, they are trivial when $n = 1$.) Suppose that $(a, \ldots, b, c, d, e)$ exists. In other words, there is an irreducible representation of $\mathrm{GL}(n+1,\mathbb{C})$ with lowest weight $a, \ldots, b, c, d, e$. Choose v_0, a lowest weight vector. Then

$$v_0 \otimes e_{n+1} \in (a, \ldots, b, c, d, e) \otimes (0, \ldots, 0, 1)$$

is a minimal weight vector of weight $a, \ldots, b, c, d, e+1$. If $d < e$, then, from (2.23) and the commutation relations in $\mathfrak{gl}(n+1,\mathbb{C})$, it is easy to check that

$$(e-d)v_0 \otimes e_n - x_n v_0 \otimes e_{n+1}$$

is a minimal weight vector of weight $a, \ldots, b, c, d+1, e$ (its first term guarantees that it is non-zero). It is straightforward though tedious to write down explicit formulae for all the minimal weight vectors that should occur in this tensor product according to the Littlewood–Richardson rules (4.3). The following example, for the case $b < c$, should be sufficient for the reader to deal with the general case:

$$(e-b+2)[(d-b+1)[(c-b)v_0 \otimes e_{n-2} - X_{n-2} v_0 \otimes e_{n-1}] + X_{n-1} v_0 \otimes e_n] - X_n v_0 \otimes e_{n+1}$$

where

$$\begin{aligned} X_{n-2} &= x_{n-2} \\ X_{n-1} &= (c-b+1)x_{n-1}X_{n-2} - (c-b)X_{n-2}x_{n-1} \\ X_n &= (d-b+2)x_n X_{n-1} - (d-b+1)X_{n-1}x_n. \end{aligned}$$

This is not to say that these are the *only* minimal weight vectors that occur in this tensor product but merely that minimal weight vectors of the weights predicted by (4.3) do actually occur. We may conclude that

$$\begin{aligned} &(a,b,\ldots,c,d,e) \otimes (0,\ldots,0,1) \supseteq \\ &\quad (a,b,\ldots,c,d,e+1) \oplus \underbrace{(a,b,\ldots,c,d+1,e)}_{\text{if } d<e} \oplus \cdots \oplus \underbrace{(a+1,b,\ldots,c,d,e)}_{\text{if } a<b} \end{aligned} \tag{4.4}$$

meaning that if the left hand side exists then so do all the representations on the right hand side and, moreover, that the right hand side can be realized as an invariant subspace of the left hand side.

The kth power of the determinant

$$\mathrm{GL}(n+1,\mathbb{C}) \ni A \longmapsto (\det A)^k$$

defines the one-dimensional representation $(k,k,\ldots,k)$. Tensoring with this representation has the effect of adding $k,k\ldots,k$ to each weight. By Lemma 3.4, therefore,

$$\Gamma(\mathbb{CP}_n, \mathcal{O}(a+k|b+k,\ldots,e+k)) = \Gamma(\mathbb{CP}_n, \mathcal{O}(a|b,\ldots,e)) \otimes (k,k,\ldots,k).$$

Hence, in our proof of the Borel–Weil theorem, we may assume without loss of generality, that $a=0$. We must now show that

$$\Gamma(\mathbb{CP}_n, \mathcal{O}(0|b,c,\ldots,d,e)) = (0,b,c,\ldots,d,e)$$

when $0 \leqslant b \leqslant c \leqslant \cdots \leqslant d \leqslant e$ and this we shall do by induction on the quantity $b+c+\cdots+d+e$ starting at zero for then

$$\Gamma(\mathbb{CP}_n, \mathcal{O}(0|0,\ldots,0)) = (0,0,\ldots,0)$$

is simply the statement that any holomorphic function on $\mathbb{CP}_n$ is constant.

Restricted to P, the standard representation of $\mathrm{GL}(n+1,\mathbb{C})$ preserves the line through e_1. This gives rise to the following exact sequence of representations of P:

$$0 \to (1|0,0,\ldots,0,0) \to (0,0,0,\ldots,0,1) \to (0|0,0,\ldots,0,1) \to 0. \tag{4.5}$$

(The corresponding sequence of homogeneous bundles on $\mathbb{CP}_n$ is known as the *Euler sequence* (with $\mathcal{O}(1|0,\ldots,0)$ more traditionally denoted by $\mathcal{O}(-1)$).)

Exercises 4.6 Prove the following statements. We may regard $\mathbb{CP}_n$ as the quotient of $\mathbb{C}^{n+1}\setminus\{0\}$ by the multiplicative action of the non-zero complex numbers. The total space of $\mathcal{O}(1|0,\ldots,0,0)$ with its zero section removed may be identified with $\mathbb{C}^{n+1}\setminus\{0\}$ in a manner compatible with the action of the group $\mathrm{GL}(n+1,\mathbb{C})$. If $z\in\mathbb{C}^{n+1}\setminus\{0\}$ represents a point $[z]\in\mathbb{CP}_n$, then the Euler sequence over $[z]$ may be identified with the exact sequence

$$0\to\{\text{the span of } z\}\to\mathbb{C}^{n+1}\to\mathbb{C}^{n+1}/\{\text{the span of } z\}\to 0.$$

As homogeneous bundles, $\mathcal{O}(-1|0,\ldots,0,1)$ is the tangent bundle to $\mathbb{CP}_n$.

Supposing that we have constructed $(0,b,c,\ldots,d,e)$ as in the statement of the theorem and also that we know the Littlewood–Richardson rules (4.3) for $\mathrm{GL}(n,\mathbb{C})$, we may tensor through (4.5) to obtain the exact sequence

$$\begin{gathered}0\to(1|b,c,\ldots,d,e)\to(0|b,c,\ldots,d,e)\otimes(0,0,\ldots,0,1)\to\\ \left[(0|b,c,\ldots,d,e+1)\oplus\underbrace{(0|b,c,\ldots,d+1,e)}_{\text{if } d<e}\oplus\cdots\oplus\underbrace{(0|b+1,c,\ldots,d,e)}_{\text{if } b<c}\right]\to 0.\end{gathered}\tag{4.7}$$

By Lemma 3.4, the weak form (4.4) of the Littlewood–Richardson rules, and our inductive hypothesis,

$$\begin{aligned}&\Gamma(\mathbb{CP}_n,\mathcal{O}((0|b,c,\ldots,d,e)\otimes(0,0,\ldots,0,1)))=\\ &\quad(0,b,c,\ldots,d,e)\otimes(0,0,\ldots,0,1)\supseteq\\ &\qquad(0,b,c,\ldots,d,e+1)\oplus\underbrace{(0,b,c,\ldots,d+1,e)}_{\text{if } d<e}\oplus\cdots\\ &\qquad\qquad\oplus\underbrace{(0,b+1,c,\ldots,d,e)}_{\text{if } b<c}\oplus\underbrace{(1,b,c,\ldots,d,e)}_{\text{if } 0<b}.\end{aligned}\tag{4.8}$$

If $b=0$, then $\Gamma(\mathbb{CP}_n,\mathcal{O}(1|b,c,\ldots,d,e))=0$ as we have already seen. On the other hand, if $b>0$, then

$$\begin{aligned}\Gamma(\mathbb{CP}_n,\mathcal{O}(1|b,c,\ldots,d,e))&=\Gamma(\mathbb{CP}_n,\mathcal{O}(0|b-1,\ldots,e-1))\otimes(1,1,\ldots,1)\\ &=(0,b-1,\ldots,e-1)\otimes(1,1,\ldots,1)\\ &=(1,b,c,\ldots,d,e)\end{aligned}$$

by the inductive hypothesis. The remaining term of (4.7) gives rise to

$$\Gamma(\mathbb{CP}_n,\mathcal{O}(0|b,c,\ldots,d,e+1))\oplus\cdots\oplus\underbrace{\Gamma(\mathbb{CP}_n,\mathcal{O}(0|b+1,c,\ldots,d,e))}_{\text{if } b<c},$$

a direct sum of representations of $\mathrm{GL}(n+1,\mathbb{C})$ which are either zero or irreducible with lowest weight as in the statement of the theorem. Combining all these observations, we may conclude that (4.7) gives a short exact sequence of global sections of the corresponding homogeneous bundles. (From general reasoning, only the surjectivity of the last mapping would be in doubt.) This not only gives the Borel–Weil theorem by induction but also forces equality in (4.8). In other words, we obtain the full Littlewood–Richardson rules (4.3) for use at the next stage. □

Remark 4.9 In fact, if one merely wishes to show the *existence* of the representations $(a,b,\ldots,d)$ for dominant $a,b,\ldots,d$, then the weaker form of the Littlewood–Richardson rules (namely (4.4)), suffice for an inductive argument. Theorem 4.1 is much better than this since it gives a specific realization and also proves the genuine rules (4.3) along the way. Though the rules (4.3) are not, by themselves, sufficient to compute the dimensions of the irreducible representations of $\mathrm{GL}(n+1,\mathbb{C})$, a sufficiently powerful generalization may be obtained by replacing the Euler sequence (4.5) in the proof of Theorem 4.1 by the exact sequence

$$0\to(1|0,0,\ldots,0,\underbrace{1,\ldots,1}_{k-1})\to(0,0,\ldots,0,\underbrace{1,\ldots,1,1}_{k})\to(0|0,\ldots,0,\underbrace{1,\ldots,1,1}_{k})\to 0.$$

This enables one to establish the Littlewood–Richardson rules for the splitting of

$$(a,b,\ldots,d,e)\otimes(0,\ldots,0,0,\underbrace{1,\ldots,1,1}_{k})=(a,b,\ldots,d,e)\otimes\textstyle\bigwedge^k\mathbb{C}^{n+1}$$

into irreducibles. (Rather than finding explicit minimal weight vectors as in the proof of Theorem 4.1, the crucial surjectivity follows easily from the vanishing of cohomology shown in the next section.) This is enough to construct an inductive proof that

$$\frac{[(b-a+1)(c-a+2)\cdots(e-a+n)][(c-b+1)\cdots(d-b+(n-2))(e-b+(n-1))]\cdots[e-d+1]}{n!(n-1)!\cdots 2!1!}\tag{4.10}$$

is the dimension of $(a,b,c,\ldots,d,e)$. A little more complex analysis provides an easier route to these dimensions. We shall indicate this route in the next section.

Exercise 4.11 Construct an inductive proof of (4.10) as outlined.

Remark 4.12 Realizing the irreducible representations of $\mathrm{GL}(n+1,\mathbb{C})$ inside tensor powers of the standard representation is due to Weyl (see 'Weyl's construction' in [8]). Our notation corresponds to *Young diagrams* (also explained in [8]):

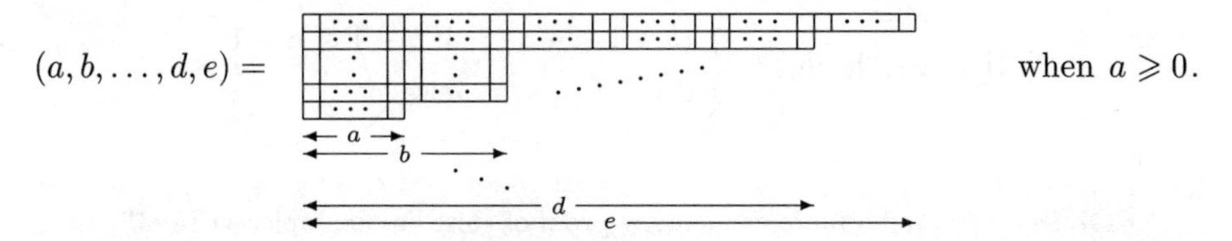

when $a \geqslant 0$.

5 Cohomological variations

These final sections will be more of a sketch. For integers $1 \leqslant i < j < \cdots < k \leqslant n$, define the *flag manifold*

$$\mathbb{F}_{i,j,\ldots,k}(\mathbb{C}^{n+1}) = \left\{ (L_i, L_j, \ldots, L_k) \text{ s.t. } \begin{matrix} L_p \text{ is a } p\text{-dimensional linear subspace} \\ \text{of } \mathbb{C}^{n+1} \text{ and } L_i \subset L_j \subset \cdots \subset L_k \end{matrix} \right\}.$$

Like $\mathbb{CP}_n = \mathbb{F}_1(\mathbb{C}^{n+1})$, these are compact complex manifolds, homogeneous under the action of $\mathrm{GL}(n+1, \mathbb{C})$. There is a natural projection $\mathbb{F}_{1,2}(\mathbb{C}^{n+1}) \xrightarrow{\nu} \mathbb{CP}_n$ given by $\nu(L_1, L_2) = L_1$. The fibres of this projection may be identified with $\mathbb{CP}_{n-1}$. There is a homogeneous bundle $\mathcal{O}(a|b|c, \ldots, d)$ defined on $\mathbb{F}_{1,2}(\mathbb{C}^{n+1})$ so that

$$\nu_* \mathcal{O}(a|b|c, \ldots, d) = \mathcal{O}(a|b, c, \ldots, d) \quad \text{if } b \leqslant c$$

where ν_* denotes direct image. This is essentially the inductive definition of $\mathcal{O}(a|b, c, \ldots, d)$. From Theorem 4.1, it follows that

$$\Gamma(\mathbb{F}_{1,2}(\mathbb{C}^{n+1}), \mathcal{O}(a|b|c, \ldots, d)) = (a, b, c, \ldots, d) \quad \text{if } a \leqslant b \leqslant c.$$

Following this procedure to its natural conclusion gives

$$\Gamma(\mathbb{F}_{1,2,3,\ldots,n}(\mathbb{C}^{n+1}), \mathcal{O}(a|b|c|\cdots|d)) = (a, b, c, \ldots, d) \quad \text{if } a \leqslant b \leqslant c \leqslant \cdots \leqslant d$$

for an appropriate homogeneous *line* bundle $\mathcal{O}(a|b|c|\cdots|d)$ on the manifold of *complete flags*. This is the classical form of the Borel–Weil theorem.

Exercise 5.1 Write the manifold of complete flags as $\mathrm{GL}(n+1, \mathbb{C})/B$ for a suitable subgroup B.

We shall now assume familiarity with sheaf cohomology. The canonical bundle on $\mathbb{CP}_1$ is $\mathcal{O}(1|-1)$. It is easy to check (for example, by looking at weights) that, as representations of $\mathrm{GL}(2, \mathbb{C})$,

$$(c, d)^* = (-d, -c).$$

Combining this with Serre duality gives

$$\mathrm{H}^1(\mathbb{CP}_1, \mathcal{O}(a|b)) = \begin{cases} (b+1, a-1) & \text{if } b+1 \leqslant a-1 \\ 0 & \text{else.} \end{cases}$$

Exercises 5.2 Find a more elementary proof of this. For example, write $\mathbb{CP}_1$ as the union of two open sets, each of which is biholomorphic to $\mathbb{C}$ and compute the cohomology by using this Čech cover with corresponding Laurent expansions. As a warm-up exercise, use the same method to compute $\Gamma(\mathbb{CP}_1, \mathcal{O}(a|b))$.

We claim that, more generally,

$$\mathrm{H}^r(\mathbb{CP}_n, \mathcal{O}(a|\underbrace{b, \ldots, c}_{r}, d, \ldots, e))$$

$$= \begin{cases} (b+1, \ldots, c+1, a-r, d, \ldots, e) & \text{if } c+1 \leqslant a-r \leqslant d \\ 0 & \text{else.} \end{cases}$$

This is the Bott–Borel–Weil theorem for complex projective space (cf. [4]). It can be proven by induction on n using the map ν together with $\mu : \mathbb{F}_{1,2}(\mathbb{C}^{n+1}) \to \mathbb{F}_2(\mathbb{C}^{n+1})$. Notice that μ has $\mathbb{CP}_1$ as fibres. If $a, b, \ldots, c, d$ is dominant, then

$$\begin{aligned}
\mathrm{H}^s(\mathbb{CP}_n, \mathcal{O}(a|b,\ldots,c,d)) &= \mathrm{H}^s(\mathbb{CP}_n, \nu_*^{n-1}\mathcal{O}(a|d+n-1|b-1,\ldots,c-1)) \\
&= \mathrm{H}^{s+n-1}(\mathbb{F}_{1,2}(\mathbb{C}^{n+1}), \mathcal{O}(a|d+n-1|b-1,\ldots,c-1)) \\
&= \mathrm{H}^{s+n-1}(\mathbb{F}_2(\mathbb{C}^{n+1}), \mathcal{O}(a,d+n-1|b-1,\ldots,c-1)) \\
&= \mathrm{H}^{s+n-1}(\mathbb{F}_2(\mathbb{C}^{n+1}), \mu_*^1\mathcal{O}(d+n|a-1|b-1,\ldots,c-1)) \\
&= \mathrm{H}^{s+n}(\mathbb{F}_{1,2}(\mathbb{C}^{n+1}), \mathcal{O}(d+n|a-1|b-1,\ldots,c-1)) \\
&= \mathrm{H}^{s+n}(\mathbb{CP}_n, \mathcal{O}(d+n|a-1,b-1,\ldots,c-1)).
\end{aligned}$$

Since $\dim_{\mathbb{C}} \mathbb{CP}_n = n$, this vanishes for $s > 0$. The general case is similar.

Exercise 5.3 Establish the general case (for complex projective space). Formulate and prove the Bott–Borel–Weil theorem for the manifold of complete flags.

The Hirzebruch–Riemann–Roch theorem [11] gives (4.10) as the value of the holomorphic Euler characteristic

$$\sum_{r=0}^{n} (-1)^r \dim \mathrm{H}^r(\mathbb{CP}_n, \mathcal{O}(a|b, c, \ldots, d, e))$$

for *any* homogeneous vector bundle $\mathcal{O}(a|b, c, \ldots, d, e)$. In particular, when the weight $a, b, c, \ldots, d, e$ is dominant, this yields the dimension of the corresponding irreducible representation. Indeed, a proof of the full Weyl character formula using, instead, the Atiyah–Bott–Lefschetz fixed point formula (and the Kodaira vanishing theorem to dispose of the higher cohomology) can be found in [1, 5]. These proofs easily extend to more general groups. In the case of complex projective space, however, a more elementary argument can be made as follows. The line through e_1 in $\mathbb{C}^{n+1}$ provides a basepoint in $\mathbb{CP}_n$ which can be defined by n homogeneous functions $f_1, f_2, \ldots, f_n$ of degree 1. (For example, we could take all but the first element of the basis dual to the standard basis.) In traditional notation, each f_j is a section of the bundle $\mathcal{O}(1)$ on $\mathbb{CP}_n$. The *Koszul complex* (using the notation of [17, especially p. 95])

$$0 \to \mathcal{O}^{\overbrace{[ij\cdots kl]}^{n}}(-n) \xrightarrow{f_l} \cdots \longrightarrow \mathcal{O}^{[ij]}(-2) \xrightarrow{f_j} \mathcal{O}^{i}(-1) \xrightarrow{f_i} \mathcal{O}$$

provides a resolution of the constant sheaf supported at the basepoint. Tensoring this complex with $\mathcal{O}(0|b, c, \ldots, e)$ gives a resolution of the sheaf with fibre the vector space $(0|b, c, \ldots, e)$ supported at the basepoint. We now proceed by induction on n assuming we know the dimension of this vector space. Each of the resolving sheaves is a direct sum of sheaves of the form

$$\mathcal{O}(r|b, c, \ldots, e) \quad \text{for } 0 \leqslant r \leqslant n.$$

Standard cohomological procedures and the Bott–Borel–Weil theorem for complex projective space now allow an induction on $b + c + \cdots + e$ to compute the dimension of $(0, b, c, \ldots, e)$. The details are left to the reader.

Exercise 5.4 Fill in the details of this inductive proof.

6 Other groups

The usual general form of the Borel–Weil and Bott–Borel–Weil theorems applies to homogeneous spaces G/B where G is a complex semisimple Lie group and B is what is known as a *Borel* subgroup. In fact, this generalizes the manifold $\mathbb{F}_{1,2,\ldots,n}(\mathbb{C}^{n+1})$ of complete flags as a homogeneous space for $\mathrm{SL}(n+1, \mathbb{C})$ rather than $\mathrm{GL}(n+1, \mathbb{C})$ but the distinction is minor: in Theorem 2.18 and the discussion which precedes it, two weights become equivalent if they differ by a repeated integer $k, k, \ldots, k$. In other words, only the differences $b-a, \ldots, d-c$ are significant. Dominance becomes the statement that all these differences are non-negative.

Complex semisimple Lie groups are like $\mathrm{SL}(n+1, \mathbb{C})$ because they are built out of copies of $\mathrm{SL}(2, \mathbb{C})$ much like the blocks of (2.17). In general, the blocks are put together in a slightly different way and there is a convenient way of specifying how. It is called a *Dynkin diagram* and the classification of complex semisimple Lie groups (or, more usually, their corresponding Lie algebras) is normally given in terms of these diagrams. The Dynkin diagram of $\mathrm{SL}(n+1,\mathbb{C})$ is

$$\bullet\!\!-\!\!\!-\!\!\bullet\!\!-\!\!\!-\!\!\bullet \cdots \!-\!\!\!-\!\!\bullet\!\!-\!\!\!-\!\!\bullet \quad \text{with } n \text{ nodes.}$$

These constitute the *A-series* of such diagrams. Each node represents an $\mathrm{SL}(2,\mathbb{C})$ building block. The lines between neighbouring nodes encode the interaction between the corresponding Lie subalgebras. There being no other lines means that non-neighbouring subalgebras don't interact at all (as is clear from (2.17)).

The *D-series* of semisimple Lie algebras has Dynkin diagrams of the form

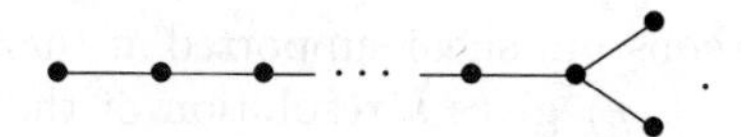

With n nodes, this is D_n and the corresponding simply connected complex Lie group is $\mathrm{Spin}(2n,\mathbb{C})$, the double cover of the complex orthogonal group $\mathrm{SO}(2n,\mathbb{C})$. The irreducible finite-dimensional complex representations of $\mathrm{Spin}(2n,\mathbb{C})$ are conveniently recorded by attaching a non-negative integer to each node of the Dynkin diagram. This exactly parallels the A-series where $b-a,\ldots,d-c$ would be attached. Up to conjugation, there is only one Borel subgroup and all its irreducible representations are one-dimensional. The Bott–Borel–Weil theorem computes the analytic cohomology of the corresponding homogeneous line bundles on G/B.

Having G built out of copies of $\mathrm{SL}(2,\mathbb{C})$ is useful, not only in developing the theory of weights for the general semisimple Lie algebra, but also in fashioning a proof of the Bott–Borel–Weil theorem using the case of $\mathbb{CP}_1$ as the basic ingredient. Such a proof may be found in [3]. This book also treats the case of G/P where P is a *parabolic* subgroup. Complex projective space falls into this category.

We remarked earlier that the holomorphic finite-dimensional representation theory of $\mathrm{GL}(n+1,\mathbb{C})$ is equivalent to the finite-dimensional representation theory of $\mathrm{U}(n+1)$. Similar remarks are valid for $\mathrm{SL}(n+1,\mathbb{C})$ versus $\mathrm{SU}(n+1)$ and $\mathrm{Spin}(2n,\mathbb{C})$ versus $\mathrm{Spin}(2n)$. This generalizes to the case of a complex semisimple Lie group versus its *compact real form*. Alternatively, one can start with a compact Lie group and pass to its *complexification* as in [6, 13]. The Bott–Borel–Weil theorem is probably most often stated for compact groups. Often it is regarded as a model, ripe for generalization. See [14] for an exposition from this point of view.

There is also a purely algebraic version of the Bott–Borel–Weil theorem due to Kostant [16]. For a discussion of this and further developments in the algebraic setting see [15]. For an exposition of the structure of complex semisimple Lie algebras and their finite-dimensional representations see [12].

An Instructional Conference was held at the International Centre for Mathematical Sciences in Edinburgh in 1996. Its proceedings [2] contain many fine expository articles suitable for further study.

References

1. M.F. Atiyah and R. Bott: A Lefschetz fixed-point formula for elliptic complexes II: applications, *Ann. Math.* **88** (1968), 451–491.
2. T.N. Bailey and A.W. Knapp (eds.): 'Representation Theory and Automorphic Forms', Proc. Symp. Pure Math. Volume 61, Amer. Math. Soc. (1997).
3. R.J. Baston and M.G. Eastwood: 'The Penrose Transform: Its interaction with Representation Theory', Oxford University Press (1989).
4. R. Bott: Homogeneous vector bundles, *Ann. Math.* **60** (1957), 203–248.
5. R. Bott: On induced representations, in 'The Mathematical Heritage of Hermann Weyl,' Proc. Symp. Pure Math. Volume 48, Amer. Math. Soc. (1988), 1–13.
6. T. Bröcker and T. tom Dieck: 'Representations of Compact Lie Groups', Springer (1985).
7. H. Cartan and J.-P. Serre: Un théorème de finitude concernant les variétés analytiques compactes, *C. R. Acad. Sci. Paris* **237** (1953), 128–130.
8. W. Fulton and J. Harris: 'Representation Theory: A First Course', Springer (1991).
9. P.B. Gilkey: 'The Index Theorem and the Heat Equation', Publish or Perish (1974).
10. R.C. Gunning: 'Introduction to Holomorphic Functions of Several Variables', Wadsworth & Brooks/Cole (1990).
11. F. Hirzebruch: 'Topological Methods in Algebraic Geometry', Springer (1966).
12. J.E. Humphreys: 'Introduction to Lie Algebras and Representation Theory', Springer (1972).
13. A.W. Knapp: 'Representation Theory of Semisimple Groups: An Overview Based on Examples', Princeton University Press (1986).
14. A.W. Knapp: Introduction to representations in analytic cohomology, in 'The Penrose Transform and Analytic Cohomology in Representation Theory,' Cont. Math. Volume 154, Amer. Math. Soc. (1993), 1–19.
15. A.W. Knapp and D.A. Vogan Jr: 'Cohomological Induction and Unitary Representations', Princeton University Press (1995).
16. B. Kostant: Lie algebra cohomology and the generalized Borel–Weil Theorem, *Ann. Math.* **74** (1961), 329–387.
17. R. Penrose and W. Rindler: 'Spinors and Space-time', Volume 2, Cambridge University Press (1986).
18. J.-P. Serre: Représentations linéaires et espaces homogènes Kählériens des groupes de Lie compacts, Séminaires Bourbaki, n° 100, Institut Henri Poincaré 1954, Benjamin (1966).
19. R.O. Wells Jr: 'Differential analysis on complex manifolds', Springer (1980).

5
Morse theory in the 1990s

MARTIN A. GUEST

Introduction

Since the publication of Milnor's book [47] in 1963, Morse theory has been a standard topic in the education of geometers and topologists. This book established such high standards for clarity of exposition and mathematical influence that it has been reprinted several times, and it is still the most popular introductory reference for the subject.

Morse theory is not merely a useful technique. It embodies a far-reaching *idea*, which relates analysis, topology, and (most recently) physics. This no doubt is responsible for the resilience of Morse theory over the past several decades: despite the essential simplicity of the idea, it seems to re-emerge every few years to play a crucial role in some major new mathematical development. The title of Bott's article [13] was no doubt inspired by this resilience – here is a subject which appears to be completely 'worked out', yet which time after time has come back to yield something unexpected. Characteristically, a few years after the appearance of [13], Morse theory is again at the forefront of mathematics, as a motivating example of a 'topological field theory'.

This article is based loosely on four lectures given at a Graduate School in Differential Geometry, held at the University of Durham in September 1996. The purpose of the lectures was to give a topical introduction to Morse theory, for postgraduate students in the general area of geometry. In order to save time and yet provide a concrete focus, I omitted most of the standard proofs, and used Morse functions on *Grassmannian manifolds* as a fundamental collection of examples to illustrate the theorems. As well as introducing some basic aspects (such as Schubert varieties) of these important manifolds, this gave the opportunity of discussing a link between Morse theory and Lie theory. A second feature was that I emphasized from the start the fundamental role played by the *gradient flow lines* of a Morse function. With the benefit of hindsight – see §4 – this is a very natural point of view to take. It also fits well with the Grassmannian examples, where the flow lines are known explicitly.

In late 1997 I gave a more leisurely series of lectures for advanced undergraduate students at Tokyo Metropolitan University, and I took this opportunity to expand greatly my original notes. As in the earlier lectures, I emphasized the Grassmannians and the role of the gradient flow lines, but this time I went 'beyond homology groups' in order to illustrate the real power of Morse theory, and to give some idea of the developments since [13]. I have also tried to give an account of the 'toric' point of view, the full significance of which is only just beginning to be appreciated.

Before getting started, a few historical comments are appropriate. After the pioneering work of Morse, the 'modern' period of Morse theory began with Bott's work in the 1950s on the homology and homotopy groups of compact symmetric spaces. One of the main achievements of the Morse-theoretic approach was the extension of this work to the loop space of a symmetric space; this led to the discovery of the (Bott) Periodicity theorem and ultimately to K-theory. Although the role of Morse theory in this area was quickly taken over by the new machinery of algebraic topology, the geometrical nature of Bott's proof of the Periodicity theorem still retains great appeal. During the 1960s, Morse theory was used most prominently to investigate the topology of manifolds, and most prominently of all in the work of Smale, which led to a proof of the Poincaré conjecture in dimensions greater than four. Following several dormant years in the early 1970s, Morse theory returned as a guiding force in the development of mathematical Yang–Mills theory, in which the critical points of the Yang–Mills functional are studied. In the 1980s, having been pushed out of the limelight by the rapidly developing analytic and algebraic geometrical aspects of gauge theory, Morse theory found a dramatic new role. This time the primary motivation was a new approach to Morse theory due to Witten, together with an extension of these ideas by Floer. In this approach the gradient flow lines play the central role, and this laid the groundwork for the 'field-theoretic' point of view pioneered by Cohen–Jones–Segal, Betz–Cohen, and Fukaya.

The main emphasis of these notes is the Morse theory of compact – in particular, finite-dimensional – manifolds. I have tried to give at least some references for each aspect of finite-dimensional Morse theory, but not for the infinite-dimensional theory where the subject is much more diverse.

1 Morse functions

In this section and the next we give a brief explanation of Morse theory, referring mainly to the first 40 pages of [47] for proofs. We begin by defining Morse functions and by mentioning several non-trivial examples. We introduce the flow lines and the stable and unstable manifolds, and give some examples to illustrate these fundamental concepts.

1.1 A basic question

Let us agree that it is important to study manifolds. One way to study a manifold might be to study all possible real-valued functions on it. Presumably, different types of manifolds will possess different types of functions.

Every manifold admits real-valued functions, e.g. constant functions. But it is not immediately obvious how to write down explicit formulae for non-trivial real-valued functions on a given manifold. For example, consider the Grassmannians

$$Gr_k(\mathbb{R}^n) = \{\text{real } k\text{-dimensional linear subspaces of } \mathbb{R}^n\}$$
$$Gr_k(\mathbb{C}^n) = \{\text{complex } k\text{-dimensional linear subspaces of } \mathbb{C}^n\},$$

which are compact manifolds of (real) dimensions $k(n-k)$, $2k(n-k)$ respectively. To give a function $f : M \to \mathbb{R}$, where $M = Gr_k(\mathbb{R}^n)$ or $Gr_k(\mathbb{C}^n)$, we must associate to each k-plane a real number. How can we do this in a natural and non-trivial way?

As a much easier example, consider the manifold S^1 consisting of complex numbers $e^{2\pi i\theta}$ of unit length (i.e. the circle). The angle θ defines a function $S^1 \to \mathbb{R}/\mathbb{Z}$, which is not quite what we want, but we can obtain a real-valued function by using $\cos 2\pi\theta$. It seems reasonable to regard this as the 'simplest' kind of non-trivial real-valued function on S^1. Note that this function may be interpreted as the first coordinate of the embedding $S^1 \to \mathbb{R}^2$, $e^{2\pi i\theta} \mapsto (\cos 2\pi\theta, \sin 2\pi\theta)$. This suggests a useful source of functions on a general manifold M: first embed M into a euclidean space, then take a coordinate function. (But in order to find nice functions, one has to find a nice embedding.)

As a slight modification of the previous example, we could take the torus: $T = S^1 \times S^1$. For a fixed choice of $(a, b) \in \mathbb{R}^2$, we have a 'coordinate function' $(e^{2\pi ix}, e^{2\pi iy}) \mapsto a\cos 2\pi x + b\cos 2\pi y$. This is a coordinate function for an embedding of T in $\mathbb{R}^4$. We could instead use the familiar embedding of T in $\mathbb{R}^3$, and take a coordinate function there (page 1 of [47]). But we shall see later that this function is slightly less satisfactory, for the purposes of present-day Morse theory.

The basic question which Morse theory addresses is: *what is the relation between the properties of a manifold and the properties of its real-valued functions?* By 'properties' we mean global properties, as all manifolds of the same dimension have the same local properties. Thus, Morse theory aims to relate topological properties of M with analytical properties of real-functions on M.

1.2 Morse functions

First we recall a standard definition:

Definition 1.2.1 *Let M be a (smooth)[1] manifold, and let $f : M \to \mathbb{R}$ be a (smooth) function. A point $m \in M$ is called a critical point of f if $Df_m = 0$.*

The derivative Df_m at m is a linear functional on the tangent space T_mM; thus m is critical if and only if this derivative is the zero linear functional.

In terms of a local coordinate chart $\phi : U \to \mathbb{R}^n$, where U is an open neighbourhood of m in M and $\phi(m) = 0$, f corresponds to the function

$$f \circ \phi^{-1} : U \to \mathbb{R},$$

[1] The word 'smooth', meaning 'infinitely differentiable', will usually be omitted in future.

and Df_m is represented by the $1 \times n$ matrix

$$\left(\frac{\partial(f \circ \phi^{-1})}{\partial x_1}(0), \ldots, \frac{\partial(f \circ \phi^{-1})}{\partial x_n}(0) \right).$$

It will simplify notation if we just write f instead of $f \circ \phi^{-1}$. Using this convention, Taylor's theorem may be written as

$$f(x) - f(0) = \sum a_i x_i + \tfrac{1}{2} \sum a_{ij} x_i x_j + \text{remainder}$$

where $a_i = (\partial f / \partial x_i)|_0$ and $a_{ij} = (\partial^2 f / \partial x_i \partial x_j)|_0$.

When m is a critical point, i.e. $\sum a_i x_i = 0$, it can be shown that the matrix (a_{ij}) defines a symmetric bilinear form on the tangent space $T_m M$. (This bilinear form is called the Hessian of f.) Hence it is diagonalizable, and the rank and nullity do not depend on the choice of ϕ. This leads to two more definitions:

Definition 1.2.2 *Let m be a critical point of $f : M \to \mathbb{R}$. The index of m is defined to be the index of (a_{ij}), i.e. the number of negative eigenvalues of (a_{ij}).*

Definition 1.2.3 *Let m be a critical point of $f : M \to \mathbb{R}$. We say that m is a non-degenerate critical point if and only if the nullity of (a_{ij}), i.e. the dimension of the 0-eigenspace of (a_{ij}), is zero.*

Since any function on a compact manifold has critical points (e.g. maxima and minima), we cannot get very far by considering functions without critical points. In other words, it is unreasonable to insist that the first term in the Taylor series be a non-degenerate linear functional at every point. The next most favourable condition is that a function has no degenerate critical points, i.e. that (at each critical point) the quadratic term in the Taylor series be non-degenerate:

Definition 1.2.4 *Let M be a (smooth) manifold, and let $f : M \to \mathbb{R}$ be a (smooth) function. We say that f is a Morse function if and only if every critical point of f is non-degenerate.*

To be a Morse function is in some sense a weak condition, as it can be shown that the space of Morse functions is dense in the space of functions. But in another sense[2] it is a strong condition, as Morse functions have a very special local canonical form:

Lemma 1.2.5 [The Morse lemma] *Let $f : M \to \mathbb{R}$ be a Morse function. Then, for any $m \in M$, there exists a local chart ϕ at m such that*

$$f(x) - f(0) = -\sum_{i=1}^{\lambda} x_i^2 + \sum_{i=\lambda+1}^{n} x_i^2.$$

□

[2] This somewhat paradoxical situation is indicative of a 'good' definition, perhaps.

Note that the remainder term has disappeared. It follows from this formula that the index of any local maximum point is n, and the index of any local minimum point is 0.

Example 1.2.6 [Height functions on the torus] In the diagram below we have two embeddings of the torus T^2 in $\mathbb{R}^3$. By taking the z-coordinate function, we obtain two real-valued functions on T^2. The critical points are marked with crosses.

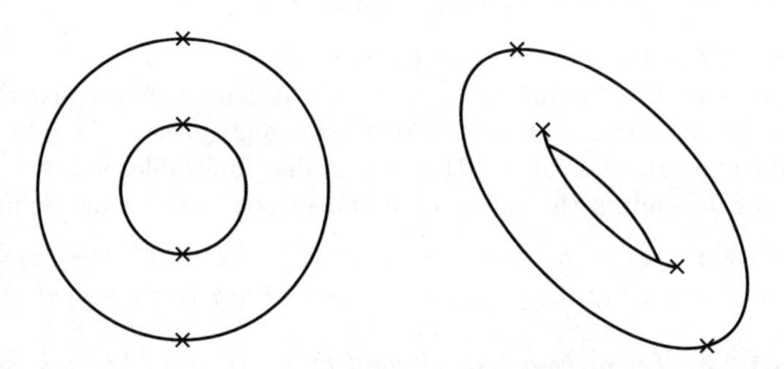

It is instructive to write down explicit formulae for these functions on T^2, and to verify that the critical points are all non-degenerate, with indices $2, 1, 1, 0$. □

Example 1.2.7 [General theory of height functions] Let M be a compact submanifold of $\mathbb{R}^N$. For any $v \in \mathbb{R}^N$ we may define the 'height function' h^v and the 'distance function' L^v on M by

$$h^v(m) = \langle m, v \rangle, \qquad L^v(m) = \langle m - v, m - v \rangle$$

where $\langle \; , \; \rangle$ is the standard inner product. (Note that these functions are essentially the same if M is embedded in the sphere $S^{N-1} \subseteq \mathbb{R}^N$.) It can be shown that, for almost all values of v, the functions h^v and L^v are Morse functions. The critical point theory of these functions is intimately related to the Riemannian geometry of M (with its induced metric). A brief treatment of L^v is given in [47]; the general theory is developed in [53]. □

1.3 Geometry

To obtain geometrical information from a Morse function $f : M \to \mathbb{R}$ it is useful to consider the gradient vector field ∇f of f. To define the gradient vector field we assume that a Riemannian metric $\langle \; , \; \rangle$ has been chosen on M, and we define $(\nabla f)_m$ by $\langle (\nabla f)_m, X \rangle = (Df)_m(X)$ (for all $X \in T_m M$). For many purposes (although not all!) the particular choice of Riemannian metric is unimportant.

By the theorem of local existence of solutions to first-order ordinary differential equations, there exists an integral curve γ of the vector field $-\nabla f$ through any point of M. (We introduce the minus sign because we want to consider γ as flowing 'downwards'.) It should be noted that *explicit formulae* for these integral curves are not readily available in general.

If M is compact, then the domain of any such integral curve is $\mathbb{R}$, and $\mathbb{R}$ acts on M as a group of diffeomorphisms. We shall denote this action by $m \mapsto t \cdot m$, for $m \in M$, $t \in \mathbb{R}$. In other words, $t \cdot m = \gamma(t)$, where γ is the solution of the o.d.e. $\gamma'(t) = -(\nabla f)_{\gamma(t)}$ such that $\gamma(0) = m$. There are only a finite number of critical points in this case (as critical points are isolated, by the Morse lemma). Each integral curve γ 'begins' and 'ends' at critical points, i.e. $\lim_{t\to\pm\infty} \gamma(t)$ are critical points. Evidently these integral curves are constrained by the global nature of M, and we shall see that they are a very useful tool for investigating M.

Example 1.3.1 Let $M = S^2$, and embed S^2 in $\mathbb{R}^3$ as the unit sphere. The z-coordinate function $f : S^2 \to \mathbb{R}$, $f(x, y, z) = z$ is a Morse function and it has precisely two critical points, the north pole and the south pole. It is the restriction of $F : \mathbb{R}^3 \to \mathbb{R}$, $F(x, y, z) = z$, and $-\nabla f$ is the component of $-\nabla F = (0, 0, -1)$ which is tangential to the sphere. Hence each integral curve of $-\nabla f$ is a line of longitude running from the north pole to the south pole (with a certain parametrization). □

Example 1.3.2 [Height functions on the torus, continued] Consider the two height functions on T^2 from Example 1.2.6. The integral curves are illustrated below, with respect to the induced Riemannian metric from $\mathbb{R}^3$. In each case the torus is represented by $\mathbb{R}^2/\mathbb{Z}^2$, and a fundamental domain is shown. The maximum point is at the centre of the square; the minimum point is represented by the four corners.

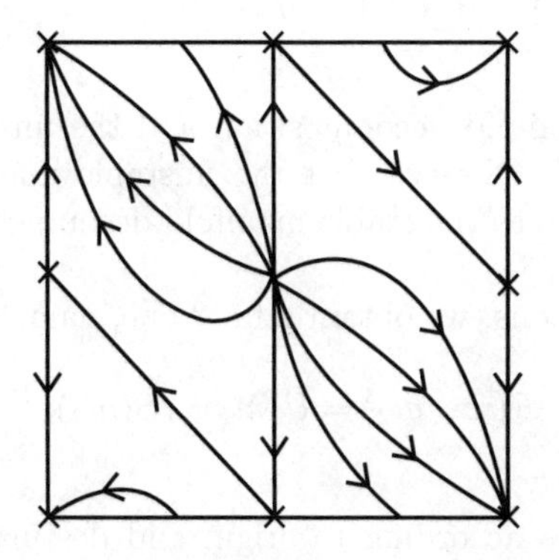

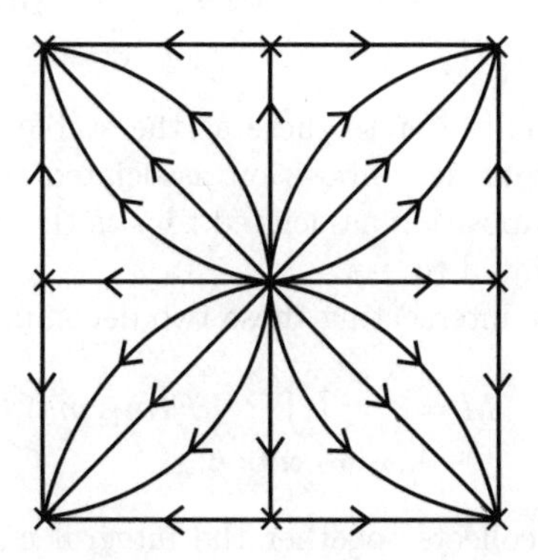

Example 1.3.3 It seems obvious that the least number of isolated critical points of any function on S^2 is 2 (proof?). Example 1.3.1 provides such a function. But what is the least number of critical points of any function on the torus T^2? In the case of a Morse function, we shall see later that the answer is 4. But if

arbitrary (smooth) functions are allowed, the answer is 3. (For an example, see [56], Example 1, page 19).) □

1.4 Stable and unstable manifolds

Integral curves, or flow lines, may be assembled into the following important objects:

Definition 1.4.1 *Let M be a compact manifold, let $f : M \to \mathbb{R}$ be a Morse function, and let m be a critical point of f. The stable manifold $S(m)$ of m is the set of points which flow 'down' to m, i.e.*

$$S(m) = \left\{ x \in M \mid \lim_{t \to \infty} t \cdot x = m \right\}.$$

The unstable manifold $U(m)$ of m is the set of points which flow 'up' to m, i.e.

$$U(m) = \left\{ x \in M \mid \lim_{t \to -\infty} t \cdot x = m \right\}.$$

The next result can be proved from the Morse lemma.

Proposition 1.4.2 *Let the index of m be λ. Then $S(m), U(m)$ are homeomorphic (respectively) to $\mathbb{R}^{n-\lambda}, \mathbb{R}^{\lambda}$.* □

It follows that a Morse function f on M provides two decompositions of M into disjoint 'cells':

$$M = \bigcup_{m \text{ critical}} S(m) = \bigcup_{m \text{ critical}} U(m).$$

We shall refer to these as the stable manifold decomposition and the unstable manifold decomposition associated to f. Observe that the unstable manifold decomposition associated to f is the same as the stable manifold decomposition associated to $-f$.

By intersecting these two decompositions, we obtain a finer one, namely

$$M = \bigcup_{m_1, m_2 \text{ critical}} F(m_1, m_2), \qquad F(m_1, m_2) = U(m_1) \cap S(m_2).$$

This collects together the integral curves according to origin and destination: $F(m_1, m_2)$ consists of all integral curves which go from m_1 to m_2.

Example 1.4.3 Consider the Morse function of Example 1.3.1, on $M = S^2$. The stable manifold decomposition has two pieces, i.e. one copy of each of $\mathbb{R}^0$ and $\mathbb{R}^2$. The unstable manifold decomposition is similar. The intersection of these decompositions has three pieces – two points and one copy of $\mathbb{R}^2 - \{0\}$. □

Example 1.4.4 [Height functions on the torus, continued] Consider again the two height functions on T^2 from Examples 1.2.6 and 1.3.2. In each case there are four unstable manifolds: a 0-cell, two 1-cells, and a 2-cell. The same is true for the stable manifolds. However, the decompositions $T^2 = \cup_{m_1,m_2 \text{ critical}} F(m_1, m_2)$ are quite different. □

The behaviour of the stable and unstable manifolds is particularly nice if we impose the condition that they intersect transversely:

Definition 1.4.5 *A Morse function $f : M \to \mathbb{R}$ on a Riemannian manifold M is said to be a Morse–Smale function if $U(m_1)$ is transverse to $S(m_2)$ for all critical points m_1, m_2 of f.*

For the meaning of transversality, we refer to Chapter 3 of [41] or Chapter 3 of [40], or other texts on differential topology. This concept was introduced into Morse theory by Smale – see [60], and the historical discussion in [12].

The transversality condition implies that $U(m_1) \cap S(m_2)$ is a manifold, and that

$$\operatorname{codim} U(m_1) \cap S(m_2) = \operatorname{codim} U(m_1) + \operatorname{codim} S(m_2)$$

whenever $U(m_1) \cap S(m_2)$ is non-empty. Since $F(m_1, m_2) = U(m_1) \cap S(m_2)$, and $\dim U(m_i) = \lambda_i$, we have

$$n - \dim F(m_1, m_2) = (n - \lambda_1) + \lambda_2$$

and hence

$$\dim F(m_1, m_2) = \lambda_1 - \lambda_2.$$

In particular, if there exists a flow line from m_1 to m_2, then we must have $\lambda_1 > \lambda_2$.

Example 1.4.6 Of the two Morse functions on the torus in Example 1.2.6 (see also Example 1.3.2), only one satisfies the Morse–Smale condition. (Which one? Note that, by the previous paragraph, the existence of a flow line connecting two critical points of the same index is not possible for a Morse–Smale function.) □

The concepts introduced so far will help us to address the question 'what configurations of critical points and flow lines are possible?' for a Morse (or Morse–Smale) function on a given manifold M. This is a more precise version of our original question 'what kind of smooth functions are possible?' on M.

Finally, we should mention that the behaviour of a smooth function is much less predictable without the Morse condition, i.e. in the presence of degenerate critical points. For example, critical points are not necessarily isolated, and flow lines do not necessarily converge to critical points. For an example of the latter phenomenon, see page 14 of [54].

2 Topology

Morse theory gives a fundamental relation between topology and analysis. This is traditionally expressed by the 'Morse inequalities'. We describe various forms of this relation, and its generalizations.

2.1 Topology and analysis

The examples in §1 clearly suggest that there is a relation between the topology of M and the critical point data of a function $f : M \to \mathbb{R}$. We list four specific examples below, following [12]. In each case, topological information predicts the existence of critical points of f.

(i) It is an elementary fact of topology that, if M is compact, *then f has maximum and minimum points, and these are critical points.*

(ii) Assume that f has a finite number k of critical points (but is not necessarily a Morse function). Then we still have $M = \cup_{m \text{ critical}} S(m)$, and the $S(m)$ are a finite number of contractible sets (but not necessarily cells). It can be shown that this implies the following cohomological condition: if $m > k$, then the product of any m cohomology classes (of positive dimension) on M must be zero. We shall not make any use of this idea, which is part of the theory of Lyusternik–Schnirelmann category, so we refer to Lecture 2 of [12] for further information. However, we note that it gives a stronger version of the prediction of (i): *if M has i cohomology classes (of positive dimension) whose product is non-zero, then any function on M must have at least $i+1$ critical points.*

For example, consider the torus T^2. There are two one-dimensional cohomology classes whose product is non-zero, so any function on T^2 must have at least three critical points. It is not possible to find three cohomology classes (of positive dimension) whose product is non-zero, so we cannot improve the estimate beyond three. This is just as well, in view of Example 1.3.3.

(iii) The 'minimax principle' – see [12].

(iv) Finally we come to our main example, the Morse inequalities. We shall consider this matter in detail later on, but the basic fact is easily stated: if f is a Morse function on a compact manifold M, *then the number of critical points of index k is greater than or equal to the kth Betti number b_k of M.* This is a considerable generalization of (i), but in fact it is only a hint of the power of Morse theory, as we shall see.

2.2 The main theorem of Morse theory

Recall from §1.4 that a Morse function f gives a decomposition

$$M = \bigcup_{m \text{ critical}} U(m)$$

of a compact manifold M, where $U(m)$ is homeomorphic to $\mathbb{R}^{\lambda_m}$, λ_m being the index of m. The main theorem of Morse theory gives information about how these pieces fit together:

Theorem 2.2.1 *Let M be a compact manifold, and let $f : M \to \mathbb{R}$ be a Morse function on M. Then M has the homotopy type of a cell complex, with one cell of dimension λ for each critical point of index λ.*

To be precise, this theorem means that M is homotopy equivalent to a topological space of the form

$$X_r = ((D^{\lambda_1} \cup_{f_1} D^{\lambda_2}) \cup_{f_2} D^{\lambda_3}) \cup_{f_3} \cdots$$

where $0 = \lambda_1, \lambda_2, \lambda_3, \ldots, \lambda_r = m$ are the indices of the critical points of f (listed in increasing order, with repetitions where necessary), and $f_1, f_2, f_3, \ldots, f_{r-1}$ are certain continuous 'attaching maps', with

$$f_i : \partial D^{\lambda_{i+1}} \to X_i.$$

To simplify the notation, we shall drop the parentheses in future and simply write $X_r = D^{\lambda_1} \cup_{f_1} D^{\lambda_2} \cup_{f_2} \cdots \cup_{f_{r-1}} D^{\lambda_r}$.

In §2.4 we sketch the main steps in the proof of this theorem. First, however, we give some simple examples.

Example 2.2.2 The height function on S^1 defined by the illustrated embedding of S^1 in $\mathbb{R}^2$ has three local minima A, B, C and three local maxima D, E, F.

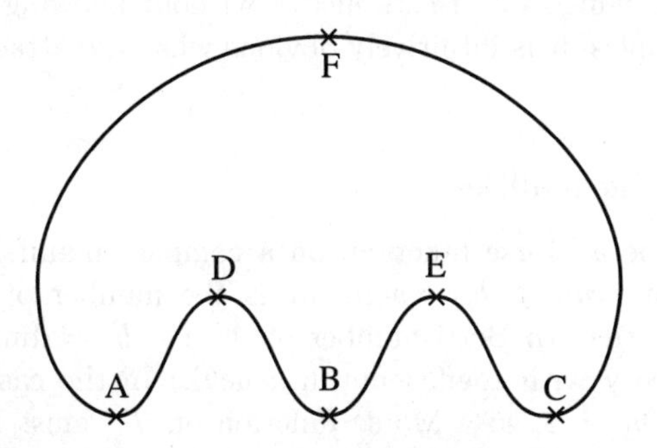

We have $S^1 \simeq \{A, B, C\} \cup_f D^1 \cup_g D^1 \cup_h D^1$. Each of the attaching maps f, g, h is an injective map from $\{-1, 1\}$ to $\{A, B, C\}$. Contemplation of this example suggests that one should be able to say more than 'there is at least one minimum point and at least one maximum point'. It seems plausible, for example, that the number of local maxima must always be *equal* to the number of local minima for this manifold. We shall soon see that this is correct (Corollary 2.3.2). □

Example 2.2.3 The height function on S^2 defined by the illustrated embedding of S^2 in $\mathbb{R}^3$ has one local minimum, one critical point of index 1, and two local maxima.

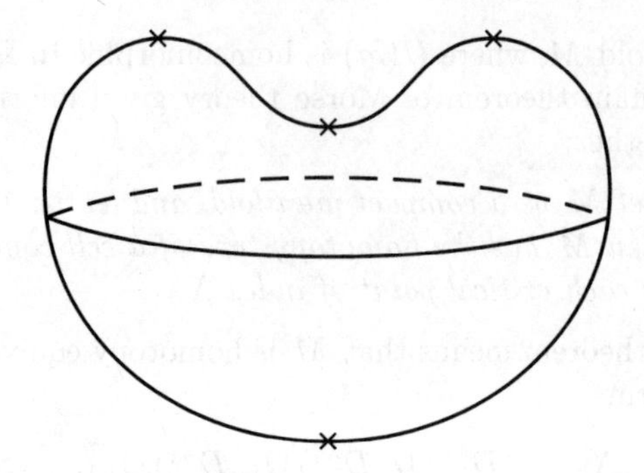

We have $S^2 \simeq D^0 \cup_f D^1 \cup_g D^2 \cup_h D^2$. Here, $D^0 \cup_f D^1$ is a copy of S^1, and the maps g, h serve to attach two hemispheres to this circle. □

Example 2.2.4 [Height functions on the torus, continued] The two height functions on T^2 in Example 1.2.6 give two cell decompositions $T^2 \simeq D^0 \cup_f D^1 \cup_g D^1 \cup_h D^2$, However, the attaching maps behave quite differently in each case, as is clear from Examples 1.3.2 and 1.4.4. □

Theorem 2.2.1 does not tell us anything about the attaching maps, other than the fact that they exist. This appears to be a serious deficiency. However, we can deduce quite a lot of information on the topology of M – for example, the Morse inequalities, which we discuss next – without knowing anything further. In the above examples, it is intuitively obvious what the attaching maps are.

2.3 The Morse inequalities

Let $f : M \to \mathbb{R}$ be a Morse function, on a compact manifold M. The Morse inequalities say that $m_i \geqslant b_i$, where m_i is the number of critical points of index i, and b_i is the ith Betti number of M, i.e. $b_i = \dim H_i(M)$. (We use any homology theory with coefficients in a field.) In the case of T^2, we have $b_0 = b_2 = 1$ and $b_1 = 2$, so a Morse function on T^2 must have at least four critical points. We have already given examples where this happens.

In terms of the *Morse polynomial*

$$M(t) = \sum_{i=0}^{n} m_i t^i$$

and the *Poincaré polynomial*

$$P(t) = \sum_{i=0}^{n} b_i t^i$$

we may express the Morse inequalities symbolically as $M(t) \geqslant P(t)$. Using this convenient notation, there is a stronger form of the Morse inequalities:

Theorem 2.3.1 *Let $f : M \to \mathbb{R}$ be a Morse function on a compact manifold M. Then $M(t) - P(t) = (1+t)Q(t)$, for some polynomial $Q(t)$ such that $Q(t) \geqslant 0$.*

We shall sketch the proof in the next section. Before that, we give some simple consequences and some examples. To begin with, we put $t = -1$ in the theorem to obtain an expression for the Euler characteristic of M:

Corollary 2.3.2 *Let $f : M \to \mathbb{R}$ be a Morse function on a compact manifold M. Then $\sum_{i=0}^{n}(-1)^i m_i = \sum_{i=0}^{n}(-1)^i b_i$.* □

(This is a special case of the Hopf index theorem on vector fields, namely for vector fields of the particular form ∇f.)

Corollary 2.3.3 *Let $f : M \to \mathbb{R}$ be a Morse function on a compact manifold M. If $M(t)$ contains only even powers of t, then $M(t) = P(t)$.*

Proof Assume that $M(t)$ contains only even powers of t. Then $M(t) \geqslant P(t) + (1+t)Q(t)$, so neither $P(t)$ nor $(1+t)Q(t)$ can contain odd powers of t. But this is possible only if $Q(t)$ is the zero polynomial. □

(Various generalizations of this result can be obtained by similar reasoning – the basic principle is that a 'gap' in the sequence $m_0, m_1, m_2, \ldots$ forces a relation between $M(t)$ and $Q(t)$. This is the 'lacunary principle' of Morse.)

Example 2.3.4 For our height functions on the torus T^2, we have $M(t) = P(t) = 1 + 2t + t^2$. □

Example 2.3.5 Let $M = \mathbb{CP}^n$, i.e. n-dimensional complex projective space. This may be defined as $\mathbb{C}^{n+1} - \{0\}/\mathbb{C}^*$, where $\mathbb{C}^*$ acts by multiplication on each coordinate of $\mathbb{C}^{n+1}$. It may be identified with the space of all complex lines in $\mathbb{C}^{n+1}$. The equivalence class of $(z_0, \ldots, z_n)$ will be denoted by the standard notation $[z_0; \ldots; z_n]$. A Morse function $f : \mathbb{CP}^n \to \mathbb{R}$ is given on page 26 of [47]. It may be defined by the formula

$$f([z_0, \ldots, z_n]) = \sum_{i=0}^{n} c_i |z_i|^2 / \sum_{i=0}^{n} |z_i|^2$$

where $c_0 < \cdots < c_n$ are fixed real numbers. The critical points of f are the coordinate axes $L_0 = [1; 0; \ldots; 0]$, $L_1 = [0; 1; \ldots; 0]$, $\ldots$, $L_n = [0; 0; \ldots; 1]$. The index of L_i is $2i$. Since all indices are even, Corollary 2.3.4 applies, and we deduce that the Poincaré polynomial of $\mathbb{CP}^n$ is $1 + t^2 + t^4 + \cdots + t^{2n}$. □

It is already clear from these results and examples that Morse theory works 'both ways': (1) the homology groups of a manifold impose conditions on the critical points of any Morse function, and (2) the critical point data of a Morse function sometimes permits the computation of the homology groups.

If $f : M \to \mathbb{R}$ is a Morse function such that $M(t) = P(t)$, we say that f is *perfect.* The question naturally arises: does every compact manifold posess a perfect Morse function? This matter is discussed in detail in [56]; the answer, essentially, is negative. One reason is illustrated by the next example.

Example 2.3.6 Let $\mathbb{RP}^n$ be n-dimensional real projective space, with $n \geqslant 2$. Consider the function $f : \mathbb{RP}^n \to \mathbb{R}$ defined by the formula of Example 2.3.5. Is this a Morse function? (Answer: yes.) Is it a perfect Morse function? (Answer: no, if the coefficient field is $\mathbb{R}$; yes, if the coefficient field is $\mathbb{Z}/2\mathbb{Z}$.) □

2.4 Sketch proofs of the main theorems

Theorem 2.2.1 is a consequence of two fundamental results. Following [12] we call these Theorem A and Theorem B. They describe how the structure of the space

$$M^t = \{m \in M \mid f(m) \leqslant t\}$$

changes when t changes.

Theorem A *If $f^{-1}([a,b])$ contains no critical point of f, then M^b is diffeomorphic to M^a.*

The integral curves of (a slight modification of) $-\nabla f$ give the required diffeomorphism. See [47], pages 12–13.

Theorem B *If $f^{-1}([a,b])$ contains a single critical point of f, then M^b is homotopy equivalent to $M^a \cup_f D^\lambda$, for some $f : \partial D^\lambda \to M^a$, where λ is the index of the critical point.*

The proof is given in [47], pages 14–19 (see also the 'Proof by picture' in [12], pages 339–340). By Theorem A, it suffices to consider the situation in a small neighbourhood of the critical point. The example $f : \mathbb{R}^2 \to \mathbb{R}$, $f(x,y) = -x^2 + y^2$, in a neighbourhood of the critical point $(0,0)$, is instructive.

To deduce Theorem 2.2.1 from Theorems A and B, an induction argument is needed (see [47] again, pages 20–24).

There are various ways of proving (and expressing) the Morse inequalities, but all of them are based on the fact that Theorem B allows us to compute the relative homology groups $H_*(M^b, M^a)$ (in the situation of Theorem B, $H_*(M^b, M^a)$ is isomorphic to the coefficient field when $* = \lambda$, and is zero otherwise). A rather formal argument is given in [47], pages 28–31; the proof in [56] is perhaps more transparent. Later expositions of the Morse inequalities and the necessary background material may be found in Chapter 8 of [41] and Chapter 6 of [40], for example.

A more intuitive version of the proof of Theorem B appears in [12]; the idea is to consider how the Morse and Poincaré polynomials $M^a(t)$, $P^a(t)$ of M^a change when we pass from a to b. Clearly we have

$$M^b(t) = M^a(t) + t^\lambda.$$

For the Poincaré polynomial there are two possibilities, as the λ-cell D^λ either introduces a new homology class in dimension λ or bounds a homology class of M^a in dimension $\lambda - 1$. (More precisely, the connecting homomorphism $H_\lambda(M^b, M^a) \to H_{\lambda-1}(M^a)$ in the long exact homology sequence is either zero or injective.) Thus we have

$$P^b(t) = P^a(t) + (t^\lambda \text{ or } -t^{\lambda-1}).$$

We are interested in the difference $M^b(t) - P^b(t)$, and for this we have

$$M^b(t) - P^b(t) = M^a(t) - P^a(t) + (0 \text{ or } t^\lambda + t^{\lambda-1}).$$

By induction, this gives the Morse inequalities in the form $M^b(t) \geqslant P^b(t)$. It also gives immediately the stronger result $M^b(t) - P^b(t) = (1+t)Q(t)$ with $Q(t) \geqslant 0$.

2.5 Generalization: Morse–Bott functions

Since Morse functions necessarily have isolated critical points, Morse theory immediately disqualifies many 'natural' functions. Constant functions provide trivial examples of this phenomenon, but there are many non-trivial ones, such as functions which are equivariant with respect to the action of a Lie group – here the orbit of a critical point consists entirely of critical points. We begin with a simple example:

Example 2.5.1 Consider the sphere $S^2 = \{(x,y,z) \mid x^2 + y^2 + z^2 = 1\}$. Define $f : S^2 \to \mathbb{R}$ by $f(x,y,z) = -z^2$. It is easy to verify that the critical points are (1) the north pole $(0,0,1)$, (2) every point of the equator $z = 0$, and (3) the south pole $(0,0,-1)$. It is also easy to verify that (1) and (3) are non-degenerate critical points of index zero. To understand what is happening at the equator, let us choose local coordinates

$$\left(\sqrt{1-u^2-v^2}, u, v\right) \mapsto (u,v)$$

around the point $(1,0,0)$. Then, with the notational conventions of §1, we have

$$f(u,v) = -v^2, \quad f(0,0) = 0.$$

Comparing this with the form of the Morse lemma, we see that degeneracy is indicated here by the absence of $\pm u^2$. Now, the u-direction is precisely the direction of the equator, and along the equator f is constant, so degeneracy 'in the u-direction' is inescapable. (If $f : M \to \mathbb{R}$ is a smooth function and N is a connected submanifold of M consisting of critical points of f, then f is constant on N.) In the v-direction, however, we are in the situation of the Morse lemma. The integral curves of $-\nabla f$ (with respect to the standard induced metric on the sphere) are easy to imagine in this situation, and we appear to have the generalized cell decomposition

$$S^2 \simeq (\text{north pole} \cup \text{south pole}) \cup_g \left(\text{equator} \times \left[-\tfrac{1}{2}, \tfrac{1}{2}\right]\right)$$

where

$$g : \text{equator} \times \{-\tfrac{1}{2}, \tfrac{1}{2}\} \to \text{north pole} \cup \text{south pole}$$

is the map which sends 'equator $\times\{-\frac{1}{2}\}$' to the south pole and 'equator $\times\{\frac{1}{2}\}$' to the north pole. (If we use $-f$ instead of f, the decomposition would involve attaching the northern and southern hemispheres to the equator.) □

The generalization of Morse theory to such examples was developed by Bott in [10]. This depends on the following definition:

Definition 2.5.2 *Let V be a connected submanifold of M, such that every point of V is a critical point of $f : M \to \mathbb{R}$. We say that V is a non-degenerate critical manifold of f (or NDCM, for short) if, for every $v \in V$, T_vV is equal to the null-space N_v of the bilinear form $(\partial^2 f/\partial x_i \partial x_j)$ on T_vM.*

With the hypotheses of the definition, it is obvious that T_vV is contained in N_v; for an NDCM they are required to be equal. Since V is assumed to be connected, each critical point $v \in V$ has the same index, and we refer to this number as the index of V.

Definition 2.5.3 *A function $f : M \to \mathbb{R}$ is a Morse–Bott function if every critical point of M belongs to a non-degenerate critical manifold.*

Any Morse function is a Morse–Bott function, of course. A Morse–Bott function is a Morse function if and only if each NDCM is a single point.

Bott's generalization of the main theorem of Morse theory is based on the fact that each stable and unstable 'cell' has the structure of a vector bundle over an NDCM (see [10]). In the above example, the unstable manifold of the equator is in fact a trivial vector bundle (of rank 1), but in general the bundle could be non-trivial. This vector bundle is usually called the *negative bundle* of the NDCM.

Theorem 2.5.4 *Let M be a compact manifold, and let $f : M \to \mathbb{R}$ be a Morse–Bott function on M. Then M has the homotopy type of a 'cell-bundle complex'. Each non-degenerate critical manifold V of index λ contributes a cell-bundle $D^\lambda(V)$ of rank λ, i.e. a fibre bundle over V with fibre D^λ.* □

This means

$$M \simeq D^{\lambda_1}(V_1) \cup_{f_1} D^{\lambda_2}(V_2) \cup_{f_2} D^{\lambda_3}(V_3) \cup_{f_3} \cdots \cup_{f_{r-1}} D^{\lambda_r}(V_r)$$

where $0 = \lambda_1, \lambda_2, \ldots, \lambda_r \leqslant m$ are the indices of the non-degenerate critical manifolds $V_1, V_2, \ldots, V_r$ of f, and $f_1, f_2, f_3, \ldots, f_{r-1}$ are certain maps, with

$$f_i : \partial D^{\lambda_{i+1}}(V_{i+1}) \to D^{\lambda_1}(V_1) \cup_{f_1} D^{\lambda_2}(V_2) \cup_{f_2} \cdots \cup_{f_{i-1}} D^{\lambda_i}(V_i).$$

(The boundary $\partial D^\lambda(V)$ of a cell-bundle $D^\lambda(V)$ is a sphere-bundle, i.e. a fibre bundle over V with fibre $S^{\lambda-1}$.)

What would be an appropriate generalization of the Morse inequalities? Well, the natural generalization of Theorem B of §2.4 leads one to consider the relative homology groups $H_*(D^\lambda(V), \partial D^\lambda(V))$. By the Thom isomorphism theorem we have

$$H_*(D^\lambda(V), \partial D^\lambda(V)) \cong H_{*-\lambda}(V; \theta)$$

where θ is the 'orientation sheaf' of the negative bundle. There are two commonly occurring situations where this orientation sheaf is constant: (1) if we use homology with coefficients in the field $\mathbb{Z}/2\mathbb{Z}$, then $\theta = \mathbb{Z}/2\mathbb{Z}$; (2) if M and V are complex manifolds and the negative bundle is a complex vector bundle, then $\theta = F$ is constant for any coefficient field F (or even for integer coefficients). If the negative bundle is trivial (as in Example 2.5.1), then θ is certainly constant.

For notational simplicity we shall assume from now on that we are in the favourable situation where θ is constant, so that $H_*(D^\lambda(V), \partial D^\lambda(V)) \cong H_{*-\lambda}(V)$. Following the argument of §2.4, we see that the contribution of V to the Poincaré polynomial is either $t^\lambda P^V(t)$ or $-t^{\lambda-1}P^V(t)$, where $P^V(t)$ is the Poincaré polynomial of V. Let us define the *Morse–Bott polynomial* of f to be

$$MB(t) = \sum_{i=1}^{r} P^{V_i}(t) t^{\lambda_i}.$$

Then the argument of §2.4 gives the following analogue of Theorem 2.3.1, which we refer to as the Morse–Bott inequalities:

Theorem 2.5.5 *Let $f : M \to \mathbb{R}$ be a Morse–Bott function on a compact manifold M. Then $MB(t) - P(t) = (1+t)Q(t)$, where $Q(t) \geqslant 0$.* □

In particular we have $MB(t) \geqslant P(t)$. The comments in §2.3 on 'gaps' apply equally to this situation: if $MB(t)$ contains no odd power of t, then $MB(t) = P(t)$, i.e. f is a perfect Morse–Bott function.

Constant functions are now (satisfyingly) incorporated into the theory, because a constant function on M has one NDCM, namely M itself, of index zero, and we have $MB(t) = P(t)$. Somewhat more interesting are functions of the form $\pi \circ f$, where $\pi : E \to M$ is a fibre bundle and $f : M \to \mathbb{R}$ is a Morse–Bott function. Any function of this form is a Morse–Bott function; the NDCMs are of the form $\pi^{-1}(V)$, where V is an NDCM of f, and the index of $\pi^{-1}(V)$ is equal to the index of V.

Bott explains the following intuition behind Theorem 2.5.5: if a Morse–Bott function is deformed slightly to give a Morse function, each NDCM V of index λ breaks up into a finite set of isolated critical points, and for each Betti number b_i of V there are b_i critical points of index $i + \lambda$. In the case of Example 2.5.1, the equator contributes $tP^{S^1}(t) = t(1+t)$, which is equivalent to having two isolated critical points of indices $1, 2$. The way in which the equator can break up into two isolated critical points is illustrated in the diagrams below:

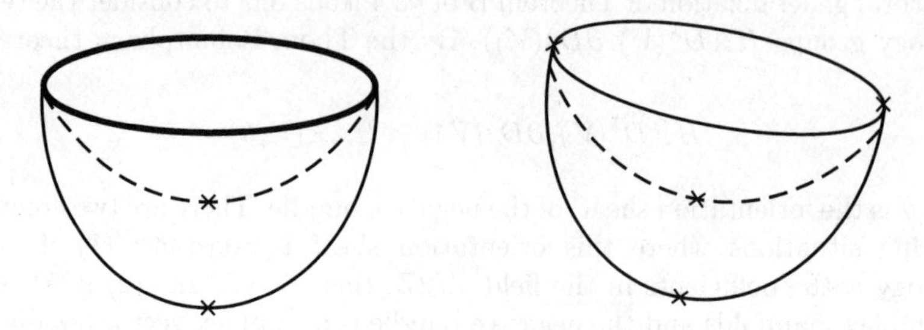

Let us look at some further concrete examples.

Example 2.5.6 In Example 2.3.5 we examined the Morse function

$$f([z_0,\ldots,z_n]) = \sum_{i=0}^{n} c_i|z_i|^2 / \sum_{i=0}^{n} |z_i|^2$$

on $\mathbb{CP}^n$, where $c_0 < \cdots < c_n$. If we allow *arbitrary* real numbers $c_0 \leqslant \cdots \leqslant c_n$ then the same formula defines a Morse–Bott function. Regarding $c_0, \ldots, c_n$ as the eigenvalues of a diagonal matrix $C = \operatorname{diag}(c_0, \ldots, c_n)$, we can describe the NDCMs as the submanifolds $\mathbb{P}(V_1), \ldots, \mathbb{P}(V_k)$ of $\mathbb{CP}^n$, where $V_1, \ldots, V_k$ are the eigenspaces of C. For example, in the extreme case where $c_0 = 0$, $c_1 = \cdots = c_n = 1$, we have two NDCMs, namely

$$\begin{aligned} &\{[1;0;\ldots;0]\} = \mathbb{CP}^0 \quad \text{(an isolated critical point)} \\ &\{[0;*;\ldots;*] \mid * \in \mathbb{C}\} \cong \mathbb{CP}^{n-1}. \end{aligned}$$

Clearly the isolated point is an absolute minimum point, so it has index zero. The other NDCM consists of absolute maxima, so it has index 2 (i.e. $\dim \mathbb{CP}^n - \dim \mathbb{C}P^{n-1}$). Hence the Morse–Bott polynomial is $MB(t) = 1 + t^2(1 + t^2 + \cdots + t^{2(n-1)})$. This is equal to the Poincaré polynomial of $\mathbb{CP}^n$, so we have a perfect Morse–Bott function. (It is easy to see that, for any choice of $c_0 \leqslant \cdots \leqslant c_n$, we obtain a perfect Morse–Bott function, in fact.) □

Example 2.5.7 Let us consider a new height function on the torus T^2, defined by the following embedding of T^2 in $\mathbb{R}^3$:

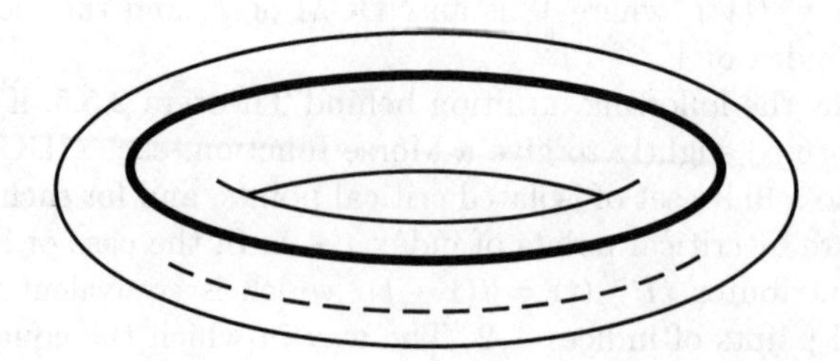

There are two NDCMs, each being a copy of S^1, with indices $0, 1$. If we tilt this embedding slightly, we obtain one of our old Morse functions on T^2 (which one?). □

2.6 Generalization: non-compactness and singularities

For a Morse function on a non-compact manifold M, there are two immediate difficulties. First, flow lines do not necessarily exist for all time (consider the manifold obtained by removing a non-critical point from a compact manifold). Second, even the flow lines which exist for all time do not necessarily converge to critical points (consider the result of removing a critical point!).

If the function $f : M \to \mathbb{R}$ is bounded below and proper, then the sets $f^{-1}(-\infty, a]$ are compact, and there is essentially no difficulty in doing Morse theory. For example, it is possible to construct Morse functions on $\mathbb{CP}^\infty = \cup_{n \geqslant 0} \mathbb{CP}^n$ by extending Example 2.3.5 in an obvious way. A more interesting example, perhaps, is the example which was fundamental for Morse himself, namely the space of paths on a Riemannian manifold (see [47]). We shall say a little more about infinite-dimensional examples such as these in §4.

Morse theory can be extended in another direction, to functions on singular spaces. Different approaches can be found in [35], [42], and section 3.2 of [45].

2.7 Generalization: equivariant Morse theory

Let G be a compact Lie group acting (smoothly) on a compact manifold M. Assume that $f : M \to \mathbb{R}$ is G-equivariant, i.e. $f(gm) = f(m)$ for all $g \in G$, $m \in M$. Then, as we have already pointed out, f is unlikely to be a Morse function as the G-orbit of any critical point consists entirely of critical points. Morse–Bott theory is designed to cope with this kind of situation, but there is a further extension of Morse–Bott theory which applies to equivariant functions.

The basic idea (see [4] and [12]) is that one would like to relate the Morse theory of an equivariant Morse–Bott function $f : M \to \mathbb{R}$ to the 'Morse theory' of the induced map $f : M/G \to \mathbb{R}$. If G acts freely on M, then M/G is a compact manifold and $M \to M/G$ is a locally trivial bundle, so such a relation exists by the remarks following Theorem 2.5.4. But if the action of G is not free, one must find another way to 'take account of the symmetry due to G'. The method introduced by Atiyah and Bott is to replace M/G by the homotopy quotient

$$M//G = M \times_G EG$$

where $EG \to BG$ is a universal bundle for G. The (contractible) space EG is not a finite-dimensional manifold (if G is non-trivial), but it can be constructed as a limit of compact manifolds $(EG)_n$, and f extends naturally to a function on each compact manifold $M \times_G (EG)_n$. The following standard example ([4]) illustrates this construction.

Example 2.7.1 Let f be the height function on $M = S^2$ as in Example 1.3.1, and let $G = S^1$ act on S^2 by 'rotation about the vertical axis'. Thus, the two critical points are isolated orbits, and all other orbits are circles. Clearly ordinary Morse theory does not work on the quotient space $S^2/S^1 \cong [-1,1]$, so we consider instead

$$S^2//S^1 = S^2 \times_{S^1} S^\infty = \lim_{n\to\infty} S^2 \times_{S^1} S^{2n+1}.$$

Now, $S^2 \times_{S^1} S^{2n+1}$ is a manifold on which (the extension of) f has two nondegenerate critical manifolds, each homeomorphic to $\{\text{point}\} \times_{S^1} S^{2n+1} \cong \mathbb{CP}^n$. The indices of these NDCMs are 0 and 2, and so we have a perfect Morse–Bott function with Morse–Bott polynomial

$$t^0(1+t^2+\cdots+t^{2n}) + t^2(1+t^2+\cdots+t^{2n}).$$

Taking the limit $n \to \infty$, we obtain the Poincaré polynomial of $S^2//S^1$ as $(1+t^2)\sum_{i\geqslant 0} t^{2i}$ (conveniently abbreviated as $(1+t^2)/(1-t^2)$). This agrees with the well-known fact that $S^2//S^1$ is homotopy equivalent to $\mathbb{CP}^\infty \vee \mathbb{CP}^\infty$. The Poincaré polynomial of $S^2 \times_{S^1} S^{2n+1}$ could also be obtained by using the fact that $S^2 \times_{S^1} S^{2n+1}$ is a bundle over $\mathbb{CP}^n$ with fibre S^2; if we choose any perfect Morse function on $\mathbb{CP}^n$ then we obtain a perfect Morse–Bott function on $S^2 \times_{S^1} S^{2n+1}$ by the remarks following Theorem 2.5.4. □

The next instructive example is also taken from [4].

Example 2.7.2 Let $f : S^2 \to \mathbb{R}$ be the function $-z^2$ of Example 2.5.1. This Morse–Bott function is not perfect, and neither is the extended Morse–Bott function on $S^2 \times_{S^1} S^{2n+1}$. The NDCMs of the latter are

$$\begin{array}{rl}\{\text{point}\} \times_{S^1} S^{2n+1} \cong \mathbb{CP}^n & (\text{index } 2)\\ \{\text{equator}\} \times_{S^1} S^{2n+1} \cong S^{2n+1} & (\text{index } 0)\\ \{\text{point}\} \times_{S^1} S^{2n+1} \cong \mathbb{CP}^n & (\text{index } 2)\end{array}$$

so the Morse–Bott polynomial is

$$t^0(1+t^{2n+1}) + 2t^2(1+t^2+\cdots+t^{2n}),$$

which has a superfluous term t^{2n+1}. But as $n \to \infty$, the effect of this term disappears, and we obtain a perfect 'Morse–Bott function' on $S^2//S^1$. □

In general, a G-equivariant Morse–Bott function $f : M \to \mathbb{R}$ is converted by the above procedure into a 'Morse–Bott function' on $M//G$; each NDCM N in M of index λ is converted into an NDCM $N//G$ of index λ. Atiyah and Bott give a criterion for the new Morse–Bott function to be perfect (even when f is not), the 'self-completion principle'.

It is important to understand what this extension of Morse theory does *not* do: it does not give additional information about the homology groups of M, but rather about those of the (much larger) auxiliary space $M//G$.

Sometimes it is useful to associate a *smaller* auxiliary space to the original Morse-theoretic data, especially when dealing with infinite-dimensional manifolds. One example of this is provided by the space of 'broken geodesics' in Morse's original application to the space $M = \Omega X$ of closed smooth paths on a manifold X, or (analogously) the space of 'algebraic loops' when X is a Lie group (see [57]). Another famous example is due to Floer [24].

In all of these situations, one may regard the homology groups of the auxiliary construction as generalized homology groups of the original space M; from equivariant Morse theory one obtains equivariant homology groups, and from Floer's theory Floer homology groups.

In the case of equivariant homology, it is, roughly speaking, the *coefficient ring* that is being generalized, from $H^*(\text{point})$ to $H^*(BG)$. The equivariant (co)homology groups $H^*_G(M)$ are defined by $H^*_G(M) = H^*(M//G)$. This is a module over the ring $H^*_G(\text{point}) = H^*(BG)$, but it is not in general isomorphic to the free module $H^*(M) \otimes H^*(BG)$; indeed, it is the $H^*(BG)$-module structure (and the multiplicative structure) of $H^*_G(M)$ which reflects the nature of the action of G on M.

A second lesson from equivariant Morse theory is that one can sometimes 'do Morse theory without a Morse function': the space $M//G$ is not a manifold in general, yet we have constructed a Morse–Bott polynomial and the Morse–Bott inequalities are satisfied.

3 Morse functions on Grassmannians

We examine in detail a special but very important example, namely the Morse theory of a family of functions on the complex Grassmannian manifold $Gr_k(\mathbb{C}^n)$. Our main tool is an explicit formula for the integral curves. This will give information on the homology and cohomology of $Gr_k(\mathbb{C}^n)$, and on the behaviour of its Schubert subvarieties.

3.1 Morse functions on Grassmannians

The (complex) Grassmannian

$$Gr_k(\mathbb{C}^n) = \{\text{complex } k\text{-dimensional linear subspaces of } \mathbb{C}^n\}$$

provides a good 'test case' for the theory of the previous two sections. In addition, $Gr_k(\mathbb{C}^n)$ is an important manifold which appears in many parts of mathematics. We shall study it in detail in this section.

To produce a 'nice' real-valued function on $Gr_k(\mathbb{C}^n)$, we use the embedding

$$\begin{aligned} Gr_k(\mathbb{C}^n) &\subseteq \text{SkewHerm}_n = \{n \times n \text{ complex matrices } A \mid A^* = -A\} \\ V &\mapsto \sqrt{-1}\,\pi_V \end{aligned}$$

where $\pi_V : \mathbb{C}^n \to \mathbb{C}^n$ denotes 'orthogonal projection on V' with respect to the standard Hermitian inner product on $\mathbb{C}^n$'. The (real) vector space $\mathrm{SkewHerm}_n$ has an inner product $\langle\langle\ ,\ \rangle\rangle$, defined by $\langle\langle A, B\rangle\rangle = \operatorname{trace} AB^* = -\operatorname{trace} AB$. We can obtain real-valued functions on $Gr_k(\mathbb{C}^n)$ by taking height functions with respect to this embedding. Specifically, we shall choose real numbers $a_1, \ldots, a_n$ and consider the function

$$f : Gr_k(\mathbb{C}^n) \to \mathbb{R}, \quad V \mapsto \langle\langle \sqrt{-1}\,\pi_V, \sqrt{-1}\,D\rangle\rangle = \operatorname{trace} \pi_V D$$

where $D = \operatorname{diag}(a_1, \ldots, a_n)$ denotes the diagonal matrix whose diagonal entries are $a_1, \ldots, a_n$.

In the special case $k = 1$ (so that $Gr_k(\mathbb{C}^n) = \mathbb{CP}^{n-1}$), we may write

$$V = \begin{bmatrix} z_1 \\ \vdots \\ z_n \end{bmatrix} = [z] \in \mathbb{CP}^{n-1}$$

and we have

$$\pi_V = \frac{zz^*}{|z|^2}.$$

So our real-valued function $f : \mathbb{CP}^{n-1} \to \mathbb{R}$ is given by

$$f([z]) = \sum_{i=1}^{n} a_i |z_i|^2 / \sum_{i=1}^{n} |z_i|^2.$$

This is our old friend from Example 2.3.5.

Let $V_i = \mathbb{C}e_i$, i.e. the line spanned by the ith basis vector of $\mathbb{C}^n$. We shall prove:

Theorem 3.1.1 *Assume that $a_1 > \cdots > a_n \geqslant 0$. Then f is a perfect Morse function. An element V of $Gr_k(\mathbb{C}^n)$ is a critical point of f if and only if $V = V_{i_1} \oplus \cdots \oplus V_{i_k}$ for some $i_1, \ldots, i_k$ with $1 \leqslant i_1 < \cdots < i_k \leqslant n$.*

One way (and indeed the conventional way) to prove this theorem would be by direct calculation, using a local chart. We shall take a slightly different point of view here, by concentrating on the integral curves of $-\nabla f$. Remarkably, there is an *explicit formula* for these integral curves, which greatly simplifies the Morse-theoretic analysis of f:

Lemma 3.1.2 *The integral curve γ of $-\nabla f$ through a point $V \in Gr_k(\mathbb{C}^n)$ is given by $\gamma(t) = \operatorname{diag}(e^{-a_1 t}, \ldots, e^{-a_n t})V = e^{-tD}V$.*

Remark It follows immediately from this lemma that the critical points of f are the special k-planes $V_{\mathbf{u}} = V_{u_1} \oplus \cdots \oplus V_{u_k}$, as stated in the above theorem.

Proof It is easy to verify that the tangent space of $Gr_k(\mathbb{C}^n)$ is given as a subspace of $\mathrm{SkewHerm}_n$ by

$$T_V Gr_k(\mathbb{C}^n) = \{T - T^* \mid T \in \mathrm{Hom}(V, V^\perp)\},$$

and that the orthogonal projection of any $X \in \mathrm{SkewHerm}_n$ on this tangent space is $\pi_V X \pi_V^\perp + \pi_V^\perp X \pi_V$. Since f is the restriction of the linear function $X \mapsto \langle\langle X, \sqrt{-1}\, D\rangle\rangle$ on $\mathrm{SkewHerm}_n$, the gradient ∇f_V is the projection of $\sqrt{-1}\, D$ on $T_V Gr_k(\mathbb{C}^n)$, so we have

$$-\nabla f_V = -\sqrt{-1}\,(\pi_V D \pi_V^\perp + \pi_V^\perp D \pi_V).$$

(This formula shows that V is a critical point of f if and only if $DV \subseteq V$, $DV^\perp \subseteq V^\perp$, i.e. if and only if V is of the form $V_{i_1} \oplus \cdots \oplus V_{i_k}$.)

To find $\dot\gamma(t)$, we first write $e^{-tD} = U_t P_t$, where U_t is unitary and P_t is an invertible complex $n \times n$ matrix such that $P_t V = V$ (such a factorization may be accomplished by the Gram–Schmidt orthogonalization process). Using this we have

$$\begin{aligned}
\dot\gamma(t) &= \frac{d}{dt}\sqrt{-1}\,\pi_{e^{-tD}V} = \frac{d}{dt}\sqrt{-1}\,\pi_{U_tV} = \frac{d}{dt}\sqrt{-1}\,U_t \pi_V U_t^{-1} \\
&= \sqrt{-1}\,\{\dot U_t \pi_V U_t^{-1} - U_t \pi_V U_t^{-1} \dot U_t U_t^{-1}\} \\
&= \sqrt{-1}\,U_t\{U_t^{-1}\dot U_t \pi_V - \pi_V U_t^{-1}\dot U_t\}U_t^{-1} \\
&= \sqrt{-1}\,U_t\{\pi_V^\perp U_t^{-1}\dot U_t \pi_V - \pi_V U_t^{-1}\dot U_t \pi_V^\perp\}U_t^{-1}.
\end{aligned}$$

From the identity

$$-D = \frac{d}{dt}(e^{-tD})(e^{-tD})^{-1} = \frac{d}{dt}(U_tP_t)(U_tP_t)^{-1}$$

we find that $U_t^{-1}\dot U_t = -U_t^{-1} D U_t - \dot P_t P^{-1}$. From the fact that $P_t V = V$ we have $\pi_V^\perp P_t \pi_V = 0$ and hence

$$0 = \pi_V^\perp \dot P_t \pi_V = \pi_V^\perp \dot P_t P_t^{-1} \pi_V.$$

Combining these we obtain

$$\pi_V^\perp U_t^{-1}\dot U_t \pi_V = -\pi_V^\perp U_t^{-1} D U_t \pi_V,$$

so the first term of $\dot\gamma(t)$ is

$$\sqrt{-1}\,U_t(-\pi_V^\perp U_t^{-1} D U_t \pi_V)U_t^{-1} = -\sqrt{-1}\,\pi_{U_tV}^\perp D \pi_{U_tV},$$

which agrees with the first term of $-\nabla f_{\gamma(t)}$.

To deal with the second term of $\dot\gamma(t)$, we use the factorization $e^{-tD} = (e^{-tD})^* = P_t^* U_t^* = Q_t U_t^{-1}$, where $Q_t = P_t^*$. From the identity

$$-D = (e^{-tD})^{-1}\frac{d}{dt}(e^{-tD}) = U_t Q_t^{-1}\frac{d}{dt}(Q_t U_t^{-1})$$

we find that $U_t^{-1}\dot{U}_t = Q_t^{-1}\dot{Q}_t + U_t^{-1}DU_t$. Since $Q_tV^\perp = V^\perp$, we obtain $\pi_V Q_t^{-1}\times \dot{Q}_t\pi_V^\perp = 0$. Hence the second term of $\dot{\gamma}(t)$ is

$$-\sqrt{-1}\,U_t(\pi_V U_t^{-1}DU_t\pi_V^\perp)U_t^{-1} = -\sqrt{-1}\,\pi_{U_tV}D\pi_{U_tV}^\perp,$$

and this agrees with the second term of $-\nabla f_{\gamma(t)}$. □

The lemma allows us to identify the stable and unstable manifolds. We begin with the case $k = 1$. Observe that

$$\lim_{t\to\infty}\begin{pmatrix} e^{-a_1t} & & \\ & \ddots & \\ & & e^{-a_nt}\end{pmatrix}\begin{bmatrix} * \\ \vdots \\ * \\ 1 \\ 0 \\ \vdots \\ 0\end{bmatrix} = \begin{bmatrix} 0 \\ \vdots \\ 0 \\ 1 \\ 0 \\ \vdots \\ 0\end{bmatrix} = V_u.$$

Hence the stable manifold S_u of V_u contains all points of the form

$$[*\cdots*10\cdots0]^t.$$

As u varies from 1 to n, such points account for all of $\mathbb{CP}^{n-1}$ – so we deduce that

$$S_u \cong \left\{\begin{pmatrix} * \\ \vdots \\ * \\ 1 \\ 0 \\ \vdots \\ 0\end{pmatrix} \text{ with } * \in \mathbb{C}\right\} \cong \mathbb{C}^{u-1}.$$

Similarly we have

$$U_u \cong \left\{\begin{pmatrix} 0 \\ \vdots \\ 0 \\ 1 \\ * \\ \vdots \\ *\end{pmatrix} \text{ with } * \in \mathbb{C}\right\} \cong \mathbb{C}^{n-u}.$$

It is clear from this description that S_u and U_u meet tranversally at V_u, and that f must be a Morse function. Finally, since the stable and unstable manifolds are even-dimensional, f is perfect.

We use similar notation in the case of $Gr_k(\mathbb{C}^n)$ for general k. First, we represent the critical point $V_{\mathbf{u}} = V_{u_1} \oplus \cdots \oplus V_{u_k}$ by an $n \times k$ matrix:

$$\begin{bmatrix} \vdots & & \vdots \\ V_{u_1} & \cdots & V_{u_k} \\ \vdots & & \vdots \end{bmatrix} = \begin{bmatrix} 1 & & \\ & \ddots & \\ & & 1 \end{bmatrix}.$$

Then we observe that

$$\lim_{t\to\infty} \begin{pmatrix} e^{-a_1 t} & & \\ & \ddots & \\ & & e^{-a_n t} \end{pmatrix} \begin{bmatrix} * & & * \\ \vdots & & \vdots \\ * & & \vdots \\ 1 & & \vdots \\ & \ddots & * \\ & & 1 \\ & & \end{bmatrix} = \begin{bmatrix} 1 & & \\ & \ddots & \\ & & 1 \end{bmatrix}.$$

But any $n \times k$ matrix may be brought into the 'echelon form'

$$\begin{bmatrix} * & & * \\ \vdots & & \vdots \\ * & & \vdots \\ 1 & & \vdots \\ & \ddots & * \\ & & 1 \\ & & \end{bmatrix}$$

by 'column operations', so we have found the stable manifold $S_{\mathbf{u}}$ of $V_{\mathbf{u}}$.

Now, an element of $S_{\mathbf{u}}$ may be represented by more than one matrix in echelon form. To obtain a unique representation, we should bring the matrix into 'reduced echelon form', so that it looks (for example) like this:

$$\begin{pmatrix} * & * & * \\ * & * & * \\ 1 & 0 & 0 \\ & 1 & 0 \\ & & * \\ & & 1 \\ & & 0 \\ & & 0 \end{pmatrix}.$$

It follows that

$$\begin{aligned}\dim_{\mathbb{C}} S_{\mathbf{u}} &= (u_1 - 1) + (u_2 - 2) + (u_3 - 3) + \cdots + (u_k - k) \\ &= \sum_{i=1}^{k} u_i - \tfrac{1}{2}k(k+1)\end{aligned}$$

(and $S_{\mathbf{u}}$ may be identified with a complex vector space – or cell – of this dimension).

There is a similar description of the unstable manifold $U_{\mathbf{u}}$. We have

$$\lim_{t\to-\infty} \begin{pmatrix} e^{-a_1 t} & & \\ & \ddots & \\ & & e^{-a_n t} \end{pmatrix} \begin{bmatrix} 1 & & \\ * & \ddots & \\ \vdots & & 1 \\ \vdots & & * \\ \vdots & & \vdots \\ * & & * \end{bmatrix} = \begin{bmatrix} 1 & & \\ & \ddots & \\ & & 1 \end{bmatrix},$$

from which we obtain a description of $U_{\mathbf{u}}$ of the form

$$U_{\mathbf{u}} = \begin{pmatrix} & & \\ & & \\ 1 & & \\ 0 & 1 & \\ * & * & 0 \\ 0 & 0 & 1 \\ * & * & * \\ * & * & * \end{pmatrix}.$$

This is diffeomorphic to a complex vector space whose dimension is

$$\begin{aligned}\dim_{\mathbb{C}} U_{\mathbf{u}} &= (n-(u_1-1)-k)+(n-(u_2-1)-(k-1))+\cdots+(n-(u_k-1)-1) \\ &= k(n-k) - \dim_{\mathbb{C}} S_{\mathbf{u}} \\ &= \dim_{\mathbb{C}} Gr_k(\mathbb{C}^n) - \dim_{\mathbb{C}} S_{\mathbf{u}}.\end{aligned}$$

This identification of the stable and unstable manifolds leads to a proof of Theorem 3.1.1 for general k, as in the case $k = 1$. Note that the indices of the critical points are twice the complex dimensions of the unstable manifolds.

Remark From Lemma 3.1.2, we see that the integral curves of $-\nabla f$ 'preserve' the submanifold $Gr_k(\mathbb{R}^n)$ of $Gr_k(\mathbb{C}^n)$. In fact, the above analysis works equally well for the restriction of f to $Gr_k(\mathbb{R}^n)$.

There is an interesting group-theoretic interpretation of $S_{\mathbf{u}}$ and $U_{\mathbf{u}}$. The group $GL_n\mathbb{C}$ of invertible complex $n \times n$ matrices acts naturally on $Gr_k(\mathbb{C}^n)$ (by multiplying a column vector on the left). We have the usual exponential map

$$\exp : \mathfrak{gl}_n\mathbb{C} \to GL_n\mathbb{C}, \quad X \mapsto I + \frac{X}{1!} + \frac{X^2}{2!} + \frac{X^3}{3!} + \cdots$$

where $\mathfrak{gl}_n\mathbb{C}$ denotes the Lie algebra of $GL_n\mathbb{C}$, i.e. the vector space of all $n \times n$ complex matrices. With this notation we see that $S_{\mathbf{u}}$ consists of k-planes of the form

$$\exp\begin{pmatrix} 0 & 0 & * & * & 0 & * & 0 & 0 \\ 0 & 0 & * & * & 0 & * & 0 & 0 \\ 0 & 0 & 0 & 0 & 0 & 0 & 0 & 0 \\ 0 & 0 & 0 & 0 & 0 & 0 & 0 & 0 \\ 0 & 0 & 0 & 0 & 0 & * & 0 & 0 \\ 0 & 0 & 0 & 0 & 0 & 0 & 0 & 0 \\ 0 & 0 & 0 & 0 & 0 & 0 & 0 & 0 \\ 0 & 0 & 0 & 0 & 0 & 0 & 0 & 0 \end{pmatrix} \begin{bmatrix} 0 & 0 & 0 \\ 0 & 0 & 0 \\ 1 & 0 & 0 \\ 0 & 1 & 0 \\ 0 & 0 & 0 \\ 0 & 0 & 1 \\ 0 & 0 & 0 \\ 0 & 0 & 0 \end{bmatrix}.$$

Thus, $S_{\mathbf{u}} = \{(\exp X)V_{\mathbf{u}} \mid X \in \mathfrak{n}_{\mathbf{u}}\}$, where $\mathfrak{n}_{\mathbf{u}}$ is a *nilpotent Lie subalgebra* of $\mathfrak{gl}_n\mathbb{C}$. This means that $S_{\mathbf{u}}$ is the orbit $V_{\mathbf{u}}$ under the corresponding Lie group $N_{\mathbf{u}}$. (Note that $\exp X = I + X$ here.) In fact, since the isotropy subgroups are trivial, $S_{\mathbf{u}}$ can be *identified* with the Lie group $N_{\mathbf{u}}$ or its Lie algebra $\mathfrak{n}_{\mathbf{u}}$. There is an analogous description of $U_{\mathbf{u}}$.

The action of $A = \operatorname{diag}(e^{-a_1 t}, \ldots, e^{-a_n t})$ (which gives the integral curves of $-\nabla f$) is easy to express in term of the Lie algebra $\mathfrak{n}_{\mathbf{u}}$. For $X \in \mathfrak{n}_{\mathbf{u}}$, we have

$$\begin{aligned} A(\exp X)V_{\mathbf{u}} &= A(\exp X)A^{-1}AV_{\mathbf{u}} \\ &= A(\exp X)A^{-1}V_{\mathbf{u}} \text{ (as } V_{\mathbf{u}} \text{ is fixed by } A) \\ &= (\exp AXA^{-1})V_{\mathbf{u}}. \end{aligned}$$

The map $X \mapsto AXA^{-1}$ has the effect of multiplying the (i,j)th entry of X by $e^{-(a_i - a_j)t}$.

3.2 Morse–Bott functions on Grassmannians

By relaxing the condition $a_1 > \cdots > a_n \geqslant 0$ to $a_1 \geqslant \cdots \geqslant a_n \geqslant 0$, we obtain a Morse–Bott function on $Gr_k(\mathbb{C}^n)$. To investigate this, we introduce the notation

$$\mathbb{C}^n = E_1 \oplus \cdots \oplus E_l$$

for the eigenspace decomposition of $D = \operatorname{diag}(a_1, \ldots, a_n)$. We write b_i for the eigenvalue on E_i. Thus, the *distinct* a_i*s* are $b_1 > \cdots > b_l$.

Theorem 3.2.1 *Assume that $a_1 \geqslant \cdots \geqslant a_n \geqslant 0$. Then f is a perfect Morse–Bott function. An element V of $Gr_k(\mathbb{C}^n)$ is a critical point of f if and only if $V = V \cap E_1 \oplus \cdots \oplus V \cap E_k$, i.e. if and only if V is spanned by eigenvectors of $D = \operatorname{diag}(a_1, \ldots, a_n)$. The NDCM containing such a V is diffeomorphic to $Gr_{c_1}(E_1) \times \cdots \times Gr_{c_l}(E_l)$ where $c_i = \dim_{\mathbb{C}} V \cap E_i$.*

It may seem unnecessary to introduce Morse–Bott functions in a situation like this, where we already have a good supply of Morse functions. However, we shall see later that the special properties of Morse–Bott functions can be extremely useful.

To prove the theorem, and to identify the stable and unstable manifolds, we use the explicit formula for the integral curves of $-\nabla f$ which was given earlier. (The derivation of that formula is clearly valid for arbitrary real $a_1, \ldots, a_n$.) We shall just sketch the main points here.

First, it is immediate from the form of the integral curves that the critical points are as stated in the theorem. Note that every $V_{\mathbf{u}}$ is certainly a critical point of f, but these are not the only critical points; the others are obtained by taking the orbits of the $V_{\mathbf{u}}$'s under the product of unitary groups $U(E_1) \times \cdots \times U(E_l)$.

Let us now try to identify the stable manifold of a critical point $V_{\mathbf{u}}$. Each NDCM contains at least one point of this form. Moreover, since the negative bundles turn out to be homogeneous vector bundles on the NDCM $M_{\mathbf{u}} \cong Gr_{k_1}(E_1) \times \cdots \times Gr_{k_l}(E_l)$, the stable manifold of $V_{\mathbf{u}}$ will determine the entire negative bundle.

Let g be the Morse function on $Gr_k(\mathbb{C}^n)$ corresponding to distinct real numbers $c_1 > \cdots > c_n \geqslant 0$, as in the previous section. The stable manifold $S^f_{\mathbf{u}}$ of $V_{\mathbf{u}}$ for f is related to the stable manifold $S^g_{\mathbf{u}}$ of $V_{\mathbf{u}}$ for g, as we shall illustrate by considering the case of $Gr_3(\mathbb{C}^8)$, with $b_1 = a_1 = a_2 = a_3 = a_4$ and $b_2 = a_5 = a_6 = a_7 = a_8$. Recall that the stable manifold of $V_{\mathbf{u}} = V_3 \oplus V_4 \oplus V_6$ (for example) for g is obtained by considering integral curves of the form:

$$\begin{pmatrix} e^{-c_1 t} &&&&&&& \\ & e^{-c_2 t} &&&&&& \\ && e^{-c_3 t} &&&&& \\ &&& e^{-c_4 t} &&&& \\ &&&& e^{-c_5 t} &&& \\ &&&&& e^{-c_6 t} && \\ &&&&&& e^{-c_7 t} & \\ &&&&&&& e^{-c_8 t} \end{pmatrix} \begin{bmatrix} * & * & * \\ * & * & * \\ 1 & 0 & 0 \\ 0 & 1 & 0 \\ & & * \\ & & 1 \\ & & 0 \\ & & 0 \end{bmatrix}.$$

The NDCM $M_{\mathbf{u}}$ is diffeomorphic to $Gr_2(\mathbb{C}^4) \times Gr_1(\mathbb{C}^4)$. Let us consider the integral curves of $-\nabla f$ through the points of $S^g_{\mathbf{u}}$:

$$\begin{pmatrix} e^{-b_1 t} &&&&&&& \\ & e^{-b_1 t} &&&&&& \\ && e^{-b_1 t} &&&&& \\ &&& e^{-b_1 t} &&&& \\ &&&& e^{-b_2 t} &&& \\ &&&&& e^{-b_2 t} && \\ &&&&&& e^{-b_2 t} & \\ &&&&&&& e^{-b_2 t} \end{pmatrix} \begin{bmatrix} * & * & * \\ * & * & * \\ 1 & 0 & 0 \\ 0 & 1 & 0 \\ & & * \\ & & 1 \\ & & 0 \\ & & 0 \end{bmatrix}.$$

Clearly these integral curves do not necessarily approach $V_{\mathbf{u}}$ as $t \to \infty$; but they do if we set the components denoted below by # equal to zero:

$$\begin{bmatrix} \# & \# & * \\ \# & \# & * \\ 1 & 0 & 0 \\ 0 & 1 & 0 \\ & & \# \\ & & 1 \\ & & 0 \\ & & 0 \end{bmatrix}.$$

Thus, we have identified a 2-dimensional cell in $S^f_{\mathbf{u}}$ which is contained in $S^g_{\mathbf{u}}$. It can be shown that this cell is *precisely* $S^f_{\mathbf{u}}$. In general, the procedure by which we identify $S^f_{\mathbf{u}}$ as a subspace of $S^g_{\mathbf{u}}$ is that we delete those coordinates which correspond to the NDCM $M_{\mathbf{u}}$. It is possible to give a more systematic description of these stable (and unstable) manifolds, as orbits of certain subgroups of $GL_n\mathbb{C}$.

3.3 Homology groups of Grassmannians

In §3.1 we described a perfect Morse function $f : Gr_k(\mathbb{C}^n) \to \mathbb{R}$. It follows that the homology groups of $Gr_k(\mathbb{C}^n)$ may be read off from the Morse polynomial of f. For $k = 1$ this is easy, but for general k it is a little harder. In this section we shall carry out the calculation, in two quite different ways.

Recall that f has $\binom{n}{k}$ critical points, namely the k-planes $V_{\mathbf{u}} = V_{u_1} \oplus \cdots \oplus V_{u_k}$, where $1 \leqslant u_1 < \cdots < u_k \leqslant n$. From 3.2 the (complex) dimension of the stable manifold $S_{\mathbf{u}}$ is $\sum_{i=1}^k (u_i - i)$, so the index of $V_{\mathbf{u}}$ as a critical point of $-f$ is $2\sum_{i=1}^k (u_i - i)$.

If we define

$$a_d = \left| \left\{ \mathbf{u} \mid \sum_{i=1}^k (u_i - i) = d \right\} \right|$$

then the Morse polynomial of $-f$ is $M(t) = \sum_{d \geqslant 0} a_d t^{2d}$. Our task, therefore, is to calculate a_d.

There is a one to one correspondence between

$$\left\{ (u_1, \ldots, u_k) \mid 1 \leqslant u_1 < \cdots < u_k \leqslant n, \ \sum_{i=1}^k (u_i - i) = d \right\}$$

and

$$\left\{ (p_1, \ldots, p_k) = (u_1 - 1, \ldots, u_k - k) \mid 0 \leqslant p_1 \leqslant \cdots \leqslant p_k \leqslant n - k, \ \sum_{i=1}^k p_i = d \right\}.$$

Thus, a_d is equal to 'the number of partitions of d into at most k integers, where each such integer is at most $n - k$'. Unfortunately there is no simple formula for

a_d, but there is a formula for the generating function $G(t) = \sum_{d \geqslant 0} a_d t^d$ (and this is exactly what we want, since $M(t) = G(t^2)$).

Before stating this formula, we recall the well-known fact that, if b_d denotes the number of (unrestricted) partitions of d, then

$$\sum_{d \geqslant 0} b_d t^d = \frac{1}{(1-t)(1-t^2)(1-t^3)\cdots}$$

This means that b_d is equal to the coefficient of t^d in the formal expansion of the right hand side. The formula for $G(t)$ is:

Proposition 3.3.1 *The generating function $G(t) = \sum_{d \geqslant 0} a_d t^d$ is given by*

$$G(t) = \frac{(1-t^{n-k+1})(1-t^{n-k+2})\cdots(1-t^n)}{(1-t)(1-t^2)\cdots(1-t^k)}.$$

This is a purely combinatorial statement, which may be proved directly. However, we shall give an indirect proof, by making use of a Morse–Bott function on $Gr_k(\mathbb{C}^n)$. In doing so, we shall find a simple inductive formula for $G(t)$ as well.

Let $g : Gr_k(\mathbb{C}^n) \to \mathbb{R}$ be the Morse–Bott function corresponding to the choice of real numbers

$$a_1 = 1, \ a_2 = \cdots = a_n = 0.$$

Amongst the non-constant Morse–Bott functions of §3.2, this is the 'crudest', in the sense that it has the least number of critical manifolds. (Morse functions are at the opposite extreme; they have the largest number of critical manifolds.)

From Theorem 3.2.1, the critical manifolds are as follows:

$$\begin{aligned} M_{\max} &= \{V \in Gr_k(\mathbb{C}^n) \mid V = V_1 \oplus W, W \subseteq V_2 \oplus \cdots \oplus V_n\} \cong Gr_{k-1}(\mathbb{C}^{n-1}) \\ M_{\min} &= \{V \in Gr_k(\mathbb{C}^n) \mid V \subseteq V_2 \oplus \cdots \oplus V_n\} \cong Gr_k(\mathbb{C}^{n-1}). \end{aligned}$$

The unstable manifold of $M_{\max}$ must be the complement of $M_{\min}$, so the index of $M_{\max}$ for g is $2(\dim Gr_k(\mathbb{C}^n) - \dim Gr_{k-1}(\mathbb{C}^{n-1}))$, i.e. $2(n-k)$. The index of $M_{\min}$ is of course zero. So the Morse–Bott polynomial of g is

$$MB(t) = P_{k,n-1}(t) + t^{2(n-k)} P_{k-1,n-1}(t)$$

where $P_{i,j}(t)$ denotes the Poincaré polynomial of $Gr_i(\mathbb{C}^j)$.

Since g is a perfect Morse–Bott function, this gives an inductive formula for $P_{k,n}(t) = MB(t)$:

Proposition 3.3.2 *The Poincaré polynomial $P_{k,n}(t)$ of $Gr_k(\mathbb{C}^n)$ satisfies the relation*

$$P_{k,n}(t) = P_{k,n-1}(t) + t^{2(n-k)} P_{k-1,n-1}(t).$$

□

For example: $P_{2,4}(t) = P_{2,3}(t) + t^4 P_{1,3}(t) = (1 + t^2 + t^4) + t^4(1 + t^2 + t^4) = 1 + t^2 + 2t^4 + t^6 + t^8$.

We can also use this to give a proof of Proposition 3.3.1, which is equivalent to the following slightly more symmetrical statement:

Proposition 3.3.3 *The Poincaré polynomial $P_{k,n}(t)$ of $Gr_k(\mathbb{C}^n)$ is given by*

$$P_{k,n}(t) = \frac{\prod_{i=1}^{n}(1 - t^{2i})}{\prod_{i=1}^{k}(1 - t^{2i}) \prod_{i=1}^{n-k}(1 - t^{2i})}.$$

Proof Denote the right hand side by $B_{k,n}(t)$. It is easy to verify that this satisfies the same recurrence relation as $P_{k,n}(t)$, i.e. $B_{k,n}(t) = B_{k,n-1}(t) + t^{2(n-k)} B_{k-1,n-1}(t)$. Since the recurrence relation determines $B_{k,n}(t)$ or $P_{k,n}(t)$ inductively, and these agree when $k = 1$, they must be equal. □

3.4 Schubert varieties

In §3.1 we described explicitly the stable manifold $S_{\mathbf{u}}$ of a critical point $V_{\mathbf{u}} = V_{u_1} \oplus \cdots \oplus V_{u_k}$ of a Morse function $f : Gr_k(\mathbb{C}^n) \to \mathbb{R}$. It has the form

$$S_{\mathbf{u}} = \left\{ V = \begin{bmatrix} * & * & * \\ * & * & * \\ 1 & 0 & 0 \\ & 1 & 0 \\ & & * \\ & & 1 \\ & & 0 \\ & & 0 \end{bmatrix} \in Gr_k(\mathbb{C}^n) \;\middle|\; * \in \mathbb{C} \right\}.$$

From that section it is clear that such V are characterized *geometrically* by the following conditions:

$$\dim V \cap \mathbb{C}^i = \begin{cases} 0 & \text{if } 1 \leqslant i \leqslant u_1 - 1 \\ 1 & \text{if } u_1 \leqslant i \leqslant u_2 - 1 \\ \cdots & \\ k & \text{if } u_k \leqslant i \leqslant n. \end{cases}$$

In other words, we have

$$S_{\mathbf{u}} = \{V \in Gr_k(\mathbb{C}^n) \mid \dim V \cap \mathbb{C}^i = \dim V_{\mathbf{u}} \cap \mathbb{C}^i \text{ for all } i\}.$$

The condition on $\dim V \cap \mathbb{C}^i$ is called a 'Schubert condition'. It can be specified either by listing $v_i = \dim V \cap \mathbb{C}^i$, i.e.

$$v_1 = \cdots = v_{u_1 - 1} = 0,\ v_{u_1} = \cdots = v_{u_2 - 1} = 1, \quad \ldots, \quad v_{u_k} = \cdots = v_n = k$$

or, more economically, by listing those i such that $\dim V \cap \mathbb{C}^i = \dim V \cap \mathbb{C}^{i-1} + 1$, i.e.

$$u_1, u_2, \ldots, u_k.$$

The k-tuple $\mathbf{u} = (u_1, \ldots, u_k)$ is referred to as a 'Schubert symbol', and the set $S_{\mathbf{u}}$ is called the 'Schubert cell' associated to $\mathbf{u}$.

Similarly, the unstable manifold $U_{\mathbf{u}}$ is characterized by these conditions:

$$\dim V \cap (\mathbb{C}^{n-i})^{\perp} = \begin{cases} 0 & \text{if } 1 \leqslant i \leqslant n - u_k \\ 1 & \text{if } n - u_k + 1 \leqslant i \leqslant n - u_{k-1} \\ \ldots & \\ k & \text{if } n - u_1 + 1 \leqslant i \leqslant n. \end{cases}$$

Example 3.4.1 For the Morse function $f : Gr_2(\mathbb{C}^4) \to \mathbb{R}$ (of §3.1), there are six Schubert cells. We give the matrix representations below, followed by the sequence $\dim V \cap \mathbb{C}, \ldots, \dim V \cap \mathbb{C}^4$, the Schubert symbol, and the dimension of the cell.

$$\begin{pmatrix} * & * \\ * & * \\ 1 & 0 \\ 0 & 1 \end{pmatrix} \quad 0,0,1,2 \quad (3,4) \quad \dim_{\mathbb{C}} = 4$$

$$\begin{pmatrix} * & * \\ 1 & 0 \\ 0 & * \\ 0 & 1 \end{pmatrix} \quad 0,1,1,2 \quad (2,4) \quad \dim_{\mathbb{C}} = 3$$

$$\begin{pmatrix} * & * \\ 1 & 0 \\ 0 & 1 \\ 0 & 0 \end{pmatrix} \quad 0,1,2,2 \quad (2,3) \quad \dim_{\mathbb{C}} = 2$$

$$\begin{pmatrix} 1 & 0 \\ 0 & * \\ 0 & * \\ 0 & 1 \end{pmatrix} \quad 1,1,1,2 \quad (1,4) \quad \dim_{\mathbb{C}} = 2$$

$$\begin{pmatrix} 1 & 0 \\ 0 & * \\ 0 & 1 \\ 0 & 0 \end{pmatrix} \quad 1,1,2,2 \quad (1,3) \quad \dim_{\mathbb{C}} = 1$$

$$\begin{pmatrix} 1 & 0 \\ 0 & 1 \\ 0 & 0 \\ 0 & 0 \end{pmatrix} \quad 1,2,2,2 \quad (1,2) \quad \dim_{\mathbb{C}} = 0.$$

It follows that the Poincaré polynomial of $Gr_2(\mathbb{C}^4)$ is $1+t^2+2t^4+t^6+t^8$. □

Definition 3.4.2 *The Schubert variety $X_{\mathbf{u}}$ associated to the Schubert symbol* $\mathbf{u}$ *is the closure of $S_{\mathbf{u}}$ (with respect to the usual topology of $Gr_k(\mathbb{C}^n)$), i.e.*

$$X_{\mathbf{u}} = \overline{S}_{\mathbf{u}} = \{V \in Gr_k(\mathbb{C}^n) \mid \dim V \cap \mathbb{C}^i \geqslant v_i \text{ for all } i\}.$$

It is easy to show that $X_{\mathbf{u}}$ is an algebraic subvariety of $Gr_k(\mathbb{C}^n)$ (see §3.5). This subvariety may have singularities. For example, in the case of $Gr_2(\mathbb{C}^4)$, we have

$$X_{(2,4)} = \{V \in Gr_2(\mathbb{C}^4) \mid \dim V \cap \mathbb{C}^2 \geqslant 1\}$$

(the v_i's are given by $(v_1, v_2, v_3, v_4) = (0, 1, 1, 2)$, but the conditions $\dim V \cap \mathbb{C} \geqslant 0$, $\dim V \cap \mathbb{C}^3 \geqslant 1$, $\dim V \cap \mathbb{C}^4 \geqslant 2$ are automatically[3] satisfied). The point $V = \mathbb{C}^2$ is a singular point of $X_{(2,4)}$, but $X_{(2,4)} - \{\mathbb{C}^2\}$ is smooth, having the structure of a fibre bundle over $\mathbb{CP}^1$ with fibre $\mathbb{CP}^2 - \{\text{point}\}$.

Observe that

$$X_{(2,4)} = S_{(2,4)} \cup S_{(2,3)} \cup S_{(1,4)} \cup S_{(1,3)} \cup S_{(1,2)}.$$

It is clear from the definition that, in general, $X_{\mathbf{u}}$ is a disjoint union of Schubert cells. This gives rise to a partial order on the set of Schubert symbols: we define $\mathbf{u}_1 \leqslant \mathbf{u}_2$ if and only if $\overline{S}_{\mathbf{u}_1} \supseteq S_{\mathbf{u}_2}$.

In the case of $Gr_2(\mathbb{C}^4)$, this partial order is represented in the following diagrams. The second diagram indicates the conditions which define the Schubert varieties.

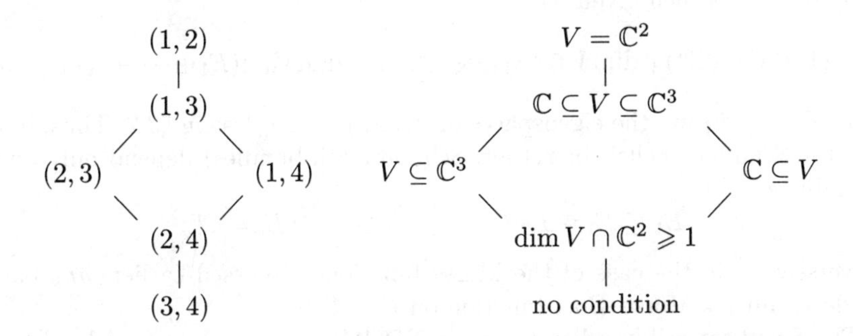

Although the partial order is a simple consequence of the definition of Schubert variety, it provides non-trivial information on the behaviour of the flow lines of $-\nabla f$. Namely, the condition $\mathbf{u}_1 \leqslant \mathbf{u}_2$ is equivalent to the existence of a flow line $\gamma(t)$ from $\mathbf{u}_2$ to $\mathbf{u}_1$, i.e. such that $\lim_{t\to-\infty} \gamma(t) = V_{\mathbf{u}_2}$, $\lim_{t\to\infty} \gamma(t) = V_{\mathbf{u}_1}$.

[3] If V, W are linear subspaces of $\mathbb{C}^n$ of dimensions k, l respectively, then we have $W/W \cap V \cong W + V/V$, and hence $\dim W \cap V + \dim(W + V) = k + l$.

(This is not immediately obvious from the condition $\overline{S}_{\mathbf{u}_1} \supseteq S_{\mathbf{u}_2}$, but it does follow from the geometrical description of $S_{\mathbf{u}_1} \cap U_{\mathbf{u}_2}$.)

We shall return to Schubert varieties in §3.6, when we discuss the cohomology ring of a Grassmannian, so we conclude this section with some further remarks.

First, since the flow lines of $-\nabla f$ 'preserve' the Schubert variety $X_{\mathbf{u}}$, we deduce that $X_{\mathbf{u}}$ inherits a decomposition into (possibly singular) 'stable manifolds' (or 'unstable manifolds').

Second, although the Morse function f depends on a choice of real numbers $a_1 > \cdots > a_n$, the Schubert varieties are independent of this choice. In fact they depend solely on the standard 'flag'

$$\mathbb{C} \subseteq \mathbb{C}^2 \subseteq \cdots \subseteq \mathbb{C}^n,$$

or on the standard ordered basis $e_1, \ldots, e_n$ of $\mathbb{C}^n$. If $a_1, \ldots, a_n$ are arbitrary distinct real numbers, we obtain a similar Morse function, possibly corresponding to a re-ordering of $e_1, \ldots, e_n$. More generally still, any choice of orthonormal basis, or equivalently any flag, corresponds to a similar Morse function. The formula for such a function is obtained by replacing the diagonal matrix A by UAU^{-1}, where U is a unitary matrix.

There is a geometrical description of the stable manifolds of the Morse–Bott functions considered in §3.2, i.e. where the real numbers $a_1 \geqslant \cdots \geqslant a_n$ are not necessarily distinct. We state the result without proof (as the easiest proof depends on the more general theory of §4). Recall that the critical manifolds are denoted $M_{\mathbf{u}}$, and that each such NDCM contains at least one point of the form $V_{\mathbf{u}}$. The stable manifold $S_{\mathbf{u}}$ of $M_{\mathbf{u}}$, i.e. the union of the stable manifolds of all points of $M_{\mathbf{u}}$, is then given by:

$$S_{\mathbf{u}} = \{V \in Gr_k(\mathbb{C}^n) \mid \dim V \cap (E_1 \oplus \cdots \oplus E_i) = \dim V_{\mathbf{u}} \cap (E_1 \oplus \cdots \oplus E_i) \text{ for all } i\},$$

where $E_1, \ldots, E_l$ are the eigenspaces of $\operatorname{diag}(a_1, \ldots, a_n)$ as in §3.2. Thus, in this case, the Schubert 'cells' (or rather, Schubert cell-bundles) depend only on the 'partial flag'

$$E_1 \subseteq E_1 \oplus E_2 \subseteq \cdots \subseteq E_1 \oplus \cdots \oplus E_l = \mathbb{C}^n.$$

Conversely, as in the case of the Morse functions discussed earlier, *any* partial flag determines a Morse–Bott function on $Gr_k(\mathbb{C}^n)$.

The Schubert cell-bundles (or their NDCMs) are parametrized by l-tuples $(c_1, \ldots, c_l)$ of non-negative integers with $c_j \leqslant \dim E_j$ and $c_1 + \cdots + c_l = k$; namely $c_j = \dim V_{\mathbf{u}} \cap E_j$. We consider $(c_1, \ldots, c_l)$ to be a 'generalized Schubert symbol'. The bundle projection map $S_{\mathbf{u}} \to M_{\mathbf{u}}$ is given explicitly by

$$V \mapsto (V_{(1)}, \ldots, V_{(l)}) \in Gr_{c_1}(E_1) \times \cdots \times Gr_{c_l}(E_l)$$

where

$$V_{(i)} = V \cap E_1 \oplus \cdots \oplus E_i + E_1 \oplus \cdots \oplus E_{i-1} \;/\; E_1 \oplus \cdots \oplus E_{i-1}.$$

The integers $w_i = c_1 + \cdots + c_i$, $i = 1, \ldots, l$, are analogous to the integers v_i in the case of a Morse function. In terms of these integers, the Schubert cell-bundle $S_{\mathbf{u}}$ is

$$S_{\mathbf{u}} = \{V \in Gr_k(\mathbb{C}^n) \mid \dim V \cap (E_1 \oplus \cdots \oplus E_i) = w_i \text{ for all } i\}.$$

By taking the closure of a Schubert cell-bundle, we obtain a generalized Schubert variety, namely

$$X_{\mathbf{u}} = \overline{S}_{\mathbf{u}} = \{V \in Gr_k(\mathbb{C}^n) \mid \dim V \cap E_1 \oplus \cdots \oplus E_i \geqslant w_i \text{ for all } i\}.$$

One of the advantages of having these explicit descriptions of Schubert cell-bundles is that we can compute easily the indices of the critical points; we shall give two examples below.

Example 3.4.3 Let us choose the partial flag $\mathbb{C} \subseteq \mathbb{C}^n$. This corresponds to a Morse–Bott function on $Gr_k(\mathbb{C}^n)$ with $a_1 > a_2 = \cdots = a_n$. We have already considered such a function in §3.3; there are two critical manifolds. The stable manifold of the maximum NDCM is just that NDCM, and the stable manifold of the minimum NDCM is the complement of the maximum NDCM.

The eigenspace decomposition of $\mathbb{C}^n$ is given by $E_1 = \mathbb{C}$, $E_2 = \mathbb{C}^\perp$, and the generalized Schubert symbols are $(c_0, c_1) = (1, k-1)$ and $(c_0, c_1) = (0, k)$. □

Example 3.4.4 Consider the Morse–Bott function on $\mathbb{CP}^6 = Gr_1(\mathbb{C}^7)$ given by $a_1 = a_2 = 2$, $a_3 = a_4 = a_5 = 1$, $a_6 = a_7 = 0$. In this case the eigenspace decomposition of $\mathbb{C}^7$ is given by

$$\mathbb{C}^7 = E_1 \oplus E_2 \oplus E_3, \quad E_1 = V_1 \oplus V_2, E_2 = V_3 \oplus V_4 \oplus V_5, E_3 = V_6 \oplus V_7.$$

We list below the generalized Schubert symbols, followed by the NDCM, and then the Schubert cell-bundle.

$$\begin{array}{lll} (c_0, c_1, c_2) = (1, 0, 0), & M_{\mathbf{u}} = \mathbb{P}(E_1), & S_{\mathbf{u}} = \mathbb{P}(E_1) \\ (c_0, c_1, c_2) = (0, 1, 0), & M_{\mathbf{u}} = \mathbb{P}(E_2), & S_{\mathbf{u}} = \mathbb{P}(E_1 \oplus E_2) - \mathbb{P}(E_1) \\ (c_0, c_1, c_2) = (0, 0, 1), & M_{\mathbf{u}} = \mathbb{P}(E_3), & S_{\mathbf{u}} = \mathbb{P}(E_1 \oplus E_2 \oplus E_3) - \mathbb{P}(E_1 \oplus E_2). \end{array}$$

In particular, the Morse indices (for the function $-f$) are $0, 4, 10$ respectively, and the Morse–Bott polynomial is $t^0(1+t^2) + t^4(1+t^2+t^4) + t^{10}(1+t^2)$. This is equal to the Poincaré polynomial of $\mathbb{CP}^6$, as expected. □

3.5 Morse theory of the Plücker embedding

There is a well-known embedding

$$Gr_k(\mathbb{C}^n) \to \mathbb{CP}^N, \quad N = \binom{n}{k} - 1$$

called the Plücker embedding. It is defined by

$$V \mapsto \wedge^k V \subseteq \wedge^k \mathbb{C}^n \cong \mathbb{C}^{N+1}.$$

If $e_1, \ldots, e_n$ are the standard basis vectors of $\mathbb{C}^n$, then the vectors $e_{\mathbf{u}} = e_{u_1} \wedge \cdots \wedge e_{u_k}$ with $1 \leqslant u_1 < \cdots < u_k \leqslant n$ constitute a basis of $\wedge^k \mathbb{C}^n$.

Let $f : Gr_k(\mathbb{C}^n) \to \mathbb{R}$ be the Morse–Bott function defined by certain real numbers $a_1, \ldots, a_n$, as in §3.2. Then by Lemma 3.1.2 the one-parameter diffeomorphism group of the vector field $-\nabla f$ is induced by the action

$$t \cdot \sum \lambda_i e_i = \sum \lambda_i e^{-ta_i} e_i$$

of $\mathbb{R}$ on $\mathbb{C}^n$.

This action is the restriction of the action

$$t \cdot \sum \lambda_{\mathbf{u}} e_{\mathbf{u}} = \sum \lambda_{\mathbf{u}} e^{-t(a_{u_1} + \cdots + a_{u_k})} e_{\mathbf{u}}$$

of $\mathbb{R}$ on $\wedge^k \mathbb{C}^n$. But this action is the one-parameter diffeomorphism group of the vector field $-\nabla F$, where $F : \mathbb{CP}^N \to \mathbb{R}$ is the Morse–Bott function defined by the $\binom{n}{k}$ real numbers $a_{u_1} + \cdots + a_{u_k}$. We conclude that the Morse–Bott theory of f on $Gr_k(\mathbb{C}^n)$ is just the 'restriction' of the (much simpler!) Morse–Bott theory of F on $\mathbb{CP}^N$:

Proposition 3.5.1 *Let $M_{\mathbf{u}}, S_{\mathbf{u}}, X_{\mathbf{u}}$ denote the NDCMs, Schubert cell-bundles, and generalized Schubert varieties for the Morse–Bott function $f : Gr_k(\mathbb{C}^n) \to \mathbb{R}$. Let $M^F_{\mathbf{u}}, S^F_{\mathbf{u}}, X^F_{\mathbf{u}}$ denote the corresponding objects for the Morse–Bott function $F : \mathbb{CP}^N \to \mathbb{R}$. Then we have $M_{\mathbf{u}} = Gr_k(\mathbb{C}^n) \cap M^F_{\mathbf{u}}$, $S_{\mathbf{u}} = Gr_k(\mathbb{C}^n) \cap S^F_{\mathbf{u}}$, and $X_{\mathbf{u}} \subseteq Gr_k(\mathbb{C}^n) \cap X^F_{\mathbf{u}}$.* □

Observe that it is possible to choose the real numbers $a_1, \ldots, a_n$ so that both f and F are Morse functions. But it may happen that f is a Morse function even when F is not.

The spaces $M^F_{\mathbf{u}}, S^F_{\mathbf{u}}, X^F_{\mathbf{u}}$ are determined by an eigenspace decomposition

$$\mathbb{C}^{N+1} = \hat{E}_1 \oplus \cdots \oplus \hat{E}_{\hat{k}}$$

in the usual way. Using this notation, the NDCMs of F are the linear subspaces $\mathbb{P}(\hat{E}_i)$, and the stable manifold of $\mathbb{P}(\hat{E}_i)$ is given explicitly as a bundle over $\mathbb{P}(\hat{E}_i)$ by

$$\mathbb{P}(\hat{E}_0 \oplus \cdots \oplus \hat{E}_i) - \mathbb{P}(\hat{E}_0 \oplus \cdots \oplus \hat{E}_{i-1}).$$

The projection map to $\mathbb{P}(\hat{E}_i)$ sends a line L in $\hat{E}_0 \oplus \cdots \oplus \hat{E}_i$ to the line $L + \hat{E}_0 \oplus \cdots \oplus \hat{E}_{i-1}$ in $\hat{E}_0 \oplus \cdots \oplus \hat{E}_i \;/\; \hat{E}_0 \oplus \cdots \oplus \hat{E}_{i-1} \cong \hat{E}_i$.

The associated Schubert variety, i.e. the closure of this stable manifold, is just

$$\mathbb{P}(\hat{E}_0 \oplus \cdots \oplus \hat{E}_i),$$

which is a *linear subspace* of $\mathbb{CP}^N$. Although $X_{\mathbf{u}}$ is not necessarily *equal* to the intersection of this space with $Gr_k(\mathbb{C}^n)$, it is in fact true that $X_{\mathbf{u}}$ is given by the intersection of *some* linear subspace with $Gr_k(\mathbb{C}^n)$. This follows from the fact that

$$S_{\mathbf{u}} = Gr_k(\mathbb{C}^n) \quad \cap \quad \mathbb{P}(\hat{E}_0 \oplus \cdots \oplus \hat{E}_i) - \mathbb{P}(\hat{E}_0 \oplus \cdots \oplus \hat{E}_{i-1})$$

(as some of the linear equations defining $\mathbb{P}(\hat{E}_0 \oplus \cdots \oplus \hat{E}_i)$ in $\mathbb{C}^{n+1}$ may become dependent in the presence of the equations defining $Gr_k(\mathbb{C}^n)$).

The fact that the Plücker embedding is compatible with the natural Morse–Bott functions on $Gr_k(\mathbb{C}^n)$ and $\mathbb{CP}^N$ may be explained group-theoretically. The key point is that the Plücker embedding is induced by an irreducible representation $U_n \to U_{N+1}$. However, it seems technically easier to work with the explicit formulae for the flow lines, as we have done in this section.

3.6 Cohomology of the Grassmannian, and the Schubert calculus

In this section we consider only Morse functions on $Gr_k(\mathbb{C}^n)$.

From a Schubert symbol $\mathbf{u} = (u_1, \ldots, u_k)$ we obtain a cell $S_{\mathbf{u}}$ in $Gr_k(\mathbb{C}^n)$ of (real) dimension $2\sum_{i=1}^k (u_i - i)$. By the Morse inequalities for the coefficient group $\mathbb{Z}$ (or by standard theory of cellular homology) it follows that $H_*(Gr_k(\mathbb{C}^n); \mathbb{Z})$ is a free abelian group with one generator for each Schubert cell. Let $x_{\mathbf{u}}$ be the homology class represented by[4] $X_{\mathbf{u}}$. Both $x_{\mathbf{u}}$ and $X_{\mathbf{u}}$ are referred to as 'Schubert cycles'.

By Poincaré duality we obtain a dual cohomology class $z_{\mathbf{u}}$ of dimension $2k(n-k) - 2\sum_{i=1}^k (u_i - i)$. Thus, $\dim z_{\mathbf{u}} = \operatorname{codim} x_{\mathbf{u}}$.

An example of particular interest is the generator of $H^2(Gr_k(\mathbb{C}^n); \mathbb{Z}) \cong \mathbb{Z}$; this corresponds to the unique codimension one Schubert cycle, i.e. to the Schubert symbol $(n-k, n-k+2, \ldots, n-1, n)$. The Schubert conditions here are

$$\dim V \cap \mathbb{C} = \cdots = \dim V \cap \mathbb{C}^{n-k-1} = 0,\ \dim V \cap \mathbb{C}^{n-k} = 1,\ \ldots,\ \dim V \cap \mathbb{C}^n = k,$$

and the Schubert variety is characterized by the single condition $\dim V \cap \mathbb{C}^{n-k} \geqslant 1$. In terms of the Plücker embedding (see §3.5), we have

$$X_{(n-k,n-k+2,\ldots,n-1,n)} = Gr_k(\mathbb{C}^n) \cap H$$

where $\mathbb{P}(H)$ is the Schubert variety for the critical point $V_{n-k} \wedge V_{n-k+2} \wedge \cdots \wedge V_{n-1} \wedge V_n$ in $\mathbb{CP}^N$. Now, since $a_1 > \cdots > a_n$, we have $a_n + \cdots + a_{n-k+1} > a_n + \cdots + a_{n-k+2} + a_{n-k} > \cdots$, so H is the *hyperplane* in $\mathbb{C}^{N+1}$ orthogonal to $V_{n-k+1} \wedge \cdots \wedge V_{n-1} \wedge V_n$. It is well known that the cohomology class dual to $\mathbb{P}(H)$ is a generator of $H^2(\mathbb{CP}^N; \mathbb{Z})$, so we deduce that the induced homomorphism $H^2(\mathbb{CP}^N; \mathbb{Z}) \to H^2(Gr_k(\mathbb{C}^n); \mathbb{Z})$ is an isomorphism.

---

[4] The precise meaning of this is explained in Appendix B of [30].

The multiplicative behaviour of $H^*(Gr_k(\mathbb{C}^n);\mathbb{Z})$ is equivalent to the behaviour of the intersections of generic representatives of homology classes (we shall make a more precise statement shortly). As a first step towards describing this, we shall need a slight generalization of Schubert varieties.

In §3.4 we pointed out that the definition of the Schubert varieties $X_{\mathbf{u}}$ depends only on the choice of the standard flag $\mathbb{C} \subseteq \mathbb{C}^2 \subseteq \cdots \subseteq \mathbb{C}^n$. If we choose a new flag $F_1 \subseteq F_2 \subseteq \cdots \subseteq F_n = \mathbb{C}^n$, denoted by F, then we obtain new objects $S^F_{\mathbf{u}}, X^F_{\mathbf{u}}, x^F_{\mathbf{u}}, z^F_{\mathbf{u}}$ defined in exactly the same way as $S_{\mathbf{u}}, X_{\mathbf{u}}, x_{\mathbf{u}}, z_{\mathbf{u}}$, but using the new flag instead of the standard flag. Since any two flags are related by an element of the unitary group U_n, however, we have $x^F_{\mathbf{u}} = x_{\mathbf{u}}$ and $z^F_{\mathbf{u}} = z_{\mathbf{u}}$. So we may regard the Schubert cycles $X^F_{\mathbf{u}}$ as a family of representatives for the homology class $x_{\mathbf{u}}$, parametrized by the space of all flags.

For example, consider the 'opposite' flag

$$(\mathbb{C}^{n-1})^\perp \subseteq (\mathbb{C}^{n-2})^\perp \subseteq \cdots \subseteq \mathbb{C}^\perp \subseteq \mathbb{C}^n;$$

let us denote the Schubert varieties with respect to this flag by $X^c_{\mathbf{u}}$. It is easy to check that

$$X^c_{\mathbf{u}} = \overline{U}_{\mathbf{u}^c}$$

where $\mathbf{u}^c = (n - u_k + 1, n - u_{k-1} + 1, \ldots, n - u_1 + 1)$. Thus, both $\overline{S}_{\mathbf{u}}$ and $\overline{U}_{\mathbf{u}^c}$ are representatives of the same homology class $x_{\mathbf{u}}$.

We now state a special case of an important general principle (see Appendix B of [30]):

Theorem 3.6.1 *If $x_{\mathbf{u}}$ and $x_{\mathbf{v}}$ are Schubert cycles with $\dim z_{\mathbf{u}} + \dim z_{\mathbf{v}} = \dim Gr_k(\mathbb{C}^n)$, then the product $z_{\mathbf{u}} z_{\mathbf{v}} \in H^{\dim Gr_k(\mathbb{C}^n)}(Gr_k(\mathbb{C}^n);\mathbb{Z}) \cong \mathbb{Z}$ of the corresponding cohomology classes is equal to the intersection number of $X_{\mathbf{u}}$ and $X_{\mathbf{v}}$.* □

This intersection number is equal to the number of points (counted with multiplicities) in $X^{F_1}_{\mathbf{u}} \cap X^{F_2}_{\mathbf{v}}$ whenever $X^{F_1}_{\mathbf{u}} \cap X^{F_2}_{\mathbf{v}}$ is a finite set.

Example 3.6.2 For any $\mathbf{u}$, we have $z_{\mathbf{u}} z_{\mathbf{u}^c} = 1$. This is because the dual homology classes are represented by $\overline{S}_{\mathbf{u}}$ and $\overline{U}_{\mathbf{u}}$ respectively, and these intersect at precisely one point, namely the critical point $V_{\mathbf{u}}$. (Of course, these homology classes may also be represented by $\overline{S}_{\mathbf{u}}$ and $\overline{S}_{\mathbf{u}^c}$. But this is of no interest to us as these cycles intersect at infinitely many points.) To see that the multiplicity of the intersection point is 1, one can use the Plücker embedding – it follows from our discussion in §3.5 that the multiplicity of *any* isolated point of intersection of two Schubert varieties is precisely 1, since we are just taking the intersection of linear subspaces in $\mathbb{CP}^N$. □

More generally, we have:

Proposition 3.6.3 *If $x_{\mathbf{u}}$ and $x_{\mathbf{v}}$ are Schubert cycles with* $\dim z_{\mathbf{u}} + \dim z_{\mathbf{v}} = \dim Gr_k(\mathbb{C}^n)$*, then*

$$z_{\mathbf{u}} z_{\mathbf{v}} = \begin{cases} 1 & \text{if } \mathbf{u} = \mathbf{v}^c \\ 0 & \text{otherwise.} \end{cases}$$

Proof We have just seen that $z_{\mathbf{u}} z_{\mathbf{u}^c} = 1$. To show that $X_{\mathbf{u}} \cap X_{\mathbf{v}} = \emptyset$ if $\mathbf{v} \neq \mathbf{u}^c$, one may use the geometrical characterization of $X_{\mathbf{u}}$ and $X_{\mathbf{v}}$ – we omit the details. □

This proposition is a manifestation of Poincaré duality, and it allows us to determine the products $z_{\mathbf{u}} z_{\mathbf{v}}$ for *arbitrary* $\mathbf{u}, \mathbf{v}$. For we may express $z_{\mathbf{u}} z_{\mathbf{v}}$ in terms of the additive Schubert basis as

$$z_{\mathbf{u}} z_{\mathbf{v}} = a_1 z_{\mathbf{u}_{(1)}} + \cdots + a_r z_{\mathbf{u}_{(r)}}$$

for some integers $a_1, \dots, a_r$, where $\dim z_{\mathbf{u}} + \dim z_{\mathbf{v}} = \dim z_{\mathbf{u}_{(i)}}$ for each i. Then we obtain a_i by multiplying the above expression by $z_{\mathbf{u}^c_{(i)}}$:

$$z_{\mathbf{u}} z_{\mathbf{v}} z_{\mathbf{u}^c_{(i)}} = a_i$$

(all other products vanish, by Proposition 3.6.3). Thus, we have to calculate all triple products $z_{\mathbf{u}} z_{\mathbf{v}} z_{\mathbf{w}}$ such that $\dim z_{\mathbf{u}} + \dim z_{\mathbf{v}} + \dim z_{\mathbf{w}} = \dim Gr_k(\mathbb{C}^n)$. Theorem 3.6.1 generalizes to this situation, so we have to calculate the corresponding triple intersections of Schubert varieties.

Example 3.6.4 We shall carry out the calculation of some triple products for $Gr_2(\mathbb{C}^4)$, and hence determine the multiplicative structure of $H^*(Gr_2(\mathbb{C}^4); \mathbb{Z})$. First we list the additive generators:

$$\begin{aligned} z_{(3,4)} &\in H^0 \\ z_{(2,4)} &\in H^2 \\ z_{(1,4)}, z_{(2,3)} &\in H^4 \\ z_{(1,3)} &\in H^6 \\ 1 = z_{(1,2)} &\in H^8. \end{aligned}$$

Proposition 3.6.3 gives the following products:

$$z_{(2,4)} z_{(1,3)} = z_{(1,4)} z_{(1,4)} = z_{(2,3)} z_{(2,3)} = 1, \; z_{(1,4)} z_{(2,3)} = 0.$$

Let us now try to compute $z_{(1,4)} z_{(2,4)} z_{(2,4)}$. We must find suitably generic representing cycles $X^F_{\mathbf{u}}$ for these classes, by choosing suitably generic flags F.

For $z_{(2,4)}$ we need two modifications of the standard representative

$$X_{(2,4)} = \{V \in Gr_2(\mathbb{C}^4) \mid \dim V \cap \mathbb{C}^2 \geqslant 1\}.$$

We shall choose the flags

$$F' : V_1 \subseteq V_1 \oplus V_4 \subseteq V_1 \oplus V_2 \oplus V_4 \subseteq \mathbb{C}^4$$
$$F'' : V_2 \subseteq V_2 \oplus V_4 \subseteq V_1 \oplus V_2 \oplus V_4 \subseteq \mathbb{C}^4.$$

The corresponding cycles are:

$$X'_{(2,4)} = \{V \in Gr_2(\mathbb{C}^4) \mid \dim V \cap V_1 \oplus V_4 \geqslant 1\}$$
$$X''_{(2,4)} = \{V \in Gr_2(\mathbb{C}^4) \mid \dim V \cap V_2 \oplus V_4 \geqslant 1\}.$$

For $z_{(1,4)}$ we shall choose the flag

$$F''' : \quad V_3 \subseteq V_2 \oplus V_3 \subseteq V_1 \oplus V_2 \oplus V_3 \subseteq \mathbb{C}^4$$

i.e. we choose

$$X'''_{(1,4)} = \{V \in Gr_2(\mathbb{C}^4) \mid V_3 \subseteq V\}.$$

It may now be verified that

$$X'_{(2,4)} \cap X''_{(2,4)} \cap X'''_{(1,4)} = \{V_3 \oplus V_4\}$$

i.e. a single point. As in Example 3.6.2, we can see that the multiplicity of this point is 1. So we conclude that $z_{(1,4)}z_{(2,4)}z_{(2,4)} = 1$. Exactly the same argument gives $z_{(2,3)}z_{(2,4)}z_{(2,4)} = 1$.

The remaining (double) products in $H^*(Gr_2(\mathbb{C}^4); \mathbb{Z})$ are

$$z_{(2,4)}z_{(2,4)} = az_{(2,3)} + bz_{(1,4)}$$
$$z_{(1,4)}z_{(2,4)} = cz_{(1,3)}$$
$$z_{(2,3)}z_{(2,4)} = dz_{(1,3)}.$$

Using the two triple products which we have just calculated, we find that $a = b = c = d = 1$. □

The same method works for $H^*(Gr_k(\mathbb{C}^n); \mathbb{Z})$, although this situation is of course more complicated. There are famous general formulae for the multiplicative structure, which constitute the 'Schubert calculus'. An elementary approach to these formulae and their traditional applications can be found in [43]; other versions can be found in [36], [39], and [30].

From the theory of Chern classes, there is a well-known 'closed formula' for the ring structure of $H^*(Gr_k(\mathbb{C}^n); \mathbb{Z})$, namely

$$\frac{\mathbb{Z}[c_1, \ldots, c_{n-k}, d_1, \ldots, d_k]}{(1 + c_1 + \cdots + c_{n-k})(1 + d_1 + \cdots + d_k) = 1}$$

where $c_i, d_i \in H^{2i}(Gr_k(\mathbb{C}^n); \mathbb{Z})$. (This is explained in detail in §23 of [15]; another good reference is [49].) It may be checked that this agrees with our description in the case of $Gr_2(\mathbb{C}^4)$. The dual of cohomology class d_i is represented by the

Schubert variety $X_{\mathbf{u}}$ with $\mathbf{u} = (n-k, n-k+1, \ldots, n-k+i-1, n-k+i+1, \ldots, n-1, n)$; these are called 'special Schubert varieties'. The cohomology class corresponding to a general Schubert variety is more complicated to describe (see [7] and [30]).

Viewing the cohomology ring of a manifold M as a collection of triple intersection numbers of (representatives of) homology classes is in fact a useful idea. The equivariant cohomology ring $H^*_G(M)$ (see §2.7) may be described in terms of $H^*(BG)$-valued triple intersections. Quantum cohomology (see §4.4) appears in a similar way.

3.7 Next steps

We have now covered the 'classical' aspects of Morse theory, and in §4 we shall turn to more recent developments. As motivation for this, we mention here a couple of points which arise from our study of Grassmannians.

An immediate question is: when can the cohomology ring be determined directly from a Morse function? It is a well-known limitation of classical Morse theory that the Morse inequalities give information only about the *additive* structure of the cohomology ring. However, we have seen that the cohomology ring of $Gr_k(\mathbb{C}^n)$ can be found from explicit knowledge of the stable manifolds of a suitable Morse function. Was this just a special trick, or is there perhaps a more general theory which works for Morse functions on arbitrary compact manifolds?

A slightly more subtle (but related) question concerns the possible configurations of flow lines of a Morse function. We saw in §2 that the possible configurations of critical points of a Morse function $f : M \to \mathbb{R}$ are limited by the topology of M. For example, it is not possible to have a Morse function on $S^1 \times S^1$ with precisely three critical points whose indices are $0, 1$, and 2. In the same way, it is not possible to have arbitrary configurations of flow lines connecting the critical points. The question arises as to how these configurations of flow lines are restricted.

These questions may be answered (for certain kinds of manifolds, at least) by generalizing Morse theory in a new way: instead of considering a single Morse function, we consider several. Indeed, when we computed triple products of cohomology classes in §3.6, we were in fact making use of three 'independent' Morse functions on $Gr_k(\mathbb{C}^n)$. The general theory underlying this calculation is what we shall study next.

4 Morse theory in the 1990s

We describe several recent applications of Morse theory, in which the gradient flow lines play a fundamental role. Although the level of discussion will be somewhat more advanced in this section, the case of a complex Grassmannian should be kept in mind as a typical example. We begin by discussing Morse functions which arise from torus actions on Kähler manifolds; these functions, which include the functions on Grassmannians in §3, have the crucial property that their gradient flow lines are explicitly identifiable. Then we describe the 'new' approach to Morse theory, due to Witten. After that we present the 'field-theoretic' Morse theory of Cohen–Jones–Segal, Betz–Cohen, and Fukaya.

4.1 Morse functions generated by torus actions

In §1 and §2 we gave a review of the 'classical' Morse theory, and then in §3 we illustrated this in detail for a particular example. We shall now focus on some contemporary aspects, which show that the power of Morse theory goes far beyond the computation of homology groups. Our starting point is an important family of examples which includes the Morse and Morse–Bott functions of §3.

Let $T \cong S^1 \times \cdots \times S^1$ be a torus, and let M be a simply-connected connected compact Kähler manifold. Assume that the group T acts on M, and that this action preserves the complex structure J and the Kähler 2-form ω of M. It follows that the action also preserves the Kähler metric $\langle\ ,\ \rangle$, as the latter is given by $\omega(A, B) = \langle A, JB \rangle$.

Let $\mathfrak{t}$ denote the Lie algebra of T. Let $X \in \mathfrak{t}$ be any generator of the torus; this means that T is the closure of its (not necessarily closed) subgroup $\exp \mathbb{R}X$. The formula

$$X^*_m = \frac{d}{dt} \exp tX \cdot m|_{t=0}$$

defines a vector field X^* on M.

Since M is simply connected, and $\omega(X^*,\ )$ is a closed 1-form, there is a function $f^X : M \to \mathbb{R}$ such that $df^X = \omega(X^*,\ )$. We shall see that f^X is a perfect Morse–Bott function, with particularly nice properties. Observe that we have

$$-\nabla f^X = JX^*,$$

from the formula $\langle \nabla f^X, A \rangle = df^X(A) = \omega(X^*, A) = -\langle JX^*, A \rangle$.

Theorem 4.1.1 [25] *For any generator X of $\mathfrak{t}$, the function f^X is a perfect Morse–Bott function on M. The critical points of f^X are the fixed points of the T-action, and they form a finite number of totally geodesic Kähler submanifolds of M.*

Sketch of the proof The fact that the critical points of f^X are the fixed points of T follows from the formula $-\nabla f^X = JX^*$.

Let m be a fixed point of T. Then there is an induced action of the Lie group T, and hence also of the Lie algebra $\mathfrak{t}$, on the vector space $T_m M$. This means that we have a Lie group homomorphism $\Theta_m : T \to Gl(T_m M)$ and a Lie algebra homomorphism $\theta_m : \mathfrak{t} \to \mathrm{End}(T_m M)$.

Since T acts by isometries (i.e. the action of T preserves the metric), $\Theta_m(t)$ is an orthogonal transformation, and $\theta_m(X)$ a skew-symmetric transformation, for each $t \in T$, $X \in \mathfrak{t}$. As T is abelian, we may put these transformations simultaneously into canonical form. This means that there exists a decomposition

$$T_m M = V_0 \oplus V_1 \oplus \cdots \oplus V_r$$

such that $\theta|_{V_0} = 0$ and

$$\theta|_{V_i} = \begin{pmatrix} 0 & w_i \\ -w_i & 0 \end{pmatrix}$$

for non-trivial linear functionals $w_1, \ldots, w_r$ on $\mathfrak{t}$. Note that the subspaces V_i for $i > 0$ are not uniquely determined in general, and that the linear functionals ω_i are determined only up to sign.

By considering geodesics through m (see [44]), it can be shown that the connected component of the fixed point set of T containing m is a submanifold of M – the geodesics through m in the direction of V_0 give a local chart (via the exponential map). This argument also shows that the submanifold is totally geodesic, with tangent space V_0 at m.

Up to this point, we have used only the fact that the action of T preserves the metric. Since T preserves the complex structure, J commutes with the linear transformations $\Theta_m(t)$ and $\theta_m(t)$. We may therefore choose the decomposition so that J preserves each subspace V_i. It follows that each connected component of the fixed point set of T is actually a Kähler manifold, and that (for $i > 0$) we may write

$$J|_{V_i} = \pm \begin{pmatrix} 0 & 1 \\ -1 & 0 \end{pmatrix}.$$

An additional consequence of the existence of J is that each V_i acquires a natural orientation, and so the linear functionals ω_i are now determined uniquely (for a given choice of subspaces V_i.

We now turn to the computation of the Hessian H of f^X. In §1.2 we gave a definition in terms of local coordinates, but this is equivalent to the following more invariant definition (see page 4 of [47]). For any vectors $V, W \in T_m M$, we have

$$H(V,W) = \tilde{V}(\tilde{W}(f^X))(m) = \tilde{V} df^X(\tilde{W})(m)$$

where $\tilde{V}, \tilde{W}$ are any extensions of V, W to local vector fields on M. We therefore have $H(V,W) = \tilde{V}\omega(X^*, \tilde{W})(m) = \tilde{V}\alpha(X^*)(m)$, where $\alpha = -i_{\tilde{W}}\omega$ (and i denotes interior product). With a suitable choice of the extensions $\tilde{V}, \tilde{W}$, the well-known formula for $d\alpha$ shows that $H(V,W) = -\omega(\tilde{W}, [\tilde{V}, X^*])$.

From the definition of θ, it follows that $\theta(X)V = [\tilde{V}, X^*]_m$. Hence we obtain the formula

$$H(V,W) = \langle W, -J\theta(X)V \rangle.$$

On each V_i with $i > 0$, it follows that H is equal to the inner product times the non-trivial linear functional ω_i. Hence the Hessian is non-degenerate on a space complementary to V_0, i.e. f^X is a Morse–Bott function.

The index of the Hessian at a critical point m is even, being twice the number of ω_i's such that $\omega_i(m) < 0$. This implies that the Morse–Bott function f^X is perfect, and so the proof of Theorem 4.1.1 is complete. □

The linear functionals $w_1, \ldots, w_r$ on $\mathfrak{t}$ which appear in this proof are called the (non-zero) weights of the action of T at the fixed point m. Theorem 4.1.1 generalizes to the case of a symplectic manifold, as was pointed out in [25]. We shall not need this extra generality, and in fact we shall make essential use of the complex structure J, so we shall only consider Kähler manifolds here.

The formula $-\nabla f^X = JX^*$ leads to a geometrical description of the flow lines of $-\nabla f^X$, as we shall explain next.

Lemma 4.1.2 *The action of $T = S^1 \times \cdots \times S^1$ on M extends to an action of the complexified torus $T^{\mathbb{C}} = \mathbb{C}^* \times \cdots \times \mathbb{C}^*$ on M. The vector $\sqrt{-1}\, X \in \mathfrak{t}^{\mathbb{C}}$ generates the vector field JX^*, i.e. $(\sqrt{-1}\, X)^* = JX^*$. The flow line γ of $-\nabla f^X$ passing through $m \in M$ is given by the action of the subgroup $\sqrt{-1}\, \mathbb{R}X$, i.e. $\gamma(t) = \exp \sqrt{-1}\, tX \cdot m$.*

Sketch of the proof The extension of the action is guaranteed by the fact that T preserves the complex structure of M (see [2]). A direct construction of the extension may be given by using the integral curves of JX^*. The fact that $(\sqrt{-1}\, X)^* = JX^*$ follows (tautologically) from this, as does the required description of the flow line γ. □

Thus, the flow lines of the gradient vector field associated to the action of T are given by the action of a subgroup of the larger group $T^{\mathbb{C}}$. *Conversely, whenever an 'algebraic torus' $\mathbb{C}^* \times \cdots \times \mathbb{C}^*$ acts complex analytically on a compact Kähler manifold then we obtain both a Morse–Bott function and its gradient flow in the above manner.* Many manifolds do admit such actions, for example the Grassmannian $Gr_k(\mathbb{C}^n)$ (which is acted upon naturally by the group of diagonal matrices with non-zero complex diagonal entries). This provides a simple explanation of the rather tricky calculation of Lemma 3.1.2, by means of which we identified the gradient flow lines for the height functions. It suffices to assume that the action of $\mathbb{C}^* \times \cdots \times \mathbb{C}^*$ preserves the complex structure, because an $S^1 \times \cdots \times S^1$-invariant Kähler metric can be obtained by averaging the given Kähler metric over the (compact) group $S^1 \times \cdots \times S^1$.

We shall now address a question which lies at the heart of Morse theory: *how are the gradient flow lines of a Morse function $f : M \to \mathbb{R}$ arranged within M?* For example, which pairs of critical points are connected by a flow line? And by how many flow lines? These questions can be answered by explicit computation of stable and unstable manifolds in the case of a height function on a Grassmannian, as we did in §3, but in general no such computation will be possible. Our main theme from now on will be to consider this question for Morse–Bott functions f^X associated to torus actions.

From the lemma it follows that the behaviour of the flow lines is related to the geometrical properties of the various orbits of the group $T^{\mathbb{C}}$. If M is an algebraic Kähler manifold, and the action of $T^{\mathbb{C}}$ is algebraic (as we shall assume from now on), the closures of such orbits are special algebraic varieties called toric varieties.[5] In general toric varieties have singularities, but they are particularly amenable to study (see [50] and [31]) because they may be characterized by purely combinatorial objects, called 'fans'. Now, in the Kähler situation at least, the fan is equivalent to a more familiar combinatorial object, namely a convex polyhedron in Euclidean space. The next theorem describes this polyhedron.

Before stating the theorem, we need to introduce the moment map associated to the action of T on M. This is the map

$$\mu : M \to \mathfrak{t}^*$$

which is determined up to an additive constant by the condition

$$d\mu(\ \ )(Y) = \omega(Y^*,\ \ )$$

for all $Y \in \mathfrak{t}$. The definition of this map comes from symplectic geometry and classical mechanics, but it has a straightforward Morse-theoretic interpretation: $\mu(\ \ )(Y)$ is the Morse–Bott function f^Y associated to the subtorus of T which is generated by Y.

The fact that our Morse–Bott function f^X is not alone, but is accompanied by a whole family of Morse–Bott functions f^Y parametrized by $Y \in \mathfrak{t}$, is significant. The moment map μ assembles these Morse–Bott functions into a single vector-valued function.

Theorem 4.1.3 [2, 38] *Let O_m denote the closure in M of the orbit of m under $T^{\mathbb{C}}$, i.e. $O_m = \overline{T^{\mathbb{C}} \cdot m}$. Then: (i) $\mu(O_m)$ is the convex hull of the finite set $\{\mu(m) \in \mathfrak{t}^* \mid m$ is a critical point of $f^X\}$, (ii) the inverse image (under μ) of each open face of $\mu(O_m)$ is a single $T^{\mathbb{C}}$-orbit in O_m, and (iii) μ induces a homeomorphism $O_m/T \to \mu(O_m)$ (although the action of T on O_m is not necessarily free).*

Proof See Theorem 2 of [2]. □

The simplest non-trivial example of this theorem is given by the action of $\mathbb{C}^*$ on $\mathbb{CP}^1 (\cong S^2)$ by $u \cdot [z_0; z_1] = [uz_0; z_1]$. The corresponding Morse–Bott function $f^X : S^2 \to \mathbb{R}$ is a height function, and there are two isolated critical points: the maximum point and the minimum point. We have $f^X = \mu$ in this situation, and $f^X(S^2)$ is obviously the line segment joining the maximum and minimum values.

We are specifically interested in the stable and unstable manifolds of f^X, and their intersections. Theorem 4.1.3 leads to the following information about these spaces. For simplicity we shall assume that M is actually a smooth projective variety, i.e. an algebraic submanifold of some complex projective space, with the induced Kähler structure. Furthermore we assume that $T^{\mathbb{C}}$ acts on M by

[5] This is essentially the definition of a toric variety.

projective transformations. From the method of §3.5 we can then deduce that the closures of the stable and unstable manifolds of our Morse functions are irreducible algebraic subvarieties.

Theorem 4.1.4 [46] *Assume that M is a smooth projective variety, and that $T^{\mathbb{C}}$ acts on M by projective transformations. Let V be an irreducible algebraic subvariety of M which is preserved by the action of $T^{\mathbb{C}}$. Then $\mu(V)$ is the convex hull of the finite set $\{\mu(m) \in \mathfrak{t}^* \mid m$ is a critical point of f^X in $V\}$.*

Sketch of the proof Let $P_1, \ldots, P_s$ be the distinct images under μ of the critical points of f^X which lie in V. (This is necessarily a finite set, as f^X is constant on any connected critical submanifold.) For any one-dimensional subalgebra $\mathbb{R}Y$ of $\mathfrak{t}$, the image of the continuous function $f^Y|_V$ is a closed finite interval in $\mathbb{R}$ (as V is connected). Moreover, since $T^{\mathbb{C}}$ preserves V, the ends of this interval (i.e. the maximum and minimum values of f^Y) are of the form $f^Y(P_i), f^Y(P_j)$, for some i, j. But $f^Y = \pi_Y \circ \mu$, where $\pi_Y : \mathfrak{t}^* \to \mathbb{R}$ is given by evaluation at Y. It follows that $\mu(V)$ is contained in the convex hull of $P_1, \ldots, P_s$.

Conversely, we must show that any point of this polyhedron is contained in $\mu(V)$. Let $P_{i_1}, \ldots, P_{i_t}$ denote the 'exterior' points of the polyhedron. For each i_j, choose some $Y_j \in \mathfrak{t}$ such that the function $f^{Y_j}|_V$ has P_{i_j} as its absolute minimum value, occurring on a critical set V_j, where $V_j = V \cap M_j$ for some connected component M_j of the fixed point set of $T^{\mathbb{C}}$ on M. (This may be done by choosing a 'generic' Y_j such that P_{i_j} is the closest point of the polyhedron to the linear functional $\langle Y_j, \ \rangle$, where $\langle\ ,\ \rangle$ is an invariant inner product on $\mathfrak{t}$.)

Let $S_j^V = V \cap S_j$, where S_j is the stable manifold of M_j (for f^{Y_j}). Since $T^{\mathbb{C}}$ preserves V, S_j^V is the stable 'manifold' of V_j (for $f^{Y_j}|_V$). Now, the stable manifold decomposition of M induces a decomposition of V. The closure (in V) of each piece of this decomposition is a subvariety of V, and precisely one such closure must be equal to V since V is irreducible. Denote this piece by $V \cap S$, where S is the corresponding stable manifold in M (for f^{Y_j}).

We claim that $S = S_j$. (This would be obvious if V were a smooth subvariety.) Assume that S and S_j are not equal, so that $S \cap S_j = \emptyset$. Then the closure of $V \cap S$ in V is disjoint from $V \cap S_j$, since the closure of S in M is disjoint from S_j. However this contradicts the defining property of $V \cap S$.

It follows that the complement of S_j in V is a subvariety of positive codimension. The complement of the intersection of all such S_j (for $j = 1, \ldots, t$) is therefore also a subvariety of positive codimension; in particular this intersection is non-empty. For any point v of the intersection, we have $P_{i_1}, \ldots, P_{i_t} \in \overline{T^{\mathbb{C}} \cdot v}$, because of the description of the flow lines as orbits of subgroups of $T^{\mathbb{C}}$. Hence $\mu(V)$, which contains $\mu(\overline{T^{\mathbb{C}} \cdot v})$, must contain the convex hull of $P_1, \ldots, P_s$, by Theorem 4.1.3. □

The case $V = M$ was stated and proved as one of the main theorems of [2], [38]. Although the proofs given there were direct, the possibility of deducing the result from Theorem 4.1.3 was in fact mentioned in [2]. Various special cases of this result (e.g. when V is a Schubert variety) were already known – we refer to [2], [38] for further information.

Theorem 4.1.4 suggests the possibility that the behaviour of the gradient flow lines of f may be encoded by combinatorial information within the polyhedron $\mu(M)$. We shall investigate this phenomenon, starting with the familiar case of height functions on Grassmannians from §3. Initial work in this direction was done in [33], where some of the results of [2], [38] were anticipated in the case of a Grassmannian.

Let $M = Gr_k(\mathbb{C}^n)$ and let T_n be the group of diagonal $n \times n$ matrices whose diagonal entries are complex numbers of unit length, and let $T_n^{\mathbb{C}}$ be its complexification, i.e. the group of diagonal $n \times n$ matrices with non-zero complex diagonal entries. We have a natural action of T_n (or $T_n^{\mathbb{C}}$) on M. The vector $X = \sqrt{-1}\,(a_1, \ldots, a_n)$ in the Lie algebra $\mathfrak{t}_n$ is a generator of T_n if and only if $\mathbb{R}(a_1, \ldots, a_n) \cap \mathbb{Z}^n = \{0\}$, i.e. the line through $(a_1, \ldots, a_n)$ has 'irrational slope'.

From the above general theory, any choice of $(a_1, \ldots, a_n)$ gives rise to (1) an action of T (the subtorus of T_n generated by $Y = \sqrt{-1}\,(a_1, \ldots, a_n)$), and (2) a Morse–Bott function f^Y, whose critical points are the fixed points of T. From Lemma 3.1.2 we see that this function is (up to an additive constant) the height function on $Gr_k(\mathbb{C}^n)$ defined in §3. The fixed points of T are of course the k-planes which can be spanned by eigenvalues of $\mathrm{diag}(a_1, \ldots, a_n)$.

From the formula for f^X in §3.1, it follows that the moment map is given explicitly by

$$\mu(V) = \text{diagonal part of } \pi_V.$$

(We identify $\mathfrak{t}^*$ with $\mathfrak{t} \cong \mathbb{R}^n$ by using the (restriction of the) standard Hermitian product on $\mathbb{C}^n$.) In particular,

$$\mu(V_{\mathbf{u}}) = e_{\mathbf{u}},$$

where $e_{\mathbf{u}} = e_{u_1} + \cdots + e_{u_k}$ (and e_i denotes the ith basis vector of $\mathbb{C}^n$). Since every critical manifold of f^X contains at least one point of the form $V_{\mathbf{u}}$, the images of the critical points of f^X under μ are precisely the points $e_{\mathbf{u}}$. Hence the general theory gives

$$\mu(Gr_k(\mathbb{C}^n)) = \text{convex hull of } \{e_{\mathbf{u}} = e_{u_1} + \cdots + e_{u_k} \mid 1 \leqslant u_1 < \cdots < u_k \leqslant n\}.$$

Example 4.1.5 Let $k = 1$, so that $Gr_k(\mathbb{C}^n) = \mathbb{CP}^{n-1}$. In this case the formula for f^X in §3.1 shows that the moment map is given even more explicitly by

$$\mu([z]) = (|z_1|^2, \ldots, |z_n|^2) / \sum_{i=0}^{n} |z_i|^2.$$

It follows immediately from this formula that $\mu(\mathbb{CP}^{n-1})$ is the convex hull of the basis vectors $e_1, \ldots, e_n$. Since the (closures of the) stable and unstable manifolds of f^X are respectively of the form $[*;\ldots;*;0;\ldots;0]$ and $[0;\ldots;0;*;\ldots;*]$ (see §3.1), their images under μ are also clear. For example, in the diagram below, we illustrate the images under μ of the closures of the stable manifolds of V_1, V_2, V_3, for a Morse function $f^X : \mathbb{CP}^2 \to \mathbb{R}$.

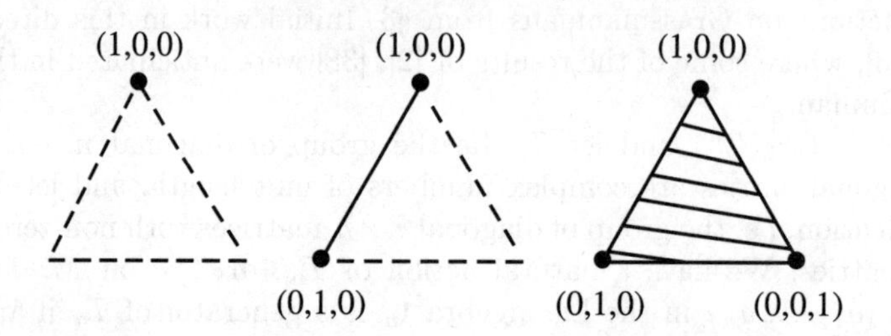

In fact, for $\mathbb{CP}^{n-1}$, all assertions of Theorems 4.1.3 and 4.1.4 may be verified directly, from the formula for μ. □

Example 4.1.6 Let us consider what the general theory says in the case of $Gr_2(\mathbb{C}^4)$. We have already investigated this space in some detail in §3 (starting with Example 3.4.1). The image of μ is the convex hull of the six points $e_i + e_j$ in $\mathbb{R}^4$ (with $1 \leqslant i < j \leqslant 4$). It may be verified that this polyhedron is a regular octahedron.

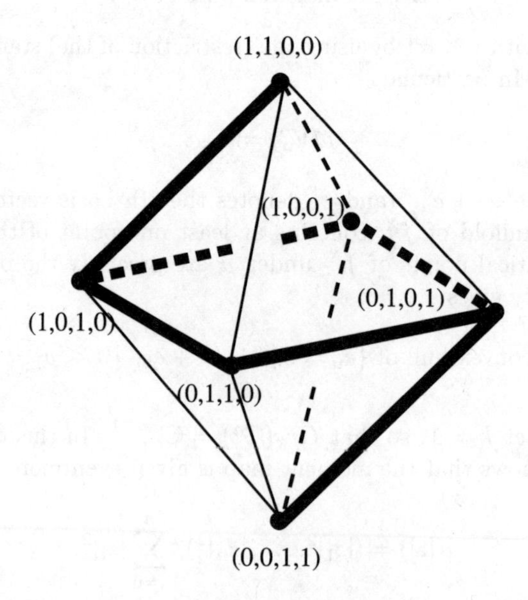

The heavy lines represent the partial order shown in the diagram following Definition 3.4.2. Combining the calculations of the stable manifolds in §3 with the statement of Theorem 4.1.4, we see that the image under μ of the closure of the stable manifold of a critical point $V_{\mathbf{u}}$ is the convex hull of those vertices which are greater than or equal to $e_{\mathbf{u}}$ in this partial order.

With a little more work, it is possible to verify the predictions of Theorem 4.1.3 in this case. First it is necessary to identify the various possible types of the closures of the orbits of the group $(\mathbb{C}^*)^4$, to which Theorem 4.1.3 will apply. Zero-dimensional orbits, i.e. points, correspond to the vertices of the octahedron. One-dimensional orbits (necessarily isomorphic to $\mathbb{CP}^1$) correspond to the edges. Two-dimensional orbits are of two types: copies of $\mathbb{CP}^2$, which correspond to faces, and copies of $\mathbb{CP}^1 \times \mathbb{CP}^1$, which correspond to squares spanned by sets of four coplanar vertices. Finally, there are two types of three-dimensional orbits, whose images under μ are either half of the octahedron or the entire octahedron. The first type is represented by the Schubert variety $X_{(2,4)}$ (see §3.4), and the second by the famous 'tetrahedral complex' (see [33]). □

Returning to the general case of $Gr_k(\mathbb{C}^n)$, we mention that the partial order on the critical points of $f^X : Gr_k(\mathbb{C}^n) \to \mathbb{R}$ (or equivalently, on the vertices of the polyhedron) may be specified purely algebraically, in terms of the action of the symmetric group Σ_n (the Weyl group of $GL_n\mathbb{C}$). This is explained in [2].

In [34] and [32], a detailed study is made of the various subpolyhedra of $\mu(Gr_k(\mathbb{C}^n))$ which arise from Schubert varieties associated to the Morse functions f^X and their intersections. These are characterized in terms of the combinatorial concept of a matroid.

By representing the stable and unstable manifolds of the function f^X as subpolyhedra of the polyhedron $\mu(M)$, we have in principle solved the problem of understanding the behaviour of the flow lines of $-\nabla f^X$. The practical value of this solution depends on being able to extract this information in an efficient manner, and this may not always be possible. However, there are two general situations where reasonably explicit results may be expected, namely for generalized flag manifolds and for toric manifolds. The essential phenomenon in both these situations is that *there exists an orbit of the complex algebraic torus $T^{\mathbb{C}}$ whose closure contains all critical points of the Morse functions associated to the torus action.* We shall call this kind of $T^{\mathbb{C}}$-action a *complete torus action.*

Example 4.1.7 A generalized flag manifold is by definition the quotient of a complex semisimple Lie group $G^{\mathbb{C}}$ by a parabolic subgroup P. Height functions on generalized flag manifolds – generalizing the Morse functions on Grassmannians in §3 – were first studied by Bott (see [11] and the article of Bott in [1]). Such functions are associated to torus actions (namely a maximal torus of $G^{\mathbb{C}}$) in the manner described at the beginning of this section. The index and, in the case of a Morse–Bott function, the nullity of a critical point may be computed in terms of the weights of the torus action (i.e. in terms of the roots of $G^{\mathbb{C}}$). Each stable or unstable manifold is an orbit of a certain subgroup of $G^{\mathbb{C}}$, as in the case $G^{\mathbb{C}}/P = Gr_k(\mathbb{C}^n)$. Indeed the decomposition of $G^{\mathbb{C}}/P$ into stable (or unstable) manifolds coincides with the well-known 'Bruhat decomposition' of $G^{\mathbb{C}}/P$, a fact which was proved in [55] as well as in various later papers (e.g. [2]). A brief summary of this theory may be found in the Appendix of [37]. The image of the moment map for the action of a maximal torus on a generalized flag manifold was worked out in [2], generalizing earlier work of Kostant. The polyhedron can be described (see [34]) as the convex hull of the *weights* of an irreducible representation of G; it is well known that $G^{\mathbb{C}}/P$ is the projectivized orbit of the maximal weight vector of a suitable representation (the generalized Plücker embedding). Schubert varieties in generalized flag manifolds have been extensively studied from the point of view of Lie theory and algebraic geometry (see [39] for an introduction and further references). The subpolyhedra obtained by taking the images of (various intersections of) Schubert varieties in generalized flag manifolds have been characterized in combinatorial terms in [34], generalizing the results mentioned earlier for Grassmannians. The homology classes represented by Schubert varieties have also been investigated thoroughly. In [7] these homology classes are related to the well-known description of the cohomology ring due to Borel, by making use of the generalized Plücker embedding. A brief explanation of the latter work can be found in [59]. □

Example 4.1.8 From the general theory of toric varieties, it is well known that a (smooth) toric variety with a Kähler metric is entirely determined by the image of its moment mapping. In particular, the behaviour of the flow lines of a Morse–Bott function associated to the given torus action is represented faithfully in the momentum polyhedron. Various geometrical and topological invariants of such manifolds have been computed explicitly in terms of this polyhedron; details can be found in [50] and [31] (see also [5] and [20]). There are relations between this and the previous example, as one may consider the toric varieties obtained as the closures of the orbits of a maximal torus of $G^{\mathbb{C}}$ acting on $G^{\mathbb{C}}/P$. These are singular varieties, in general; they have been studied in [34] and later in [23], [19], and [16]. □

We shall now change our point of view slightly by focusing on the family of Morse–Bott functions f^Y (parametrized by $Y \in \mathfrak{t}$), rather than the single Morse function f^X (corresponding to a generator X of $\mathfrak{t}$). This theme will reach maturity in §4.3, so as motivation for this we shall consider again the problem of computing the cohomology ring $H^*(Gr_2(\mathbb{C}^4))$ (cf. §3.6).

Theorem 4.1.4 gives a representation of the (image under μ of the) Schubert variety $V = X^Y_{\mathbf{u}}$ for the Morse function $f^Y : Gr_k(\mathbb{C}^n) \to \mathbb{R}$. Namely, the image under μ of V is the convex hull of those points $e_{\mathbf{u}}$ such that $V_{\mathbf{u}} \in V$. This representation also applies to the (irreducible components of the) variety $V = X^{Y_1}_{\mathbf{u}_1} \cap \cdots \cap X^{Y_k}_{\mathbf{u}_k}$. In §3.6 we saw that the problem of calculating products in cohomology can in principle be reduced to the problem of calculating intersections of (pairs or) triples of Schubert varieties $X^{Y_1}_{\mathbf{u}_1} \cap X^{Y_2}_{\mathbf{u}_2} \cap X^{Y_3}_{\mathbf{u}_3}$. 'In principle' means[6] 'providing that all necessary triple intersections are transverse'. To be more precise, we need to find all zero-dimensional triple intersections. By Theorem 4.1.4 (or by the Plücker embedding argument of §3.6), such an intersection either consists of a single point or is empty. To determine which is the case, we just need to know which cohomology classes are represented by which subpolyhedra of $\mu(Gr_k(\mathbb{C}^n))$. Let us consider two examples, $\mathbb{CP}^2$ and $Gr_2(\mathbb{C}^4)$.

Example 4.1.9 The cohomology ring $H^*(\mathbb{CP}^2)$ has additive generators in dimensions 0, 2, and 4; let us denote these respectively by 1, A, and B. They are dual to the fundamental homology classes of the Schubert varieties for a fixed Morse function f^Y. By varying Y in $\mathfrak{t}$, we arrive at the following representation of these cohomology classes on the triangular region $\mu(\mathbb{CP}^2)$:

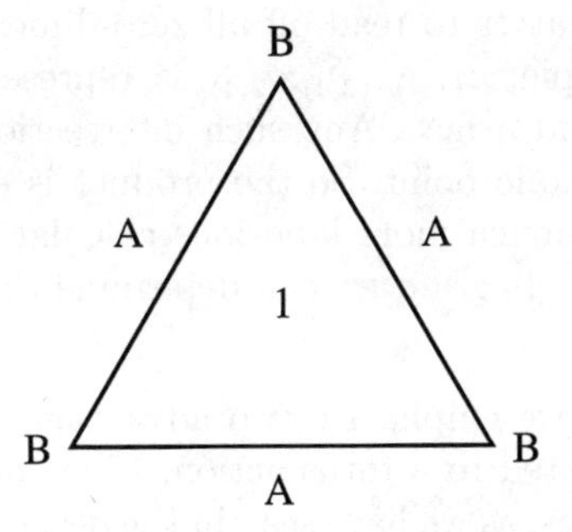

The structure of the cohomology ring $H^*(\mathbb{CP}^2)$ is determined completely by the product A^2, and from the diagram the intersection number of the dual Schubert varieties is 1. Hence $A^2 = 1B = B$, as expected. □

Example 4.1.10 The cohomology ring $H^*(Gr_2(\mathbb{C}^4))$ has the six additive generators described in §3.6. These are represented on the octahedron of Example 4.1.6 as follows:

$$\begin{aligned} z_{(3,4)} &\in H^0 : && \text{the octahedron} \\ z_{(2,4)} &\in H^2 : && \text{the half octahedra} \\ z_{(1,4)}, z_{(2,3)} &\in H^4 : && \text{alternate faces} \\ z_{(1,3)} &\in H^6 : && \text{the edges} \\ z_{(1,2)} &\in H^8 : && \text{the vertices.} \end{aligned}$$

In the diagram below, the (four) faces which represent the cohomology class $z_{(2,3)}$ are shaded.

[6] It can be shown that all intersections are indeed transverse, by using the fact that the stable manifolds are orbits of certain subgroups of $GL_n\mathbb{C}$; see [45]. This is not quite enough, for we must show in addition that there exist sufficiently many representatives of all cohomology classes. This will be obvious in the examples we consider, however.

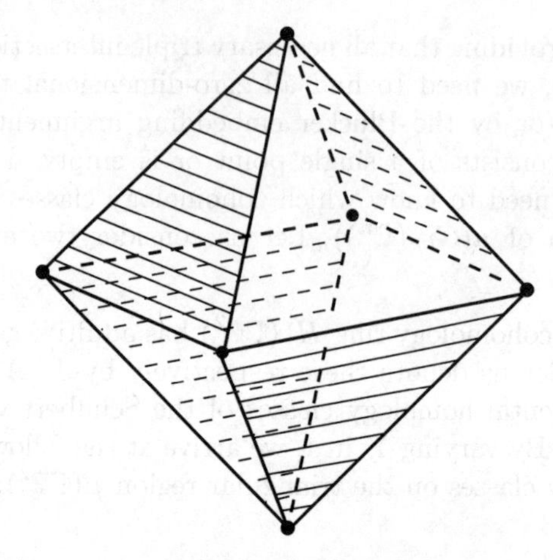

It is now a simple matter to read off all zero-dimensional double and triple intersections. For example, $z_{(1,4)}z_{(2,4)}z_{(2,4)}$ is represented by the intersection of two half-octahedra and a face. Any such intersection *giving a finite number of points* gives precisely one point. So the product is equal to the generator of H^8 – as we found by a much more laborious calculation in Example 3.6.4. By contemplating the above diagram we can determine the entire cohomology ring of $Gr_2(\mathbb{C}^4)$! □

To end this section, we emphasize two advantages of having a Morse–Bott function which is associated to a torus action. First, having a group action has computational advantages, as we have seen in the description of the gradient flow and the identification of the index (and nullity) of a critical point. Second, the torus action gives more than a single Morse–Bott function; it gives a whole family of related Morse–Bott functions, and this family can give more information than one of its members (as in the above calculation of the cohomology ring).

Additional comments (May, 2000): The theory of equivariant cohomology provides an algebraic explanation for the success of the above method of computing $H^*(M)$ from a (complete) torus action (see M. Goresky, R. Kottwitz, and R. MacPherson: Equivariant cohomology, Koszul duality, and the localization theorem, *Invent. Math.* 131 (1998), 25–83). The equivariant cohomology ring $H_T^*(M)$ (see §2.7) is in this case isomorphic as a module over $H^*(BT)$ to the tensor product $H^*(M) \otimes H^*(BT)$. The localization theorem for equivariant cohomology expresses products of equivariant cohomology classes in terms of the fixed point data of the torus action (this is another manifestation of Morse theory). As a result, it is possible to describe $H^*(M)$ explicitly as a quotient of a polynomial ring, in terms of this data.

4.2 The Witten complex

In view of the importance of the gradient flow lines of a Morse function, it is perhaps not surprising that the basic theorems of Morse theory may be developed entirely from this point of view. In fact, much stronger results (than the traditional 'Morse inequalities') are possible, as we shall see in this section and the next one.

In this section we consider the goal of computing the homology groups of a manifold M. Traditionally, this is possible only for a *perfect* Morse function. However, if we assume that $f : M \to \mathbb{R}$ is a Morse–Smale function (i.e. the stable and unstable manifolds of f intersect transversely, as in Definition 1.4.5), then the homology may be calculated whether f is perfect or not. This method became widely understood only in the 1980s, through the work of Witten and Floer (see [66] and [24]). It is easy to describe: one constructs a certain chain complex of abelian groups, the 'Witten complex', whose homology groups turn out to be the homology groups of M.

Let us assume first that M is oriented. The abelian groups are defined in terms of the critical points of the Morse–Smale function f by

$$C_i = \text{ free abelian group on the set of critical points of index } i.$$

(Since f is a Morse function and M is compact, the groups C_i have finite rank.) The boundary operators $\partial_i : C_i \to C_{i-1}$ are defined in terms of the gradient flow lines of $-\nabla f$ as follows. Let m be a critical point of f of index i (i.e. a generator of C_i). Then

$$\partial_i(m) = \sum_{\gamma} e(\gamma) m_\gamma,$$

where (1) the sum is over all flow lines γ such that

$$\lim_{t\to-\infty} \gamma(t) = m, \qquad \lim_{t\to\infty} \gamma(t) = \text{(a critical point) } m_\gamma \text{ of index } i-1,$$

and (2) $e(\gamma)$ is either 1 or -1, the choice depending on whether γ 'preserves or reverses orientation'.

Some explanation of (1) and (2) is necessary. First, since the Morse–Smale condition gives

$$\dim F(m, m_\gamma) = 1,$$

it follows that there are only finitely many such γ, hence the sum is finite. Second, to define the sign of $e(\gamma)$, we first choose arbitrary orientations of the unstable manifolds. Since M is oriented, and since the stable and unstable manifolds intersect transversely, we may then assign orientations to the stable manifolds in a consistent manner. The manifold $F(m, m_\gamma)$ itself thus acquires an orientation. We define $e(\gamma)$ to be 1 if the natural orientation of γ agrees with its orientation as a component of $F(m, m_\gamma)$; otherwise we define $e(\gamma)$ to be -1.

It can be shown that (C_*, ∂_*) is a chain complex, i.e. that $\partial_{i-1} \circ \partial_i = 0$ for all i. For this, and for the proof of the next theorem, we refer to section 2 of [6], where a detailed discussion can be found.

Theorem 4.2.1 *The homology groups of the chain complex* (C_*, ∂_*) *are isomorphic to the homology groups of* M. □

If M is orientable, as in the above definition, then homology groups with coefficients in $\mathbb{Z}$ are obtained. If M is not orientable, then some modifications to the definition are necessary. The simplest way to do this is to work over $\mathbb{Z}/2\mathbb{Z}$ instead of $\mathbb{Z}$, i.e. to replace C_i by $C_i \otimes \mathbb{Z}/2\mathbb{Z}$ and then to define $e(\gamma) = 1$ for all γ. In this case the theorem is true for homology groups with coefficients in $\mathbb{Z}/2\mathbb{Z}$.

It is possible to formulate and prove a similar theorem for the cohomology groups of M, using the dual chain complex (see [6]).

The Morse inequalities (Theorem 2.3.1) follow from the statement of the above theorem by a purely algebraic argument (using $\operatorname{rank} H_i(M) = \operatorname{rank} \operatorname{Ker}\partial_i - \operatorname{rank} \operatorname{Im}\partial_{i+1}$ and $\operatorname{rank} C_i = \operatorname{rank} \operatorname{Ker}\partial_i + \operatorname{rank} \operatorname{Im}\partial_i$). The 'lacunary principle', that f is necessarily perfect if all its critical points have even index, also follows immediately, since in this case we have $\partial_i = 0$ for all i.

Example 4.2.2 Consider the Morse function on the circle with three local maxima and three local minima depicted in Example 2.2.2.

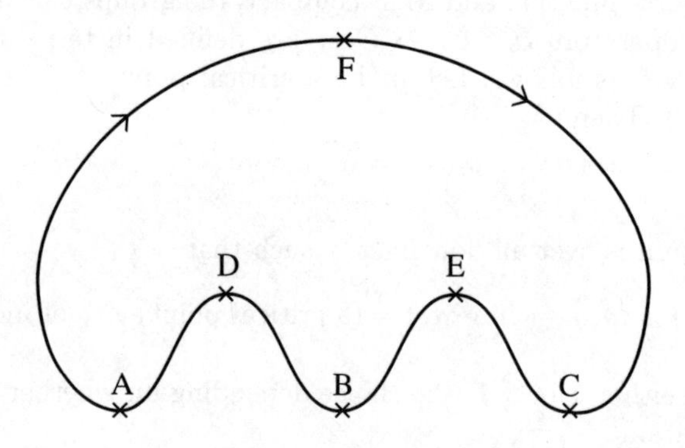

This is of course not a perfect Morse function. But the homology groups may be calculated by using the Witten complex. If we choose the 'clockwise' orientation on S^1 and on all stable manifolds, then the map $\partial_1 : C_1 \to C_0$ is given by:

$$\partial_1 F = -A + C, \quad \partial_1 D = A - B, \quad \partial_1 E = B - C.$$

The kernel and cokernel of ∂_1 are therefore both isomorphic to $\mathbb{Z}$. □

Example 4.2.3 Consider the Morse function $f : \mathbb{RP}^n \to \mathbb{R}$ defined by

$$f([x_0, \dots, x_n]) = \sum_{i=0}^{n} c_i |x_i|^2 / \sum_{i=0}^{n} |x_i|^2$$

with $c_0 > \cdots > c_n$, which we met in Example 2.3.6. Let us calculate the homology of $\mathbb{RP}^n$ by using the Witten complex. First we take coefficients in $\mathbb{Z}/2\mathbb{Z}$, to avoid the problem of dealing with orientations.

The critical points are the coordinate axes $V_0, \ldots, V_n$, and the indices are (respectively) $n, n-1, \ldots, 0$. Thus $C_i = \mathbb{Z}/2\mathbb{Z}\, V_{n-i}$ for $0 \leqslant i \leqslant n$. As in the case of $\mathbb{CP}^n$ in §3, we may identify the stable and unstable manifolds explicitly. In particular, we see that the space $F(V_i, V_{i+1})$ of points on flow lines from V_i to V_{i+1} is of the form

$$\{[0; \ldots; 0; *; *; 0; \ldots; 0] \in \mathbb{RP}^n \mid * \in \mathbb{R}\}.$$

In other words, it is a copy of $\mathbb{RP}^1 \cong S^1$, and so there are precisely two such flow lines. Thus, every homomorphism ∂_i is zero, and our Morse function is perfect.

With integer coefficients the situation is more complicated, particularly since we have not defined the Witten complex (over $\mathbb{Z}$) for a non-orientable manifold, and it is well known that $\mathbb{RP}^n$ is orientable only when n is odd. However, the Witten complex can in fact be defined for any manifold (see section 2.1 of [45]), and it turns out that for $\mathbb{RP}^n$ the maps ∂_i are given by

$$\partial_i(n) = \begin{cases} 2n & \text{if } i \text{ is even} \\ 0 & \text{if } i \text{ is odd.} \end{cases}$$

This gives the integral homology groups of $\mathbb{RP}^n$. □

Example 4.2.4 Any function associated to the standard torus action on a generalized flag manifold (see Example 4.1.7) has trivial Witten complex, since all critical points have even index. However, the real analogues of these complex manifolds (which include $\mathbb{RP}^n$ and more generally the real Grassmannians) give rise to non-trivial Witten complexes, and it may be expected that these complexes are determined by the same combinatorial information that describes the behaviour of the flow lines. This is indeed the case; full details may be found in [45]. Extensive earlier work on the Morse theory of these spaces can be found in [14], [61], [62], [26], [22]. The convexity results of [2], [38] were extended to these spaces in [21]. □

The above description of the Witten complex, phrased in the language of differential topology, is closely related to earlier work of Smale and Thom (see [27], [48], and [60]), in which the groups C_i appear as the relative homology groups of the pair (M_i, M_{i-1}), for a suitable filtration $\{M_i\}$ of M, and the maps ∂_i appear as the connecting homomorphisms in the homology exact sequence. (The space M_i is obtained by taking all cells of dimension less than or equal to i in the usual Morse decomposition of M.) Witten's original motivation was actually quite different, as it arose from quantum theory. A brief description of Witten's point of view, together with further historical information, can be found in [13].

Finally we mention that an extension of the theory to the case of Morse–Bott functions is given in [6]. Another reference where full details of the material of this section can be found is the book [58].

4.3 Morse theory as a topological field theory

In §4.1 we have seen that it can be useful to study *families* of Morse functions on a given compact manifold M. This can be taken as motivation for the 'field-theoretic' approach to Morse theory of [8], [9], [17], [18], [28, 29], so called because it is based on similar constructions in gauge theory. As such constructions tend to involve rather elaborate preparations, we shall just give an informal description here.

The basic ingredient is a certain 'moduli space' $\mathcal{M}(\Gamma)$, which is a device for counting configurations of flow lines. The definition of this space depends on M and on a choice of an oriented connected graph Γ. We assume that Γ has n_1 edges parametrized by $(-\infty, 0]$ ('incoming edges'), n_2 edges parametrized by $[0,1]$ ('internal edges'), and n_3 edges parametrized by $[0,\infty)$ ('outgoing edges'). In the example below, we have $n_1 = 2$, $n_2 = 1$, and $n_3 = 3$.

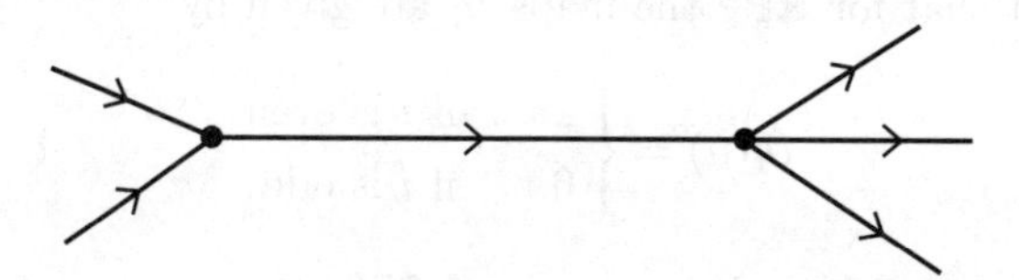

An element of $\mathcal{M}(\Gamma)$ is a 'configuration of flow lines of Morse–Smale functions on M, modelled on Γ', i.e. a continuous map $\mathcal{F} : \Gamma \to M$ such that, on each edge of Γ, $\mathcal{F}$ is (part of) a flow line of a Morse–Smale function on M. Thus, for the graph illustrated above, $\mathcal{F}(\Gamma)$ might look like the diagram below, where the white dots are critical points approached by the incoming and outgoing flow lines.

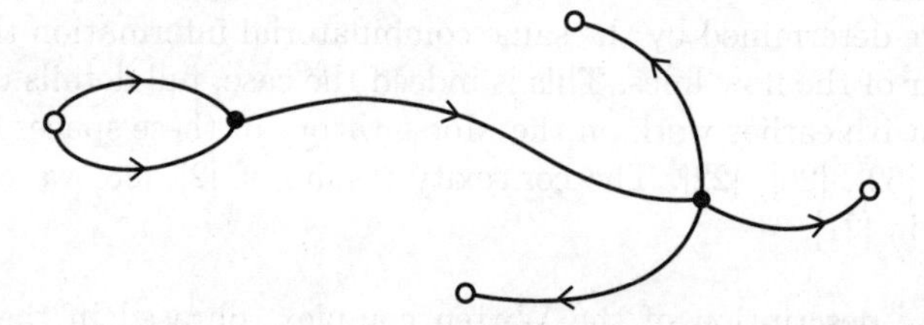

Let $f_1^{\mathcal{F}}, \dots, f_{n_1+n_2+n_3}^{\mathcal{F}}$ be the Morse–Smale functions in the definition of $\mathcal{F}$, and let $a_1^{\mathcal{F}}, \dots, a_{n_1}^{\mathcal{F}}, a_{n_1+n_2+1}^{\mathcal{F}}, \dots, a_{n_1+n_2+n_3}^{\mathcal{F}}$ be the critical points which are approached by the incoming and outgoing flow lines in that definition (the white dots in the diagram). For any $(n_1 + n_2 + n_3)$-tuple of functions $g = (g_1, \dots, g_{n_1+n_2+n_3})$ on M, we define

$$\mathcal{M}_g(\Gamma) = \{\mathcal{F} \in \mathcal{M}(\Gamma) \mid f_i^{\mathcal{F}} = g_i\}.$$

For any $(n_1 + n_3)$-tuple of points $b = (b_1, \dots, b_{n_1}, b_{n_1+n_3+1}, \dots, b_{n_1+n_2+n_3})$ of M, we define

$$\mathcal{M}_g(\Gamma; b) = \{\mathcal{F} \in \mathcal{M}_g(\Gamma) \mid a_i^{\mathcal{F}} = b_i\}.$$

(These definitions are informal versions of the precise definitions in [9].)

We shall assume from now on that all stable and unstable manifolds of $f_1^{\mathcal{F}}, \ldots, f_{n_1+n_2+n_3}^{\mathcal{F}}$ *intersect transversely;* in particular all these functions are Morse–Smale functions. Under this assumption, it can be shown that $\mathcal{M}_g(\Gamma; b)$ is a smooth manifold, that $\mathcal{M}_g(\Gamma; b)$ is oriented if M is oriented, and that

$$\dim \mathcal{M}_g(\Gamma; b) = \sum_{i=1}^{n_1} \operatorname{index} b_i - \sum_{i=1}^{n_3} \operatorname{index} b_{n_1+n_2+i} - (\dim M)(\dim H_1(\Gamma; \mathbb{R}) + n_1 - 1).$$

Example 4.3.1 Let Γ be the graph below with $n_1 = n_3 = 1$, $n_2 = 0$:

Let $g = (g_1, g_2)$, $b = (b_1, b_2)$. Then the points of $\mathcal{M}_g(\Gamma; b)$ are in one-to-one correspondence with the points of $U_{b_1}^{g_1} \cap S_{b_2}^{g_2}$, where $U_{b_i}^{g_i}$ is the unstable manifold of the critical point b_i of g_i, and $S_{b_i}^{g_i}$ is the stable manifold. From the transversality assumption we have (see §1.4)

$$\operatorname{codim} U_{b_1}^{g_1} \cap S_{b_2}^{g_2} = (\dim M - \operatorname{index} b_1) + (\operatorname{index} b_2).$$

This checks with the general formula above. □

Example 4.3.2 Let Γ be the graph below with $n_1 = 2$, $n_2 = 0$, and $n_3 = 1$:

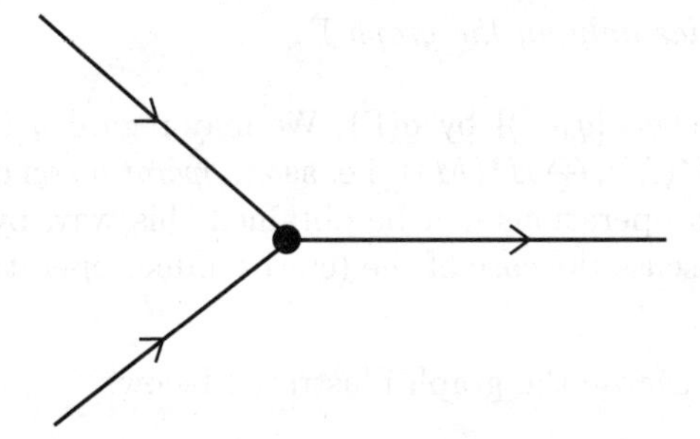

Let $g = (g_1, g_2, g_3)$, $b = (b_1, b_2, b_3)$. In this situation the points of $\mathcal{M}_g(\Gamma; b)$ correspond to points of $U_{b_1}^{g_1} \cap U_{b_2}^{g_2} \cap S_{b_3}^{g_3}$. Transversality implies that

$$\operatorname{codim} U_{b_1}^{g_1} \cap U_{b_2}^{g_2} \cap S_{b_3}^{g_3} = (\dim M - \operatorname{index} b_1) + (\dim M - \operatorname{index} b_2) + (\operatorname{index} b_3).$$

Again this is consistent with the general formula. □

We come now to the main part of the construction, which will associate to each graph Γ a topological invariant of M. This will make use of the Witten complex $(C_*(f), \partial_*)$ of a (Morse–Smale) function f, which was defined in the last section.

For a fixed choice of g (as above), we define

$$q_g(\Gamma) = \sum_{\mathcal{F},b} e(\mathcal{F})\, b_1 \otimes \cdots \otimes b_{n_1} \otimes b_{n_1+n_2+1} \otimes \cdots \otimes b_{n_1+n_2+n_3},$$

where the sum is over all $\mathcal{F}, b$ such that $\mathcal{F} \in \mathcal{M}_g(\Gamma; b)$ and $\dim \mathcal{M}_g(\Gamma; b) = 0$. If M is oriented, so is the zero-dimensional manifold $\mathcal{M}_g(\Gamma; b)$, and $e(\mathcal{F})$ is plus or minus one, according to the orientation of $\mathcal{F}$ as a point of $\mathcal{M}_g(\Gamma; b)$.

Thus, if M is oriented, we may regard $q_g(\Gamma)$ as an element of

$$\left(\bigotimes_{i=1}^{n_1} C_*(g_i) \right) \otimes \left(\bigotimes_{i=n_1+n_2+1}^{n_1+n_2+n_3} C^*(g_i) \right),$$

where $C^*(g_i)$ is the dual complex to $C_*(g_i)$. If M is not oriented, then the definition of $e(\mathcal{F})$ must be modified in the same way as the Witten complex.

The construction of $q_g(\Gamma)$ depends on the choice of the $(n_1+n_2+n_3)$-tuple of Morse–Smale functions g, of course. Nevertheless, it turns out that this choice, and all other choices necessary for a rigorous definition of $q_g(\Gamma)$, are irrelevant:

Theorem 4.3.3 ([8, 9, 28]) *The element $q_g(\Gamma)$ is annihilated by the (appropriate extension of) ∂_*, and so we obtain a class*

$$[q_g(\Gamma)] \in \left(\bigotimes_{i=1}^{n_1} H_*(M) \right) \otimes \left(\bigotimes_{i=n_1+n_2+1}^{n_1+n_2+n_3} H^*(M) \right).$$

This element depends only on the graph Γ. □

We denote the class $[q_g(\Gamma)]$ by $q(\Gamma)$. We may regard $q(\Gamma)$ as an element of the ring $\mathrm{Hom}(\bigoplus H^*(M), \bigoplus H^*(M))$, i.e. as an *operation* on cohomology classes. Various well-known operations can be obtained this way, by choosing suitable graphs. We shall discuss the case of the (cup) product operation in the following example.

Example 4.3.4 We choose the graph illustrated below:

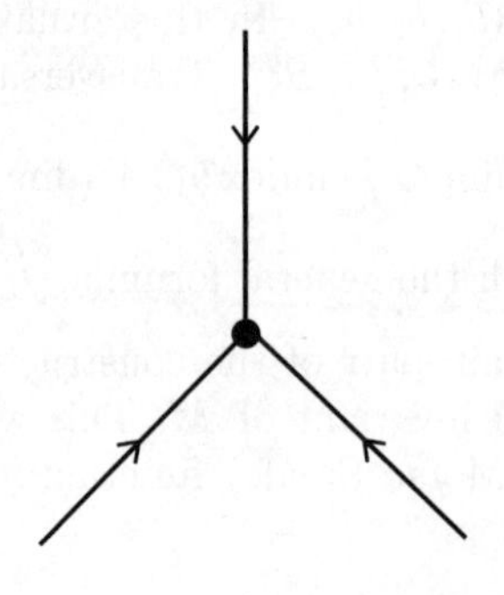

If $\mathcal{M}_g(\Gamma; b)(\cong U_{b_1}^{g_1} \cap U_{b_2}^{g_2} \cap U_{b_3}^{g_3})$ is non-empty and zero-dimensional, then by transversality we must have $\sum_{i=1}^{3}(\dim M -, b_i) = \dim M$. We obtain

$$q(\Gamma) \in \bigoplus_{i_1+i_2+i_3=\dim M} H_{\dim M - i_1}(M) \otimes H_{\dim M - i_2}(M) \otimes H_{\dim M - i_3}(M),$$

and hence also

$$q(\Gamma) \in \bigoplus_{i_1+i_2+i_3=\dim M} H^{i_1}(M)^* \otimes H^{i_2}(M)^* \otimes H^{i_3}(M)^*.$$

From our calculations of the triple product operation of $H^*(Gr_k(\mathbb{C}^n))$ in §3.6 and the end of §4.1, it is clear that $q(\Gamma)$ must be precisely that operation. This works for any orientable manifold M, because the main ingredient used in our calculation for $Gr_k(\mathbb{C}^n)$ was the existence of three functions, all of whose stable and unstable manifolds intersect transversely. In the general case, however, it is not easy to identify such 'generic' functions explicitly. The special feature of $Gr_k(\mathbb{C}^n)$, and indeed of any manifold with a complete torus action (as defined before Example 4.1.7), is that one only needs generic elements of the Lie algebra of the torus, and the existence of these is guaranteed by the convexity theorem. □

The product operation has been described by other authors in terms of the Witten complex – see [6] and [65].

In all our examples so far we had $n_2 = 0$, and the space $\mathcal{M}_g(\Gamma; b)$ was identified with a subspace of M itself. In general, $\mathcal{M}_g(\Gamma; b)$ may be identified with a subspace of the $(n_2 + 1)$-fold product $M \times \cdots \times M$.

The theory described in this section makes use only of those $\mathcal{M}_g(\Gamma; b)$ which are zero-dimensional. In [17], [18] the higher dimensional spaces are used to construct a much more complicated algebraic object, which gives correspondingly more topological information.

4.4 Origins and other directions

The study of critical points of functions on infinite-dimensional spaces (e.g. on function spaces – the calculus of variations) has been a guiding principle right from the beginning of Morse theory. Rather surprisingly, perhaps, the development of almost all of Morse theory has been prompted by infinite-dimensional examples! Since the infinite-dimensional theory is much more complicated, it is usually presented as a generalization of traditional Morse theory, and we shall continue this tradition by giving only a brief list of examples, almost as an afterthought. Nevertheless, it is very likely that future directions in Morse theory will be strongly influenced by examples like these.

One of the earliest examples was the study of geodesics as critical points of the length or energy functional on the space of paths on a Riemannian manifold. Morse's idea of using 'broken geodesics' to reduce the problem to a finite-dimensional problem is described in detail in [47]. Subsequently, a general theory of Morse functions on Hilbert manifolds was developed by Palais and Smale (see [51], [52]), under the assumption of 'Condition (C)'. This condition, a substitute for compactness, is satisfied in the case of the geodesic problem, but unfortunately not in the case of many other important examples. For example, it is not satisfied in the case of the energy functional on the space of maps from a Riemann surface into a Riemannian manifold. In this case the critical points are the harmonic maps, which are closely related to minimal surfaces.

For the Yang–Mills functional on the space of connections over a Riemann surface, a substitute for compactness was found by Atiyah and Bott [4]. This led to new developments in finite-dimensional Morse theory, namely for the functional $|\mu|^2$ where μ is the moment map for the action of a (not necessarily abelian) Lie group on a manifold M. In [42], Kirwan showed that a version of Morse theory holds for this function, even though it is not a Morse–Bott function.

A general approach to Morse theory for the Yang–Mills functional (and other functionals, such as the energy functional on maps of Riemann surfaces) has been given by Taubes (see [63] and also the survey article [64] of Uhlenbeck).

All these examples focus on the critical points of a functional, but (as we have seen) it is possible to take a different point of view by focusing on the flow lines. Floer's idea of restricting attention to certain well-behaved flow lines is such a case, and this was also motivated by an infinite-dimensional example – the 'area functional' on closed paths in a symplectic manifold X. The flow lines in this case have particular geometrical significance: they may be regarded as 'holomorphic curves' in X, where X is given an almost complex structure compatible with its symplectic structure. Such holomorphic curves arose in earlier work of Gromov, and the homology theory which is computed by the Witten complex in this situation is known as Gromov–Floer theory. (It is also closely related to quantum cohomology theory.)

Floer applied his theory to an apparently quite different example, the Chern–Simons functional on the space of connections on certain three-dimensional manifolds. Again the flow lines have a geometrical meaning: they are Yang–Mills instantons. The homology theory arising here is called Floer homology.

Acknowledgements. I would like to thank John Bolton and Lyndon Woodward for their invitation to the Graduate School at the University of Durham in 1996, and Simon Salamon for encouraging me to contribute the lecture notes to this book. I also wish to acknowledge financial support from the US National Science Foundation, as well as the financial support and extraordinary hospitality of Tokyo Metropolitan University.

My own Morse-theoretical education began at Oxford with Brian Steer in the 1970s, and I learned a lot from my fellow students Elias Micha, Andrew Pressley, Simon Salamon, Pepe Seade, and Socorro Soberon. Later on I benefited greatly from discussions about Morse theory with Haynes Miller and Bill Richter. Most of all, however, I have Brian to thank for suggesting Morse theory as a suitable direction of study. At that time I sometimes worried that Morse theory was 'too easy' a subject for serious mathematical research, and only much later did I understand the (often repeated, but usually ignored) advice that the simplest ideas are the best ones. Even then, with the applications in gauge theory still on the horizon, Brian predicted that in due course there would be tremendous developments in the subject. He was right, and for his guidance and encouragement I dedicate these notes to him, with gratitude.

References

1. M.F. Atiyah (ed.): 'Representation Theory of Lie Groups', Lond. Math. Soc. Lecture Notes 34, Cambridge Univ. Press (1979).
2. M.F. Atiyah: Convexity and commuting Hamiltonians, *Bull. Lond. Math. Soc.* **16** (1982), 1–15.
3. M.F. Atiyah: Topological quantum field theory, *Publ. Math. IHES* **68** (1989), 175–186.
4. M.F. Atiyah and R. Bott: The Yang–Mills equations over Riemann surfaces, *Phil. Trans. R. Soc. Lond.* **A308** (1982), 523–615.
5. M. Audin : 'The Topology of Torus Actions on Symplectic Manifolds', Birkhäuser (1991).
6. D.M. Austin and P.J. Braam: Morse–Bott theory and equivariant cohomology, in 'The Floer Memorial Volume' (eds:) H. Hofer, C.H. Taubes, A. Weinstein, and E. Zehnder, Progress in Math. 133, Birkhäuser (1995), pp. 123–183.
7. I.N. Bernstein, I.M. Gelfand and S.I. Gelfand: Schubert cells and cohomology of the spaces G/P, *Russian Math. Surveys* **28** (1973), 1–26.
8. M. Betz: Categorical constructions in Morse theory and cohomology operations, Ph.D. thesis, Stanford University, 1993.
9. M. Betz and R.L. Cohen: Graph moduli spaces and cohomology operations, *Turkish J. of Math.* **18** (1994), 23–41.
10. R. Bott: Nondegenerate critical manifolds, *Annals of Math.* **60** (1954), 248–261.
11. R. Bott: An application of the Morse theory to the topology of Lie groups, *Bull. Soc. Math. France* **84** (1956), 251–281.
12. R. Bott: Lectures on Morse theory, old and new, *Bull. Amer. Math. Soc.* **7** (1982), 331–358.
13. R. Bott: Morse theory indomitable, *Publ. Math. IHES* **68**, (1989), 99–114.
14. R. Bott and H. Samelson: An application of the theory of Morse to symmetric spaces, *Amer. J. Math.* **80** (1958), 964–1029.
15. R. Bott and L.W. Tu: 'Differential Forms in Algebraic Topology', GTM 82, Springer (1982).
16. J.B. Carrell and A. Kurth: Normality of torus orbits in G/P, preprint.
17. R.L. Cohen, J.D.S. Jones, and G.B. Segal: Morse theory and classifying spaces, preprint, 1991.
18. R.L. Cohen, J.D.S. Jones, and G.B. Segal: Floer's infinite dimensional Morse theory and homotopy theory, in 'The Floer Memorial Volume' (eds:) H. Hofer, C.H. Taubes, A. Weinstein, and E. Zehnder, Progress in Math. 133, Birkhäuser (1995), pp. 297–325 (also published in Surikaisekikenkyujo Kokyuroku 883, Kyoto, 1994, 68–96).

19. A. Dabrowski: On normality of the closure of a generic torus orbit in G/P, *Pacific J. Math.* **172** (1996), 321–330.
20. T. Delzant: Hamiltoniens périodiques et image convex de l'application moment, *Bull. Soc. Math. France* **116** (1988), 315–339.
21. J.J. Duistermaat: Convexity and tightness for restrictions of Hamiltonian functions to fixed point sets of an anti-symplectic involution, *Trans. Amer. Math. Soc.* **275** (1983), 417–429.
22. J.J. Duistermaat, J.A.C. Kolk, and V.S. Varadarajan: Functions, flows and oscillatory integrals on flag manifolds and conjugacy classes in real semisimple Lie groups, *Comp. Math.* **49** (1983), 309–398.
23. H. Flaschka and L. Haine: Torus orbits in G/P, *Pacific J. Math.* **149** (1991), 251–292.
24. A. Floer: Witten's complex and infinite dimensional Morse theory, *J. Diff. Geom.* **30** (1989), 207–221.
25. T. Frankel: Fixed points and torsion on Kähler manifolds, *Annals of Math.* **70** (1959), 1–8.
26. T. Frankel: Critical submanifolds of the classical groups and Stiefel manifolds, in 'Differential and Combinatorial Topology' (ed:) S.S. Cairns, Princeton University Press (1965), pp. 37–53.
27. J. Franks: Morse–Smale flows and homotopy theory, *Topology* **18** (1979), 199–215.
28. K. Fukaya: Topological field theory and Morse theory, *Sugaku Expositions* (translation from Suugaku, 46 (1994), 289–307) **10** (1997), 19–39.
29. K. Fukaya (notes by P. Seidel): Floer homology, A_∞-categories and topological field theory, in 'Geometry and Physics' (eds:) J.E. Anderson, J. Dupont, H. Pederson, A. Swann, Lecture Notes in Pure and Applied Math. 184, Marcel Dekker (1997), pp. 9–32.
30. W. Fulton: 'Young Tableaux', Cambridge University Press (1997).
31. W. Fulton: 'Introduction to Toric Varieties', Annals of Math. Stud. 131, Princeton University Press (1993).
32. I.M. Gelfand, M. Goresky, R.D. MacPherson and V.V. Serganova: Combinatorial geometries, convex polyhedra, and Schubert cells, *Adv. Math.* **63** (1987), 301–316.
33. I.M. Gelfand and R.D. MacPherson: Geometry in Grassmannians and a generalization of the dilogarithm, *Adv. Math.* **44** (1982), 279–312.
34. I.M. Gelfand and V.V. Serganova: Combinatorial geometries and torus strata on homogeneous compact manifolds, *Usp. Mat. Nauk.* **42**, (1987), 107–133 (Russian Math. Surveys 42 (1987), 133–168).
35. M. Goresky and R. MacPherson: 'Stratified Morse Theory', Springer (1988).
36. P. Griffiths and J. Harris: 'Principles of Algebraic Geometry', Wiley (1978).
37. M.A. Guest and Y. Ohnita: Group actions and deformations for harmonic maps, *J. Math. Soc. Japan.* **45** (1993), 671–704.
38. V. Guillemin and S. Sternberg: Convexity properties of the moment mapping, *Invent. Math.* **67** (1982), 491–513.
39. H. Hiller: 'Geometry of Coxeter Groups', Research Notes in Math. 54, Pitman, (1982).
40. M.W. Hirsch: 'Differential Topology', GTM 33, Springer (1976).
41. D.W. Kahn: 'Introduction to Global Analysis', Academic Press (1980).
42. F.C. Kirwan: 'Cohomology of Quotients in Symplectic and Algebraic Geometry', Math. Notes 31, Princeton University Press (1986).
43. S.L. Kleiman and D. Laksov: Schubert calculus, *Amer. Math. Monthly* **79** (1972), 1061–1082.

44. S. Kobayashi: Fixed points of isometries, *Nagoya Math. J.* **13** (1958), 63–68.
45. R.R. Kocherlakota: Integral homology of real flag manifolds and loop spaces of symmetric spaces, *Advances in Math.* **110** (1995), 1–46.
46. W. Liu, Ph.D. thesis, University of Rochester, 1998.
47. J. Milnor: 'Morse Theory', Annals of Math. Studies 51, Princeton University Press (1963).
48. J. Milnor: 'Lectures on the h-cobordism Theorem', Math. Notes 1, Princeton University Press (1965).
49. J. Milnor and J.D. Stasheff: 'Characteristic Classes', Annals of Math. Studies 76, Princeton University Press (1974).
50. T. Oda: 'Convex Bodies and Algebraic Geometry: An Introduction to the Theory of Toric Varieties', Springer (1988).
51. R.S. Palais: Morse theory on Hilbert manifolds, *Topology* **2** (1963), 299–340.
52. R.S. Palais: 'Foundations of Global Non-linear Analysis', Benjamin (1968).
53. R.S. Palais and C.-L. Terng: 'Critical Point Theory and Submanifold Geometry', LNM. 1353, Springer (1988).
54. J. Palis and W. de Melo: 'Geometric Theory of Dynamical Systems', Springer (1982).
55. G.D. Parker: Morse theory on Kähler homogeneous spaces, *Proc. Amer. Math. Soc.* **34** (1972), 586–590.
56. E. Pitcher: Inequalities of critical point theory, *Bull. Amer. Math. Soc.* **64** (1958), 1–30.
57. A.N. Pressley: The energy flow on the loop space of a compact Lie group, *J. Lond. Math. Soc.* **26** (1982), 557–566.
58. M. Schwarz: 'Morse Homology', Progress in Math. 111, Birkhäuser (1993).
59. G. Segal: An introduction to the paper Schubert cells and cohomology of the spaces G/P, in 'Representation Theory', Lond. Math. Soc. Lecture Notes 69 (eds:) I.M. Gelfand *et al.*, Cambridge University Press (1982), pp. 111–114.
60. S. Smale: Morse inequalities for dynamical systems, *Bull. Amer. Math. Soc.* **66** (1960), 43–49.
61. M. Takeuchi: Cell decompositions and Morse inequalities on certain symmetric spaces, *J. Fac. Sci. Univ. Tokyo* **12** (1965), 81–192.
62. M. Takeuchi and S. Kobayashi: Minimal embeddings of R-spaces, *J. Diff. Geom.* **2** (1968), 203–215.
63. C.H. Taubes: A framework for Morse theory for the Yang–Mills functional, *Invent. Math.* **94** (1988), 327–402.
64. K. Uhlenbeck: Applications of non-linear analysis in topology, in 'Proc. Int. Congress of Math.' Math. Soc. Japan/Springer (1991), pp. 261–279.
65. C. Viterbo: The cup-product on the Thom–Smale–Witten complex, and Floer cohomology, in 'The Floer Memorial Volume' (eds:) H. Hofer, C.H. Taubes, A. Weinstein, and E. Zehnder, Progress in Math. 133, Birkhäuser (1995), pp. 609–625.
66. E. Witten: Supersymmetry and Morse theory, *J. Diff. Geom.* **17** (1982), 661–692.

6
The Dirac operator

NIGEL HITCHIN

1 Introduction

Take a trip to London and go to Westminster Abbey. Enter by the main door, follow your way round past the clusters of tourists, the tombs of kings and queens and long-forgotten bastions of society, out into the cloisters, back into the nave, cross over past the unknown soldier to reach the backwater by the screen which is where science holds sway. There, nestling by Newton's tomb, you can just make out, if the light is right, a diamond shaped memorial stone set in the floor:

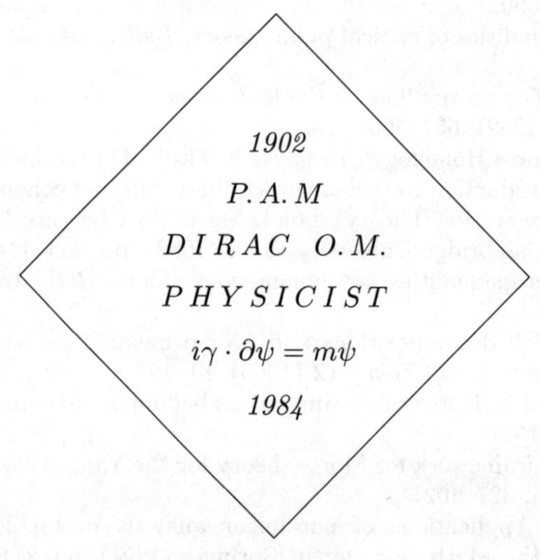

Look around and you will find no more formulae. Nothing could underline in quite the same way the importance of an equation. Yet it is not only in quantum theory that this equation appears, and indeed we can go back years before Dirac was born to find something similar. Perhaps the most appropriate place is in Miss E. Watson's notes of Clifford's *Lectures on Quaternions*, delivered in Michaelmas Term 1877 at University College, London [4]. Here we read:

> We can only give a short sketch of a part of the theory which little was done by Hamilton, but which has been worked out by Tait; the part that is which treats the operator
>
> $$\nabla = i\partial_x + j\partial_y + k\partial_z$$

Laplace's operator

$$-(\partial_x^2 + \partial_y^2 + \partial_z^2) = \nabla^2$$

is the square of

$$i\partial_x + j\partial_y + k\partial_z = \nabla$$

Hamilton laid great stress on his having thus discovered the square root of this operator, which is of importance in many physical investigations. Its properties come out in a remarkable way from the properties of ∇.

The Dirac equation has a respectable pedigree to be sure, but it has over the past 30 years or so assumed also a position of fundamental importance in the global study of manifolds, and in particular a key role in the interactions between mathematics and physics which have been so important to the development of both. My own involvement dates back to my research student days, when Brian Steer was my supervisor. He suggested, with considerable foresight, studying the null-space of the Dirac operator on a general spin manifold and seeing to what extent one could draw topological or differential geometric conclusions from studying its dimension. My thesis ultimately appeared as [10], and those original questions have been carried forward by various authors [15], [3], [14]. What I want to do here, though, after describing some general aspects of the Dirac operator, is to return to low dimensions – 1, 2, 3, and 4 – and try and convince you that the Dirac operator really does lie at the heart of a lot of mathematics, in fact most of the topics I have been associated with in the 30 years since I was a research student.

Exercises 1.1 (i) Find the Dirac memorial in Westminster Abbey (this is a non-trivial exercise especially when the Abbey has 16,000 visitors in a day as it sometimes does).

(ii) Did University College London admit women students in 1877? Did Oxford or Cambridge?

2 Clifford algebras

Having introduced W. K. Clifford, we ought to follow up on how he generalized the quaternions and how the Dirac operator in higher dimensions can be produced. Suppose we have matrices $e_1, \ldots, e_n$ such that

$$e_i^2 = -1, \qquad e_i e_j = -e_j e_i; \tag{2.1}$$

then writing

$$D = \sum_1^n e_j \frac{\partial}{\partial x_j}$$

we have an operator such that

$$D^2 = -\sum_1^n \left(\frac{\partial}{\partial x_j}\right)^2.$$

Can such matrices be found? We start first with a real vector space V with an inner product, and define an algebra $C(V)$, the *Clifford algebra*, which contains V and in which the relations (2.1) hold automatically for an orthonormal basis of V. (You might want to look for example at Lawson and Michelson [15] for more details of what follows.) This is done by fiat, taking the tensor algebra on V and quotienting by the ideal generated by the relations $v \otimes v + (v, v)1$. If we did the same for the relation $v \otimes v$ we would get the exterior algebra, and in just the same way we find a basis for $C(V)$ by taking an orthonormal basis $e_1, \ldots, e_n$ for V and forming the vectors

$$e_{i_1} e_{i_2} \ldots e_{i_k} \qquad (i_1 < i_2 < \cdots < i_k).$$

This means that $C(V)$ has finite dimension 2^n. Consider some examples:

1. When $n = 1$ we have $1, e_1$ and $e_1^2 = -1$ so $C(\mathbb{R}) = \mathbb{C}$, the complex numbers.
2. When $n = 2$ we have $1, e_1, e_2, e_1e_2$ and if we put $e_1 = i, e_2 = j, e_1e_2 = k$, then $C(\mathbb{R}^2)$ is the 4-dimensional algebra of quaternions $\mathbb{H}$, because $k^2 = e_1e_2e_1e_2 = -e_1^2e_2^2 = -1$.

On the other hand this last example is not where the quaternions appear in Clifford's 3-dimensional operator. We must consider $C(V)$ more closely to see what structure it possesses. There is an algebra involution α which is induced by the map $\alpha(v) = -v$ on V. The ± 1 eigenspaces of this are the odd and even parts of the Clifford algebra. Now take an orthonormal basis and write $\omega = e_1e_2 \ldots e_n$. Then from the basic relations (2.1)

$$\omega e_i = (-1)^{n-1} e_i \omega \qquad \omega^2 = (-1)^{n(n+1)/2}. \tag{2.2}$$

In particular, if n is odd then ω commutes with everything since it commutes with the generators, and if $n(n+1)/2$ even (i.e. $n = 4k+3$), it has eigenvalues ± 1, so we have eigenspaces $I_\pm$ which are ideals, and α interchanges them. The Clifford algebra is thus the direct sum of two isomorphic algebras. Let us look again at $n = 3$. In this case $\omega e_3 = -e_1e_2$ so $e_3 - e_1e_2 = (1+\omega)e_3$, hence *modulo* I_+, in which $(1+\omega)e_3$ lies, we have $e_1e_2 = e_3$ as well as our usual relations $e_1^1 = e_2^2 = e_3^2 = -1$. Thus the 3-dimensional Dirac operator $e_1\partial_x + e_2\partial_y + e_3\partial_z$ becomes Clifford's operator ∇ so long as we project onto one of the ideals.

The general picture is provided by the famous 8-fold periodicity

$$C(\mathbb{R}^{n+8}) \cong C(\mathbb{R}^n) \otimes C(\mathbb{R}^8)$$

so if we know the first eight Clifford algebras we know them all. This is the table, where $\mathbb{K}(n)$ denotes the algebra of $n \times n$ matrices over $\mathbb{K}$:

0	1	2	3	4	5	6	7	8
$\mathbb{R}$	$\mathbb{C}$	$\mathbb{H}$	$\mathbb{H} \oplus \mathbb{H}$	$\mathbb{H}(2)$	$\mathbb{C}(4)$	$\mathbb{R}(8)$	$\mathbb{R}(8) \oplus \mathbb{R}(8)$	$\mathbb{R}(16)$

What this table tells us is that (if we are prepared to project in dimension $4k+3$) we can realize $e_1, \ldots, e_n$ as matrices with entries in $\mathbb{R}, \mathbb{C}$, or $\mathbb{H}$ and so the operator D has something to act on: functions with values in the appropriate vector space S, the space of *spinors.*

If n is even, then from (2.2) we get

$$\omega e_i = -e_i \omega \qquad \omega^2 = (-1)^{n/2}$$

so when $n = 4k$ we can break up S into ± 1 eigenspaces of ω, denoted S_+ and S_-. Now

$$\omega D\psi = \omega \sum_1^n e_j \frac{\partial \psi}{\partial x_j} = -\sum_1^n e_j \frac{\partial \omega\psi}{\partial x_j} = -D(\omega\psi)$$

and then we have

$$D : C^\infty(S_+) \to C^\infty(S_-). \tag{2.3}$$

The vector space S is not quite canonically defined by the original orthogonal vector space V. Its symmetries are naturally defined by the Clifford algebra, but its elements not so, unlike tensors. It's a bit like the uncertainty we are faced with when we first meet complex numbers – spinors have an exotic feel to them. They also have a lot of internal structure, which is reflected in the fact that their algebra of symmetries, the Clifford algebra, does. It is not just an algebra of matrices, but has the filtered structure provided by considering products of m-tuples of elements of V. In practice, and especially in higher dimensions, it is easier for a mathematician working with spinors to simply move objects in S around with elements of the Clifford algebra rather than to lose sleep worrying about what those elements really are.

One way of moving them around is by using group elements in the connected double covering $Spin(n)$ of the special orthogonal group $SO(n)$. This sits inside the Clifford algebra by exponentiating elements of the form

$$\sum_{i,j} a_{ij} e_i e_j$$

where a_{ij} is a skew-symmetric matrix. To see this, use the basic relation $e_ie_j + e_je_i = -2\delta_{ij}$, and perform the calculation

$$\begin{aligned}\left(\sum_{i,j} a_{ij}e_ie_j\right) e_k &= -\left(\sum_{i,j} a_{ij}e_ie_ke_j\right) - 2\sum_i a_{ik}e_i \\ &= e_k\left(\sum_{i,j} a_{ij}e_ie_j\right) + 2\sum_j a_{kj}e_j - 2\sum_i a_{ik}e_i \\ &= e_k\left(\sum_{i,j} a_{ij}e_ie_j\right) - 4\sum_j a_{ik}e_i.\end{aligned}$$

It follows that if

$$a = -\sum_{i,j} a_{ij}e_ie_j/4 \tag{2.4}$$

then for $v \in V$,

$$e^{-a}ve^a = \exp(A)v$$

where $\exp(A)$ is the matrix in $SO(n)$ obtained by exponentiating the skew symmetric matrix $A = a_{ij}$. The map $\exp a \mapsto \exp A$ defines the covering homomorphism from $Spin(n)$ to $SO(n)$, with kernel ± 1.

The existence of the vector space S means that we have a natural representation space of $Spin(n)$ – the spin representation – in dimensions given by the table. Because $Spin(n)$ is compact, it always acts through a unitary, orthogonal or quaternionic unitary group, depending on the structure of the Clifford algebra in each dimension. Moreover, from (2.2), ω commutes with even products of e_i so in particular with the elements of $Spin(n)$. Hence S_+ and S_- are separately representation spaces for $Spin(n)$ when $n = 4k$.

Exercises 2.5 (i) Show that $Spin(3)$ is isomorphic under the spin representation with the group of unit quaternions $Sp(1) = \{\mathbf{q} \in \mathbb{H}: \mathbf{q}\bar{\mathbf{q}} = 1\}$.

(ii) Show by looking at S_+ and S_- that $Spin(4) \cong Sp(1) \times Sp(1)$.

(iii) Note that in eight dimensions, S_+ and S_- are real 8-dimensional vector spaces, so each spin representation describes a homomorphism from $Spin(8)$ to $SO(8)$. Since $Spin(8)$ is simply connected these lift to homomorphisms $\Delta_\pm : Spin(8) \to Spin(8)$. Prove that Δ_+ is an automorphism of order 3 and Δ_+^2 is conjugate to Δ_- (*Hint*: see what happens on a maximal torus). This outer automorphism is known as *triality*.

3 Spin structures

To make use of the Dirac operator in Riemannian geometry, we have to depart from our dependence on Euclidean space, and here the idea of manipulating spinors by using the group $Spin(n)$ is essential. Its relationship is so close – just a double covering – that we can carry over all local aspects of spinors, but there do remain the global questions which we have to take into account. Take a Riemannian manifold M and consider the principal $SO(n)$ bundle of oriented orthonormal frames. The Levi-Civita connection defines a connection on the principal bundle, and we can recover the tangent bundle by taking the associated bundle

$$P \times_{SO(n)} \mathbb{R}^n$$

or tensor bundles by taking associated bundles of the form

$$P \times_{SO(n)} (\mathbb{R}^n \otimes \mathbb{R}^n \otimes \cdots \otimes \mathbb{R}^n).$$

The connection on P induces the usual covariant derivative on sections of each bundle. What we want is a bundle whose sections can be acted on by a Dirac operator. The only problem is that while tensors are representation spaces of $SO(n)$, spinors form a representation space for $Spin(n)$. We therefore need to find a principal $Spin(n)$ bundle $\tilde{P}$ which covers P. There are obstructions to doing this. The problem is not restricted to orthonormal frame bundles for the tangent bundle, it is valid for any principal $SO(n)$ bundle P.

Think of P as determined, relative to some local trivializations over coordinate neighbourhoods U_α, by transition matrices

$$g_{\alpha\beta} : U_\alpha \cap U_\beta \to SO(n)$$

satisfying the cocycle condition

$$g_{\alpha\beta} g_{\beta\gamma} g_{\gamma\alpha} = 1 \tag{3.1}$$

on $U_\alpha \cap U_\beta \cap U_\gamma$. We need to lift these to $Spin(n)$ and if we can do so to satisfy (3.1) then we shall have our bundle $\tilde{P}$. On each $U_\alpha \cap U_\beta$ we can do this, but we have two choices, differing by multiplication by -1. Make such a choice anyway, to get $\tilde{g}_{\alpha\beta}$. Then from (3.1) we shall have

$$\tilde{g}_{\alpha\beta} \tilde{g}_{\beta\gamma} \tilde{g}_{\gamma\alpha} = \pm 1.$$

This defines a 2-cocycle with values in $\pm 1 \cong \mathbb{Z}/2$. If we change our choice of $\tilde{g}_{\alpha\beta}$, this changes the cocycle by a coboundary, and so the obstruction to being able to find $\tilde{P}$ lies in the cohomology group $H^2(M, \mathbb{Z}/2)$. It is a class found years before its association with spinors, the second Stiefel–Whitney class $w_2(M)$. If $w_2(M) = 0$ we can find a principal $Spin(n)$ bundle $\tilde{P}$. Two such choices of $\tilde{g}_{\alpha\beta}$ differ by a 1-cocycle in $\mathbb{Z}/2$, which defines a class in $H^1(M, \mathbb{Z}/2)$ and so we can say that the difference of two spin structures – liftings of P to $\tilde{P}$ – is a class in $H^1(M, \mathbb{Z}/2)$. Equivalently, we can regard the space of spin structures as an affine space over $\mathbb{Z}/2$ whose group of translations is $H^1(M, \mathbb{Z}/2)$.

One of the subtleties about spinors is the fact that two principal $Spin(n)$ bundles corresponding to two spin structures are usually C^∞ isomorphic. Another is the fact that the spin structures form an affine space – something less tangible than its model space $H^1(M, \mathbb{Z}/2)$. This is particularly relevant when one considers the action of groups of isometries on spin structures.

Suppose now that $w_2(M) = 0$ (we say then that M is a spin manifold) and we have chosen a spin structure. The principal bundle $\tilde{P}$ acquires a connection from P because it double covers P and connections are infinitesimal objects. If we now take one of the fundamental representation spaces S for $Spin(n)$ from the matrix description of Clifford algebras we can define a vector bundle with connection

$$\tilde{P} \times_{Spin(n)} S$$

and indeed a Dirac operator

$$D : C^\infty(S) \to C^\infty(S).$$

The local description of D is obtained by taking an orthonormal basis of sections $e_1, \ldots, e_n$ of the tangent bundle and writing

$$D\psi = \sum_i e_i \nabla_i \psi. \tag{3.2}$$

This is an elliptic operator and so if M is compact, it has a finite-dimensional null-space. In general (see [10], [3]) this depends on the metric. It is normal to consider S as a complex vector bundle with the extra structure required to make it real or quaternionic according to the table of Clifford algebras. Given a complex vector bundle E with connection, we can also define a coupled Dirac operator:

$$D : C^\infty(S \otimes E) \to C^\infty(S \otimes E).$$

The formula is just the same as in (3.2) except that ∇ is now the connection on $S \otimes E$ induced by the Levi-Civita connection on S and the given connection on E.

There are two fundamental results concerning the null-space of the Dirac operator.

The first is an application of the Atiyah–Singer index theorem. In even dimensions $n = 2k$, as we have seen in (2.2),

$$\omega e_i = -e_i \omega \qquad \omega^2 = (-1)^k$$

so if we decompose S into ± 1 or $\pm i$ eigenspaces of ω, the Dirac operator maps sections of S_+ to sections of S_-. The index theorem tells us that

$$\dim \ker D|_{S_+} - \dim \ker D|_{S_-} = \hat{A}(M)$$

where

$$\hat{A} = 1 - \tfrac{1}{24} p_1 + \tfrac{1}{45.2^7}\left(-4p_2 + 7p_1^2\right) + \cdots$$

is a certain universal polynomial in Pontryagin classes. These classes only exist in degrees a multiple of 4, so non-zero indices for the Dirac operator on its own occur just in these dimensions. However, the coupled Dirac operator has an index

$$\dim\ker D|_{S_+} - \dim\ker D|_{S_-} = \mathrm{ch}(E)\hat{A}(M) \tag{3.3}$$

where

$$\mathrm{ch}(E) = \mathrm{rk}\, E + c_1(E) + \tfrac{1}{2}\left(c_1^2 - 2c_2\right)(E) + \cdots$$

is the Chern character of E. Since the Chern classes exist in all even degrees we can get non-zero indices for the coupled Dirac operator for any even-dimensional spin manifold.

The second result is the vanishing theorem of Lichnerowicz, or more precisely the Weitzenböck decomposition that goes with it. This consists of a rewriting of the spinor Laplacian D^2 in terms of what is generally known as the *rough Laplacian* $\nabla^*\nabla$ and a curvature term.

The rough Laplacian is just the composition of the covariant derivative $\nabla : C^\infty(S) \to C^\infty(S \otimes T^*)$ with its formal adjoint $\nabla^* : C^\infty(S \otimes T^*) \to C^\infty(S)$. If $\nabla^2\psi \in C^\infty(S \otimes T^* \otimes T^*)$ is the covariant derivative of $\nabla\psi$ then the rough Laplacian is

$$\nabla^*\nabla\psi = -\,\mathrm{tr}(\nabla^2\psi)$$

where tr denotes contracting the two T^* components using the metric.

The Dirac operator D is the composition $C\nabla$ where $C : S \otimes T^* \to S$ is the vector bundle homomorphism defined by multiplication of a spinor by a vector in the Clifford algebra, so the spinor Laplacian

$$D^2\psi = C^2\nabla^2\psi.$$

But now the basic identity $e_ie_j + e_je_i = -2\delta_{ij}$ tells us that the symmetric part of C^2 is $-\,\mathrm{tr}$. The skew-symmetric part of ∇^2 is just the curvature of the Levi-Civita connection on S, so put together this gives, in local terms

$$D^2\psi = \nabla^*\nabla\psi + \sum_{ijkl} R_{ijkl}e_ie_j\Big(-\tfrac{1}{4}e_ke_l\Big)\psi$$

where we have used (2.4) to get the expression for the action of a skew-symmetric matrix on spinors. This last term needs simplifying. You can do it directly but can also see its form by appealing to invariant theory – it has to be a multiple of the scalar curvature because it represents an invariant linear map from the space of Riemann tensors to the Clifford algebra. The direct calculation gives the coefficient and this is in Exercise 3.4(iv). In fact the Weitzenböck decomposition turns out to be

$$D^2 = \nabla^*\nabla + \tfrac{1}{4}R$$

and this, in some measure, is the curved version of the property that so impressed Hamilton, of finding a ‘square root’ of the Laplacian.

The Lichnerowicz result is a vanishing theorem. If we assume $D\psi = 0$, then integrating over M we get

$$0 = \int_M \langle \nabla^*\nabla\psi + \tfrac{1}{4}R\psi, \psi\rangle = \int_M \langle \nabla\psi, \nabla\psi\rangle + \tfrac{1}{4}R\langle\psi,\psi\rangle$$

so if $R > 0$, $\psi = 0$. This tells us that the null-space of the Dirac operator on Riemannian manifolds like spheres or compact symmetric spaces is zero. If $R = 0$ then we must have $\nabla\psi = 0$.

Exercises 3.4 (i) Use the fact that the Clifford algebra in dimensions $8k+4$ is quaternionic to deduce that for a spin manifold M of dimension $8k+4$, $\hat{A}(M)$ is an even integer.

(ii) Find the original Weitzenböck formula (for the Hodge Laplacian $dd^* + d^*d$ rather than the spinor Laplacian) in [19].

(iii) While you have the book [19] out of the library to do Exercise 2 look at the Foreword and find an acrostic (this is not recommended for Francophiles).

(iv) Use the first Bianchi identity $R_{ijkl} + R_{iljk} + R_{iklj} = 0$ to prove directly that

$$\sum_{ijkl} R_{ijkl} e_i e_j e_k e_l = 2R.$$

4 The Dirac operator in dimensions 1 and 2

Consider first the 1-dimensional situation: M is the circle S^1 with metric $d\theta^2$. In this case the tangent bundle is trivial – $d/d\theta$ is a unit vector field – so S^1 is certainly a spin manifold. On the other hand $H^1(S^1, \mathbb{Z}/2) \cong \mathbb{Z}/2$ so there are two spin structures, two double coverings of the principal bundle of orthonormal frames, which is the circle itself. What are the two Dirac operators?

Take first the trivial double covering $\tilde{P} = S^1 \times \pm 1$. The spinor bundle is then

$$S = S^1 \times \mathbb{C}$$

so a section of S is just a complex valued function $\psi : S^1 \to \mathbb{C}$ and the Dirac operator is

$$D\psi = e_1 \frac{d\psi}{d\theta} = i\frac{d\psi}{d\theta}$$

since e_1 acts as i on the spinor space $\mathbb{C}$. The null-space is clearly one-dimensional, consisting of the constants.

The other spin structure is the non-trivial double covering, so represent it by the map $S^1 \to S^1$ defined by $e^{i\phi} \mapsto e^{2i\phi} = e^{i\theta}$. The spin bundle here is

$$S = S^1 \times_{\mathbb{Z}/2} \mathbb{C}.$$

As a C^∞ complex vector bundle this is still trivial, but we should be keeping careful track of the bundle considered as a spinor bundle in order to get the Dirac operator. To this end we don't choose an arbitrary trivialization, but instead represent a section of S by a function $\psi : S^1 \to \mathbb{C}$ on the connected double covering of $M = S^1$ satisfying

$$\psi(-e^{i\phi}) = -\psi(e^{i\phi}) \tag{4.1}$$

The Dirac operator is then

$$D\psi = i\frac{d\psi}{d\phi}.$$

In this case the constant solutions clearly cannot satisfy the constraint (4.1), so the null-space is zero.

There is a refinement of the index theorem which sets this in a more general context, and works in dimensions $8k+1$. From the table of Clifford algebras we see that $C(\mathbb{R}^{8k+1})$ is a complex matrix algebra. But remember that it also has more internal structure. In particular, looking at (2.2) in this dimension, we have

$$\omega e_i = e_i \omega \qquad \omega^2 = -1.$$

What this tells us is that ω acts as i in this algebra. We also have $\alpha(\omega) = -\omega$ since ω is a product of an odd number of e_i's, so α is an antilinear automorphism of the algebra and the fixed point set of α is a real matrix algebra. We can then regard the spinor bundle S naturally as the complexification of a real vector bundle. Since ω commutes with e_i it commutes with D also. Moreover $\omega^* = -\omega$, so ωD is a real skew-adjoint operator. For the circle it is just $\omega D = -d/d\theta$.

It is an interesting fact (see [1]) that the dimension modulo 2 of the null-space of a real skew-adjoint Fredholm operator is a deformation invariant, and for this Dirac operator, the Atiyah–Singer index theorem describes it as a KO-theory characteristic number (see [15]). Clearly in the basic situation of the circle, this number distinguishes the two spin structures, since one has an even-dimensional null-space and the other an odd one.

Something similar happens in $8k+2$ dimensions. Here

$$\omega e_i = -e_i \omega \qquad \omega^2 = -1$$

and the bundle S is complex with multiplication by i given by ω. Since ω now anticommutes with the e_i, the Dirac operator is antilinear and skew-adjoint. There is a mod 2 index theorem in this case too – the dimension modulo 2 of the null-space of a skew antilinear Fredholm operator is also a deformation invariant and again for the Dirac operator is given by a KO-theory characteristic number.

Let us consider the simplest case of a 2-dimensional compact orientable Riemannian manifold – a surface. In this case we know that the choice of a Riemannian metric defines the structure of a complex 1-dimensional manifold on M: the Hodge star operator $* : T^* \to T^*$ satisfies $*^2 = -1$ and this is the almost complex structure. Because of this, the Stiefel–Whitney class $w_2(M)$ has an interpretation as the mod 2 reduction of the first Chern class $c_1(M)$ of the tangent bundle considered as a complex line bundle. Since $H^2(M,\mathbb{Z}) \cong \mathbb{Z}$ this is just a number, the Euler characteristic of M. For an oriented surface of genus g this is the even number $2-2g$ so $w_2(M)=0$ and the manifold is always a spin manifold. On the other hand since $H^1(M,\mathbb{Z}/2) \cong (\mathbb{Z}/2)^{2g}$ there are many spin structures.

In this case the holomorphic structure comes to the rescue in distinguishing different spin structures, as follows. The principal $SO(2)$-bundle of oriented orthonormal frames is a circle bundle because $SO(2) \cong U(1)$ thus its double covering $Spin(2)$ is also a circle. The choice of a spin structure therefore involves the choice of a line bundle L such that L^2 is isomorphic to the complex tangent bundle (or more conveniently for us its dual, the canonical bundle K). The Levi-Civita connection on the principal bundle $\tilde{P}$ gives L a connection and this in particular defines on L a holomorphic structure. Now any two choices of L differ by a holomorphic line bundle U with U^2 holomorphically trivial. Holomorphic line bundles are classified up to equivalence by the sheaf cohomology group $H^1(M,\mathcal{O}^*)$. If we consider the short exact sequence of sheaves

$$1 \to \pm 1 \to \mathcal{O}^* \stackrel{z\mapsto z^2}{\longrightarrow} \mathcal{O}^* \to 1$$

then the long exact cohomology sequence gives

$$\to H^0(M,\mathcal{O}^*) \to H^0(M,\mathcal{O}^*) \to H^1(M,\mathbb{Z}/2) \to H^1(M,\mathcal{O}^*) \to .$$

When M is compact $H^0(M,\mathcal{O}^*) \cong \mathbb{C}^*$ since the only global holomorphic functions are constants and then since squaring is a surjective map from $\mathbb{C}^*$ to itself, $H^1(M,\mathbb{Z}/2) \to H^1(M,\mathcal{O}^*)$ is injective which tells us that the holomorphic structures of any two line bundles with $L^2 \cong K$ differ by an element of $H^1(M,\mathbb{Z}/2)$, just like their corresponding spin structures. Consequently the holomorphic structure of the line bundle L encodes the corresponding spin structure.

We have seen that the spinor bundle S can be thought of as a complex vector bundle with complex structure ω. More conventionally we complexify S and decompose it as eigenspaces of ω. Since the complexification of a complex vector space V is canonically $V \oplus \overline{V}$, what we are saying in $8k+2$ dimensions in general is that $S_+ \cong \overline{S}_-$. In our 2-dimensional situation, if we write $K^{1/2}$ for the holomorphic line bundle L with $L^2 \cong K$ then

$$S_+ = K^{1/2}, \qquad S_- = \overline{K}^{1/2}.$$

How then do we write the Dirac operator $D : C^\infty(S_+) \to C^\infty(S_-)$? A little calculation shows that it is just the composition of

$$\bar{\partial} : C^\infty(K^{1/2}) \to C^\infty(K^{1/2}\overline{K})$$

with the isomorphism

$$h^{-1/2} : C^\infty(K^{1/2}\overline{K}) \to C^\infty(\overline{K}^{1/2})$$

where $h \in C^\infty(K\overline{K})$ is the Riemannian metric and $h^{1/2} \in C^\infty(K^{1/2}\overline{K}^{1/2})$ its positive square root.

We note two features of this description. The first is that the null-space of D consists of the global holomorphic sections of the holomorphic line bundle $K^{1/2}$. This just depends on the conformal structure of M and not on the choice of Riemannian metric in the conformal class. This is actually true in full generality – the null-space the Dirac operator on M^n has, if given what is called conformal weight $(n-1)/2$, a conformally invariant interpretation. The second observation is that the skew antilinear interpretation of the Dirac operator gives a reason why the index should vanish in dimension $8k+2$. This is not just because Pontryagin classes exist only in dimensions a multiple of 4, but instead because of a particular mechanism – complex conjugation – which interchanges the null-space of $D|_{S_+}$ with $D|_{S_-}$.

The mod 2 problem raises some interesting questions for Riemann surfaces (see [2]). We have noted that the dimension of the null-space of D mod 2 is a deformation invariant. We call a spin structure $K^{1/2}$ on the Riemann surface M odd/even if the dimension of its space of holomorphic sections is odd/even. Unravelling the KO-theory interpretation of this mod 2 invariant for compact surfaces as in [2] one finds that if $N_0, N_1 = 2^{2g} - N_0$ are the numbers of even and odd spin structures respectively, then

$$N_0 = \tfrac{1}{2}2^g(2^g+1), \qquad N_1 = \tfrac{1}{2}2^g(2^g-1).$$

We consider a few low-genus examples next. When $g = 0$ the null-space is zero by the Lichnerowicz theorem since S^2 has positive curvature. When $g = 1$, an elliptic curve, then the canonical bundle is trivial so we have one trivial square root with a 1-dimensional space of holomorphic sections and three non-trivial ones with none, i.e. $N_0 = 3, N_1 = 1$ as predicted.

In general, one can get bounds on dimension. For example, the product of two holomorphic sections of $K^{1/2}$ is a section of K, so this defines a holomorphic map of projective spaces

$$P(H^0(M, K^{1/2})) \times P(H^0(M, K^{1/2})) \to P(H^0(M, K)) = P^{g-1}.$$

Since the zero set of a section of K can be divided into two subsets in only a finite number of ways, this says the inverse image of any point is finite, so we get the estimate

$$\dim H^0(M, K^{1/2}) \leqslant \tfrac{1}{2}(g+1).$$

This enables us to deal with $g = 2$, because the dimension can only be 0 or 1 and so by the formula there must be $10 = 2^2(2^2+1)/2$ even spin structures and $6 = 2^2(2^2-1)/2$ odd.

When $g = 3$ the estimate allows the possibility that the dimension might be 2, and then parity will not distinguish spin structures with 0- and 2-dimensional null-spaces. However, if $K^{1/2}$ has a 2-dimensional space of sections, this defines a holomorphic map $p : M \to P(H^0(M, K^{1/2})^*) = P^1$ of degree $\deg K^{1/2} = g-1 = 2$ and this means that M is hyperelliptic. If M of genus 3 is not hyperelliptic then the canonical map

$$M \to P(H^0(M, K)^*) = P^2$$

embeds M as a quartic curve. So in this case there are $28 = 2^3(2^3 - 1)/2$ spin structures with one section. What are they?

Any line in P^2 intersects the quartic M in the zero set of a section of K, because we are using the canonical embedding, so the square of a section of $K^{1/2}$ corresponds to a line which meets M tangentially at two points – a bitangent. We have recovered here through counting solutions to the Dirac equation $D\psi = 0$ the classical fact that a quartic curve has 28 bitangents.

Exercises 4.2 (i) Show that for a real periodic function u

$$L = \frac{d^3}{d\theta^3} + 4u\frac{d}{d\theta} + 2u'$$

defines a real skew-adjoint differential operator on the circle. Prove that its null-space is of dimension 1 or 3 and calculate the dimension if u is a constant.

(ii) Draw the 28 bitangents to the quartic curve $x^4 + y^4 - 6x^2 - 6y^2 + 10 = 0$, or even better read the paper [7].

(iii) Let M be a Riemann surface of genus 4 such that there are no spin structures with $\dim H^0(M, K^{1/2}) = 2$. Show that in its canonical embedding $M \subset P(H^0(M, K)^*) = P^3$ there are 120 tritangent planes – planes which meet M tangentially at three points.

5 The Dirac operator in dimension 4

Four dimensions is the natural home of the Dirac operator, and not only because Dirac formulated it in 4-dimensional Minkowski space. In recent years the study of the null-space of the Dirac operator on a compact Riemannian 4-manifold, coupled to a line bundle with connection, has given a profound insight into the structure of 4-manifolds through Seiberg–Witten theory. I don't want to go into all that here, merely to point out some of the basic features of the Dirac operator in this dimension.

The starting point is the result of Exercise 2.5(ii) of Section 2

$$Spin(4) \cong Sp(1) \times Sp(1).$$

If M is a 4-manifold with a spin structure, then this gives us two spinor bundles S_+, S_- whose internal structure is completely described by saying that their structure group is $Sp(1)$, the unit quaternions, or equivalently $SU(2)$. They are therefore rank 2 complex vector bundles with a unitary structure and a fixed non-degenerate skew form. With such a description, they are in some sense more primitive than the geometrical structure of the Riemannian metric itself and can be used to give a different viewpoint on Riemannian geometry. For example, the (complexified) bundle of Clifford algebras associated to the tangent bundle is on the one hand the bundle of endomorphisms of $S_+ \oplus S_-$ and on the other the bundle of exterior forms $\Lambda^* T^*$. Since the skew form identifies $S_\pm^*$ with $S_\pm$, we can compare

$$\mathrm{End}(S_+ \oplus S_-) = 2 \oplus \mathrm{Sym}^2 S_+ \oplus \mathrm{Sym}^2 S_- \oplus 2(S_+ \otimes S_-)$$

with

$$\Lambda^* T^* = 1 \oplus \Lambda^1 \oplus \Lambda^2 \oplus \Lambda^3 \oplus \Lambda^4 \cong 2 \oplus 2\Lambda^1 \oplus \Lambda^2$$

using the Hodge star isomorphism $* : \Lambda^p \to \Lambda^{4-p}$, and we find

$$T^* \otimes \mathbb{C} \cong S_+ \otimes S_-, \qquad \Lambda^2 T^* \cong \mathrm{Sym}^2 S_+ \oplus \mathrm{Sym}^2 S_-.$$

The decomposition of the bundle of 2-forms into $\mathrm{Sym}^2 S_+ \oplus \mathrm{Sym}^2 S_-$ corresponds precisely to the decomposition into self-dual and anti-self-dual 2-forms, the ± 1 eigenspaces of $* : \Lambda^2 T^* \to \Lambda^2 T^*$.

This notion of self-duality has played a highly important role in the interface between mathematics and physics over the past 20 years, most particularly through the study of vector bundles over 4-manifolds, with a connection whose curvature is anti-self-dual (however cumbersome this is the standard convention nowadays, motivated by the relationship with holomorphic bundles on complex surfaces). The Dirac operator has an especially nice feature when coupled to a bundle with an anti-self-dual connection. If we return to the Weitzenböck formula of Section 3, then when the Dirac operator is coupled to a connection A on the vector bundle E, the curvature term involves not just the scalar curvature of the manifold, but also the curvature F_A of the connection. The operation of the curvature, via a process of Clifford multiplication, is some invariant map from curvature tensors, which are sections of $\mathrm{End}\, E \otimes \Lambda^2_-$ to $\mathrm{End}(S \otimes E)$. If we restrict to S_+, then $\mathrm{End}\, S_+ \cong \mathrm{Sym}^2 S_+ \oplus 1$ has no component of $\Lambda^2_- = \mathrm{Sym}^2 S_-$, so F_A acts trivially on that side of the formula. In particular if the scalar curvature is positive we get vanishing. The vanishing argument will not hold on the other side but we can then use the Atiyah–Singer index theorem to get a precise expression for $\dim \ker D|_{E \otimes S_-}$. From (3.3) this is

$$\begin{aligned} \dim \ker D|_{E \otimes S_-} &= -\,\mathrm{ch}(E)\hat{A}(M) \\ &= -\,\mathrm{rk}\, E\, \tfrac{1}{24} p_1 - \tfrac{1}{2}(c_1^2 - 2c_2)(E). \end{aligned}$$

Now the scalar curvature assumption means that if we take the bundle E to be trivial with trivial connection, then the Dirac operator on both S_+ and S_- has zero null-space, so this tells us that $p_1(M)$ must be zero. If we suppose E to have structure group $SU(k)$ the formula gives us:

$$\dim\ker D|_{E\otimes S_-} = c_2(E)(M).$$

This formula is very useful. In particular it means that the dimension of the null-space of the Dirac operator is constant and this $c_2(E)$-dimensional space defines, as the connection varies, a bundle over the moduli space of anti-self-dual connections over M. We shall look at this *Dirac bundle* in some special situations in the rest of this article.

Exercises 5.1 (i) Show, by using the index theorem for the Dirac equation and the Hirzebruch signature formula $\mathrm{sign}(M) = p_1(M)/3$, that the signature of a compact 4-dimensional spin manifold is divisible by 16.

(ii) A quaternionic vector space can be thought of as a complex vector space V with an antilinear map $J : V \to V$ such that $J^2 = -1$. A real vector space can similarly be defined as an antilinear map T such that $T^2 = 1$ ($T =$ 'complex conjugation'). Show that, in this language, the tensor product of two quaternionic vector spaces is real.

(iii) Deduce from (ii) that if E is a vector bundle with structure group $SU(2)$ on a 4-dimensional spin manifold, then the null-space of the coupled Dirac operator is a real vector space.

6 Quaternionic aspects

Let $\mathbf{A}$ be a $(m+k)\times k$ quaternionic matrix, and write it in terms of four real matrices:

$$\mathbf{A} = A_0 + iA_1 + jA_2 + kA_3.$$

Let $\mathbf{A}^*$ denote the quaternionic adjoint $\mathbf{A}^* = A_0^T - iA_1^T - jA_2^T - kA_3^T$. We want to consider matrices which satisfy the condition that $\mathbf{A}^*\mathbf{A}$ is real.

What has this to do with the Dirac operator? On $\mathbb{R}^4$ consider the Dirac operator coupled to a connection on a bundle E:

$$D = e_0\nabla_0 + e_1\nabla_1 + e_2\nabla_2 + e_3\nabla_3.$$

Use e_0 to identify the two spinor bundles S_+ and S_- and then

$$\begin{aligned}\mathbf{D} = e_0 D &= -\nabla_0 + e_0e_1\nabla_1 + e_0e_2\nabla_2 + e_0e_3\nabla_3 \\ &= -\nabla_0 + i\nabla_1 + j\nabla_2 + k\nabla_3\end{aligned}$$

since we have the defining property of the quaternions:

$$(e_0e_1)^2 = (e_0e_2)^2 = (e_0e_3)^2 = -1 = (e_0e_1)(e_0e_2)(e_0e_3)$$

where the last relation comes from the fact that $\omega = e_0e_1e_2e_3 = 1$ on S_+. So we can think of the Dirac operator in $\mathbb{R}^4$ as an infinite dimensional quaternionic matrix. Since the covariant derivatives are formally skew adjoint as are e_0e_i, we have

$$\mathbf{D}^*\mathbf{D} = (\nabla_0 + i\nabla_1 + j\nabla_2 + k\nabla_3)(-\nabla_0 + i\nabla_1 + j\nabla_2 + k\nabla_3).$$

This is real if and only if

$$\begin{aligned} [\nabla_0, \nabla_1] + [\nabla_2, \nabla_3] &= 0, \\ [\nabla_0, \nabla_2] + [\nabla_3, \nabla_1] &= 0, \\ [\nabla_0, \nabla_3] + [\nabla_1, \nabla_2] &= 0 \end{aligned} \tag{6.1}$$

and this is precisely the condition that the connection should be anti-self-dual. In fact the property that $\mathbf{D}^*\mathbf{D}$ is real is essentially the Weitzenböck formula for the Dirac operator coupled to an anti-self-dual connection. As we have seen, this enables us to use the index theorem to determine the dimension of the kernel of $D|_{S_-\otimes E}$ or equivalently the cokernel of $D : C^\infty(S_+ \otimes E) \to C^\infty(S_- \otimes E)$. We need compactness for this really, so we should carry out the following process for the flat 4-torus T^4. Note that in this case the scalar curvature $R = 0$, so we can only use the vanishing theorem when there are no covariant constant sections of E.

The cokernel of $\mathbf{D}$ has some interesting properties which can be more readily seen from the finite-dimensional model of a quaternionic matrix. So, under the assumption that $\mathbf{A}^*\mathbf{A}$ is real, make the generic assumption that it is invertible, that $\mathbf{A}$ has zero null-space. Then define

$$\mathbf{P} = \mathbf{A}(\mathbf{A}^*\mathbf{A})^{-1}\mathbf{A}^*.$$

We have $\mathbf{P}^2 = \mathbf{A}(\mathbf{A}^*\mathbf{A})^{-1}\mathbf{A}^*\mathbf{A}(\mathbf{A}^*\mathbf{A})^{-1}\mathbf{A}^* = \mathbf{A}(\mathbf{A}^*\mathbf{A})^{-1}\mathbf{A}^* = \mathbf{P}$, $\mathbf{P} = \mathbf{P}^*$ and $\mathbf{PA} = \mathbf{A}$, so $\mathbf{P}$ is orthogonal projection onto the image of $\mathbf{A}$.

As $\mathbf{A}$ varies its cokernel, the image of $1 - \mathbf{P}$, defines a vector bundle V over the space of matrices $\mathbf{A}$ and it has a natural connection obtained by orthogonal projection, i.e. if s is a section of V then define

$$\nabla s = (1 - \mathbf{P})ds$$

where ds is just the ordinary derivative of a vector-valued function. This situation is just like the Levi-Civita connection of a surface embedded in $\mathbb{R}^3$. We have an equation, the Gauss equation, which relates the curvature of the bundle V to a quadratic expression in the second fundamental form. In our case the second fundamental form is

$$\mathbf{B}(s) = \mathbf{P}ds$$

or, using $\mathbf{P}s = 0$

$$\mathbf{B}(s) = -d\mathbf{P}s = -d(\mathbf{A}(\mathbf{A}^*\mathbf{A})^{-1}\mathbf{A}^*)s = -\mathbf{A}(\mathbf{A}^*\mathbf{A})^{-1}d\mathbf{A}^*s$$

since $\mathbf{A}^* s = 0$. The Gauss equation for the curvature of the connection on V is then $Fs = -(1-\mathbf{P})\mathbf{B}^* \wedge \mathbf{B}s$ but since $(1-\mathbf{P})\mathbf{A} = 0$,

$$(1-\mathbf{P})\mathbf{B}^* = -(1-\mathbf{P})d\mathbf{A}(\mathbf{A}^*\mathbf{A})^{-1}\mathbf{A}^*$$

and this yields

$$Fs = (1-\mathbf{P})d\mathbf{A}(\mathbf{A}^*\mathbf{A})^{-1}d\mathbf{A}^* s.$$

The point to note here is that when $\mathbf{A}^*\mathbf{A}$ is real, the curvature consists of linear combinations of 2-forms of the type

$$d\mathbf{A}_{ij} \wedge d\overline{\mathbf{A}}_{kl}.$$

These connections are very special. Indeed if $\mathbf{q} = x_0 + ix_1 + jx_2 + kx_3$ then the components of $d\mathbf{q} \wedge d\bar{\mathbf{q}}$ are anti-self-dual 2-forms. In higher dimensions we can say that they are of type $(1,1)$ with respect to each complex structure i, j, k.

We have defined here the bundle V over the subspace Z of the space of quaternionic matrices satisfying the condition $\mathbf{A}^*\mathbf{A}$ is real. Spelling this out, the condition is the vanishing of the three skew-symmetric real $k \times k$ matrices

$$\begin{aligned}
&A_0^T A_1 - A_1^T A_0 + A_3^T A_2 - A_2^T A_3, \\
&A_0^T A_2 - A_2^T A_0 + A_1^T A_3 - A_3^T A_1, \\
&A_0^T A_3 - A_3^T A_0 + A_2^T A_1 - A_1^T A_2.
\end{aligned}$$

This has an interpretation in terms of a *hyperkähler quotient.* It will take us too far afield to go into details here, so we just state the results, and refer to [13] for a general survey. There is an obvious action

$$\mathbf{A} \mapsto \mathbf{A}P \quad \text{for} \quad P \in O(k) \tag{6.2}$$

which commutes with i, j, k. The standard inner product on matrices is preserved by this action and the imaginary parts of $\mathbf{A}^*\mathbf{A}$, the skew-symmetric matrices above, are the hyperkähler moment maps, taking values in the Lie algebra of $O(k)$. It follows then that the quotient of the space Z by $O(k)$ is again a hyperkähler manifold. This is an example of a hyperkähler quotient. What is equally important in our situation is the observation that the action of P in (6.2) on the cokernel of $\mathbf{A}$ is trivial. This means that the bundle V with its connection descends to the quotient $Z/O(k)$. The property of its curvature then implies that it is of type $(1,1)$ with respect to each of the complex structures i, j, k on the hyperkähler quotient. In particular it has a holomorphic structure in each of these.

To put this procedure into effect for the Dirac operator, we need the Green's function to play the role of $(\mathbf{A}^*\mathbf{A})^{-1}$. This works by standard Hodge theory on the 4-torus. In this case the space of quaternionic matrices is interpreted as the space of Dirac operators coupled to connections on a bundle E. The space of anti-self-dual connections is then the zero set of this moment map and its quotient by the gauge transformations is the moduli space of anti-self-dual connections. We see here the Dirac operator appearing naturally within this quaternionic framework in four dimensions. The dimensional reductions of anti-self-dual connections inherit this property as we shall see next.

Exercises 6.3 (i) Take a fixed anti-self-dual connection A on T^4 and consider the above construction for the Dirac bundle on the family of anti-self-dual connections

$$A_\xi = A + i(\xi_0 dx_0 + \xi_1 dx_1 + \xi_2 dx_2 + \xi_3 dx_3)$$

where the $\xi_i \in \mathbb{R}$ are constant. Show that this defines an anti-self-dual connection on the dual torus, or read about it in Chapter 3 of Donaldson and Kronheimer [6].

(ii) For an overview of the hyperkähler quotient construction, read [13].

7 The Dirac operator for Higgs bundles

The formal relationship between quaternionic matrices and the Dirac operator on $\mathbb{R}^4$ or T^4 that we saw in the previous section extends also to dimensional reductions of the anti-self-dual Yang–Mills equations. Consider first the 2-dimensional version. In this we assume that we have an anti-self-dual connection on $\mathbb{R}^4$ which is invariant under translation in the last two variables. The equations (6.1) now become:

$$[\nabla_0, \nabla_1] + [\phi_0, \phi_1] = 0,$$
$$[\nabla_0, \phi_1] + [\nabla_1, \phi_0] = 0,$$
$$[\nabla_0, \phi_0] - [\nabla_1, \phi_1] = 0$$

for a connection A on a bundle E over $\mathbb{R}^2$ and two skew-adjoint sections ϕ_1, ϕ_2 of the endomorphism bundle of E. The equations can be written

$$[\nabla_0 + i\nabla_1, \phi_0 + i\phi_1] = 0, \qquad [\nabla_0, \nabla_1] + [\phi_0, \phi_1] = 0$$

or if we put $\Phi = (\phi_0 + i\phi_1)dz/2$ in the form

$$\bar{\partial}_A \Phi = 0, \qquad F_A + [\Phi, \Phi^*] = 0. \tag{7.1}$$

In this formalism the equations are conformally invariant and can be put on any Riemann surface M. The connection A defines a holomorphic structure on the vector bundle E and Φ, the Higgs field, is then a holomorphic section of $\operatorname{End} E \otimes K$ by the first equation. The second equation relates the curvature of a unitary connection on E to the Higgs field – its existence is a consequence of the stability, in the sense of algebraic geometry, of the pair (E, Φ) (see [11]).

These are dimensional reductions of the anti-self-duality equations. What is the dimensional reduction of the Dirac operator? In four dimensions we had

$$\mathbf{D} = -\nabla_0 + i\nabla_1 + j\nabla_2 + k\nabla_3$$

so now we have

$$\mathbf{D} = -\nabla_0 + i\nabla_1 + j\phi_0 + k\phi_1$$

or, using the Pauli matrix representation of the quaternions:

$$i = \begin{pmatrix} i & 0 \\ 0 & -i \end{pmatrix} \quad j = \begin{pmatrix} 0 & 1 \\ -1 & 0 \end{pmatrix} \quad k = \begin{pmatrix} 0 & i \\ i & 0 \end{pmatrix}$$

$$\mathbf{D} = \begin{pmatrix} -\nabla^{1,0} & \Phi \\ -\Phi^* & -\nabla^{0,1} \end{pmatrix} \quad \text{and} \quad \mathbf{D}^* = \begin{pmatrix} \nabla^{0,1} & \Phi \\ -\Phi^* & -\nabla^{1,0} \end{pmatrix}.$$

We give $\mathbf{D}^*$ a global interpretation on the Riemann surface as follows:

$$\mathbf{D}^* : \Omega^{1,0}(E) \oplus \Omega^{0,1}(E) \to \Omega^{1,1}(E) \oplus \Omega^{1,1}(E)$$

$$\mathbf{D}^*(\psi_1, \psi_2) = (\bar{\partial}_A \psi_1 - \Phi\psi_2, \partial_A \psi_2 - \Phi^* \psi_1) \tag{7.2}$$

using the covariant exterior derivative $d_A = \partial_A + \bar{\partial}_A$ on forms with values in E.

The formal statement that $\mathbf{D}^*\mathbf{D}$ is real translates into a vanishing theorem for the null-space of $\mathbf{D}$, at least if the connection is irreducible, so as in four dimensions we can use the index theorem to calculate the dimension of the null-space of $\mathbf{D}^*$. If E has an $SU(k)$ connection then we get

$$\dim \ker \mathbf{D}^* = k(2g-2). \tag{7.3}$$

Consequently we obtain a Dirac bundle over the moduli space $\mathcal{M}$ of Higgs bundles – solutions to equations (7.1) modulo gauge transformations. Moreover, the null-space of $\mathbf{D}^*$ in this formalism has an inner product using the Hodge star operator:

$$\int_M \langle \psi_1, *\psi_1 \rangle + \langle \psi_2, *\psi_2 \rangle.$$

This depends only on the conformal structure on M and the metric on E because $*$ is conformally invariant on 1-forms. This inner product, as in the previous section, defines a natural connection on the Dirac bundle. The moduli space $\mathcal{M}$ has a hyperkähler structure (see [11]) and this connection has curvature of type $(1,1)$ with respect to all complex structures on $\mathcal{M}$. We shall consider two different interpretations of this bundle.

First, a piece of homological algebra. If $A^{p,q}$ is a double complex with two differentials

$$d : A^{p,q} \to A^{p,q+1}, \qquad \delta : A^{p,q} \to A^{p+1,q}$$

satisfying $d^2 = \delta^2 = d\delta + \delta d = 0$ then $d + \delta$ is the differential of a complex

$$\cdots \to A^{*,n-*} \xrightarrow{d+\delta} A^{*,n+1-*} \to \cdots$$

whose cohomology groups are called *hypercohomology* groups $\mathcal{H}^n$. A Higgs bundle gives a simple example of this: we consider

$$d = \bar{\partial}_A : \Omega^{p,q}(E) \to \Omega^{p,q+1}(E), \qquad \delta = -\Phi : \Omega^{p,q}(E) \to \Omega^{p+1,q}(E)$$

and because there are no $(0,2)$ or $(2,0)$ forms on a Riemann surface, $d^2 = \delta^2 = 0$. Moreover, since Φ is a holomorphic 1-form with values in E,

$$\bar{\partial}_A(\Phi\alpha) = (\bar{\partial}_A\Phi)\alpha - \Phi\bar{\partial}_A\alpha = -\Phi\bar{\partial}_A\alpha$$

so $d\delta + \delta d = 0$ as required.

Now look at $\psi = (\psi_1, \psi_2) \in \Omega^{1,0}(E) \oplus \Omega^{0,1}(E)$ in the null-space of $\mathbf{D}^*$. From (7.2), the first component says that

$$(d+\delta)\psi = \bar{\partial}_A\psi_1 - \Phi\psi_2 = 0$$

so that ψ defines an element of the first hypercohomology group $\mathcal{H}^1(M,E)$. Moreover the second component $\partial_A\psi_2 - \Phi^*\psi_1 = 0$ is equivalent to $(d+\delta)^*\psi = 0$ and Hodge theory tells us that there is a unique such representative in any hypercohomology class of this particular complex. We can thus identify the null-space of $\mathbf{D}^*$ with $\mathcal{H}^1(M,E)$.

The hypercohomology group $\mathcal{H}^1(M,E)$ has a particularly concrete interpretation in the generic situation where $\det\Phi$ has simple zeros $z_1,\ldots,z_N$. Take $\psi = (\psi_1,\psi_2)$ with

$$(d+\delta)\psi = \bar{\partial}_A\psi_1 - \Phi\psi_2 = 0.$$

If at each point z_i where Φ is not invertible $\psi_1(z_i)$ lies in the $(k-1)$-dimensional image of $\Phi(z_i)$ then $\psi_0 = \Phi^{-1}\psi_1$ is a well-defined section of E and we have

$$\Phi\psi_0 = \psi_1, \qquad \bar{\partial}_A\psi_0 = \Phi^{-1}\bar{\partial}_A\psi_1 = \psi_2.$$

Hence

$$\psi = (d+\delta)\psi_0.$$

Thus the hypercohomology class of ψ is determined by its image in

$$\bigoplus_{i=1}^{N}(E\otimes K)_{z_i}/\Phi(z_i)E_{z_i} \cong \mathbb{C}^N.$$

This fits with our dimension count using the index theorem since $\deg K = (2g-2)$ so $\det\Phi \in H^0(M,K^k)$ vanishes at $N = 2k(g-1)$ points and by (7.3) this is $\dim\ker\mathbf{D}^*$.

Another interpretation comes by considering the connection

$$d_B = d_A + \Phi + \Phi^*.$$

From equations (7.1) we have

$$\begin{aligned} d_B^2 = (d_A + \Phi + \Phi^*)^2 &= d_A^2 + \bar{\partial}_A\Phi + \partial_A\Phi^* + [\Phi,\Phi^*] \\ &= F_A + [\Phi,\Phi^*] \\ &= 0 \end{aligned}$$

so this is a flat complex connection. Now take

$$\psi = (\psi_1, -\psi_2) \in \Omega^{1,0}(E) \oplus \Omega^{0,1}(E) = \Omega^1(E)$$

and suppose $\mathbf{D}^*\psi = 0$. Then

$$d_B\psi = \bar{\partial}_A\psi_1 - \partial_A\psi_2 - \Phi\psi_2 + \Phi^*\psi_1 = 0$$

from (7.2). Thus ψ represents a class in the first cohomology $H^1(M, \mathcal{A})$ of the flat vector bundle E with connection d_B – the local system $\mathcal{A}$ determined by d_B. Moreover it is easy to see that

$$d_B^*\psi = \bar{\partial}_A\psi_1 + \partial_A\psi_2 - \Phi\psi_2 - \Phi^*\psi_1$$

and this also vanishes from (7.2). By Hodge theory, there is then an isomorphism between the null-space of $\mathbf{D}^*$ and this flat cohomology group.

These two facts are part of a general picture. As we saw, the connection on the Dirac bundle is of type $(1,1)$ with respect to all complex structures on the hyperkähler moduli space $\mathcal{M}$ and so has a holomorphic structure with respect to each of them. We know however, the complex structures of the natural hyperkähler metric on $\mathcal{M}$ [11]. One is the moduli space of stable Higgs bundles, and then the hypercohomology group gives a holomorphic bundle. The others are isomorphic to the moduli space of flat $SL(k, \mathbb{C})$ connections and then the flat cohomology bundle is holomorphic.

Exercises 7.4 (i) Show that $d_\zeta = d_A + \zeta\Phi + \zeta^{-1}\Phi^*$ is a flat connection if (A, Φ) satisfy the Higgs bundle equations (7.1). Interpret the null-space of $\mathbf{D}^*$ in this case.

(ii) Show (cf. [11]) that the tangent bundle of $\mathcal{M}$ can be identified with the Dirac bundle for the Dirac operator coupled to $\operatorname{End} E$ with its connection induced from the connection A on E.

(iii) Look at [9] to see how my research student Tamas Hausel found relations in the cohomology of $\mathcal{M}$ by using the Dirac bundle.

8 Magnetic monopoles and the Dirac operator

Consider now the anti-self-duality equations in $\mathbb{R}^4$ which are invariant under translation in the x_0-direction. The equations (6.1) become

$$\begin{aligned}
[\nabla_1, \phi] &= [\nabla_2, \nabla_3], \\
[\nabla_2, \phi] &= [\nabla_3, \nabla_1], \\
[\nabla_3, \phi] &= [\nabla_1, \nabla_2]
\end{aligned}$$

and we can put these in more invariant form by regarding $\nabla_1, \nabla_2, \nabla_3$ as defining a connection on a vector bundle E over $\mathbb{R}^3$ and ϕ as a skew-adjoint section of the bundle End E. The equations are then the Bogomolny equations:

$$F = *\nabla\phi.$$

There is a vast literature on solutions to these equations (see for example [12], [18]) particularly in the case where E has rank 2. The usual boundary conditions to adopt require in particular that $\|\phi\| = 1 + m/r + \cdots$ at infinity. A solution to the Bogomolny equations with these asymptotic conditions is called a *magnetic monopole.* With this behaviour at infinity the $+i\|\phi\|$ eigenspace of ϕ is well defined at large distances and gives a line bundle whose first Chern class on a large 2-sphere is the basic invariant of the monopole, its magnetic charge k. The moduli space of charge k monopoles is a non-compact manifold of dimension $4k-1$. If we fix an isomorphism at infinity of E with $L \oplus L^*$ where L is the standard line bundle of degree k over S^2, then we obtain a $4k$-dimensional manifold with a circle action (changing the isomorphism at infinity) whose quotient is the moduli space. This space, denoted M_k, has a hyperkähler metric determined by $\mathcal{L}^2$ inner products.

What does the Dirac operator $\mathbf{D}$ look like here? We have

$$\mathbf{D} = -\phi + i\nabla_1 + j\nabla_2 + k\nabla_3$$

and this is just the coupled Dirac operator in three-dimensions plus the extra Higgs field term.

$$D - \phi : C^\infty(S \otimes E) \to C^\infty(S \otimes E)$$

In this non-compact situation, the standard index theorem cannot be used, but nevertheless analysis of this specific situation by Callias and then Taubes (see [12]) shows that there is an index theorem, and a vanishing theorem too for $\mathcal{L}^2$ solutions to the Dirac equation, and for $SU(2)$ monopoles of charge k we find

$$\dim \ker \mathbf{D}^* = k.$$

The moduli space in this case is again hyperkähler. This time all complex structures are equivalent, because the rotation group $SO(3)$ acts transitively on them. For charge k the moduli space is isomorphic as a complex manifold to the space R_k of rational maps of degree k of the form

$$f(z) = \frac{b_0 + b_1 z + \cdots + b_{k-1}z^{k-1}}{a_0 + a_1 z + \cdots + a_{k-1}z^{k-1} + z^k}.$$

To understand the Dirac bundle as a complex bundle we adopt a slightly odd point of view of rational functions, but it is one which arises naturally out of the various different ways of solving the Bogomolny equations [12].

Let B be a complex symmetric $k \times k$ matrix and w a vector such that $w, Bw, \ldots, B^{k-1}w$ is a basis for $\mathbb{C}^k$. There is an action of the complex group $O(k, \mathbb{C})$ on such pairs by

$$(B, w) \mapsto (QBQ^{-1}, Qw).$$

The quotient space is of dimension $k(k+1)/2 + k - k(k-1)/2 = 2k$ and is in fact biholomorphically equivalent to the space of rational maps, the isomorphism given explicitly by

$$f(z) = w^T(zI - B)^{-1}w.$$

From this point of view, the set of pairs (B, w) is a principal $O(k, \mathbb{C})$ bundle P over the space R_k. The Dirac bundle is then the associated vector bundle

$$P \times_{O(k,\mathbb{C})} \mathbb{C}^k.$$

We should note (Exercise 5.1(ii)) that the tensor product of two quaternionic bundles is real, so since the spinor bundle S and the bundle E with structure group $SU(2) \cong Sp(1)$ are quaternionic, $\ker \mathbf{D}^*$ is a real k-dimensional vector space. The natural $\mathcal{L}^2$ inner product therefore gives the Dirac bundle an $O(k)$ structure, preserved by the connection. We see this structure, in its complexification, in the appearance of $O(k, \mathbb{C})$ in the above description.

The Dirac bundle for a magnetic monopole has been put to use as a testing ground for the S-duality conjectures in mathematical physics [8]. To formulate them we need to examine slightly closer the moduli space M_k. There is an obvious action of translation in $\mathbb{R}^3$ and also the circle action described above. Clearly these are rather simple-minded actions but they can't quite be factored out. What happens is that there is a cyclic k-fold covering $\tilde{M}_k$ of M_k which splits isometrically as a product

$$\tilde{M}_k \cong \tilde{M}_k^0 \times S^1 \times \mathbb{R}^3.$$

The S-duality conjectures concern the hyperkähler manifold $\tilde{M}_k^0$ of dimension $4k - 4$ which has a free isometric $\mathbb{Z}/k$ action. It involves triality (see Exercise 3 of Section 2) and the group $SL(2, \mathbb{Z})$.

The first thing one does is to take the Dirac bundle V over M_k and pull it back to $\tilde{M}_k^0$. This, remember, we consider as a real vector bundle with a natural $O(k)$ connection on it. Then consider the tensor product $V \otimes \mathbb{R}^8$. This vector bundle has a spin structure even if V itself doesn't, because if we look at the total Stiefel–Whitney class

$$w = 1 + w_1 + w_2 + \cdots \in H^*(M, \mathbb{Z}/2)$$

then $w(V \otimes \mathbb{R}^8) = w(V)^8 = 1$ because the binomial coefficients are even, so

$$w_1(V \otimes \mathbb{R}^8) = 0 \qquad \text{and} \qquad w_2(V \otimes \mathbb{R}^8) = 0.$$

Since $\tilde{M}_k^0$ is simply connected there is a unique spin structure. Now take the spinor bundle of this, $S(V \otimes \mathbb{R}^8)$, and consider $\mathcal{L}^2$ solutions to the Dirac equation $\mathbf{D}^*\psi = 0$ on $\tilde{M}_k^0$ coupled to this bundle with connection. Denote this space by $\mathcal{H}_k$.

The space $\mathcal{H}_k$ has two group actions:

- the action of $Spin(8)$ induced from its action on $\mathbb{R}^8$;
- the action of the cyclic group $\mathbb{Z}/k$.

Under the second action, we can break up $\mathcal{H}_k$ into a direct sum of subspaces $\mathcal{H}_{k,l}$ on which the generator of the cyclic group acts as $e^{2\pi i l/k}$. S-duality relates to a conjectured action of $SL(2,\mathbb{Z})$ on the direct sum of all $\mathcal{H}_k$ which, if

$$\begin{pmatrix} a & b \\ c & d \end{pmatrix} \begin{pmatrix} k \\ l \end{pmatrix} = \begin{pmatrix} m \\ n \end{pmatrix}$$

takes $\mathcal{H}_{k,l}$ to $\mathcal{H}_{m,n}$.

We associate to (k,l) modulo 2 the three representations of $Spin(8)$ which are related by triality: the vector representation λ and the two spin representations Δ_+, Δ_-

$$(1,0) \Longleftrightarrow \Delta_+, \qquad (1,1) \Longleftrightarrow \Delta_-, \qquad (0,1) \Longleftrightarrow \lambda.$$

Since $SL(2,\mathbb{Z}/2)$ is the symmetric group on three elements, $SL(2,\mathbb{Z})$ permutes naturally these three representations. We use $\rho(k,l)$ to denote the representation of $Spin(8)$ corresponding in this way to (k,l) mod 2. Then the conjecture concerns the action of the group $\mathbb{Z}/k \times Spin(8)$ on $\mathcal{H}_{k,l}$:

- the subspace of $\mathcal{H}_{k,l}$ with representation $e^{2\pi i l/k} \otimes \rho(k,l)$ is 8-dimensional;
- the subspace with trivial representation is 2-dimensional;
- all other irreducible subspaces vanish.

For $k = 2$ there is a proof of these conjectures (see [17]) which uses the explicit form of the 2-monopole metric given in [12] and an ad hoc index theorem adapted to the specific asymptotic form of this metric.

It is perhaps natural that it is in the realm of physics that the Dirac operator and spinors appear predominantly, but there can be few examples as rich as this one, where we take the Dirac operator coupled to spinors of spaces of solutions to another Dirac operator.

Exercises 8.1 (i) Show that the pull-back of the Dirac bundle to $\tilde{M}_k^0$ has structure group $SO(k)$. Deduce from looking at the curvature of the connection on the Dirac bundle that there exists an anti-self-dual harmonic 2-form on $\tilde{M}_2^0$ which is invariant under $SO(3)$. This is called the *Sen* form.

(ii) Read in [16] about another implication of S-duality which involves harmonic (anti)-self-dual forms on monopole moduli spaces. The Sen form is an example.

(iii) Read in [5] about a relationship between the Dirac bundle on monopole moduli spaces and the braid group.

References

1. M.F. Atiyah and I.M. Singer: Index theory for skew-adjoint Fredholm operators, *Publ. Math. IHES* **37** (1969), 305–326.
2. M.F. Atiyah and I.M. Singer: Riemann surfaces and spin structures, *Ann. Sci. Ecole Norm. Sup.* **4** (1971), 47–62.
3. C. Bär: Metrics with harmonic spinors, *Geom. Funct. Anal.* **6** (1996), 899—942.
4. W.K. Clifford: 'Mathematical Papers' (R. Tucker, ed.), Macmillan and Co (London, 1882).
5. R. Cohen and J.D.S. Jones: Monopoles, braid groups, and the Dirac operator, *Comm. Math. Phys.* **158** (1993), 241–266.
6. S.K. Donaldson and P.B. Kronheimer: 'The Geometry of Four-Manifolds', Oxford University Press (Oxford, 1990).
7. W.L. Edge: 28 real bitangents, *Proc. Royal Soc. of Edinburgh* **124A** (1994), 729–736.
8. J.P. Gauntlett and J.A. Harvey. S-duality and the dyon spectrum in $N = 2$ super Yang–Mills theory, *Nucl. Phys. B* **463** (1996), 287–314.
9. T. Hausel: Vanishing of intersection numbers on the moduli space of Higgs bundles, *Adv. Theor. Math. Phys* **2** (1998), 1011–1040.
10. N.J. Hitchin: Harmonic spinors, *Adv. Math.* **14** (1974), 1–55.
11. N.J. Hitchin: The self-duality equations on a Riemann surface, *Proc. London Math. Soc.* **55** (1987), 59–126.
12. N.J. Hitchin and M.F. Atiyah: 'The Geometry and Dynamics of Magnetic Monopoles.' M. B. Porter Lectures. Princeton University Press, (Princeton, 1988).
13. N.J. Hitchin: Hyperkähler manifolds, Séminaire Bourbaki 748, 1991/92, *Astérisque* **206** (1992), 137–166.
14. D.Kotschick: Non-trivial harmonic spinors on generic algebraic surfaces, *Proc. Amer. Math. Soc.* **124** (1996), 2315–2318.
15. H. Blaine Lawson: Jr and Marie-Louise Michelson, 'Spin Geometry', Princeton University Press (Princeton, 1989).
16. G.B. Segal and A. Selby: The cohomology of the space of magnetic monopoles, *Comm. Math. Phys.* **177** (1996), 775–787.
17. S. Sethi, M. Stern and E. Zaslow: Monopole and dyon bound states in $N = 2$ supersymmetric Yang–Mills theories. *Nucl. Phys. B* **457** (1995), 484–510.
18. P.M. Sutcliffe: BPS monopoles, *Intern. J. Mod. Phys.* **12** (1997), 4663–4705.
19. R. Weitzenböck: 'Invariantentheorie', P. Noordhoff (Groningen, 1923).

7
Hermitian geometry

SIMON M. SALAMON

Introduction

These notes are based on graduate courses given by the author in 1998 and 1999. The main idea was to introduce a number of aspects of the theory of complex and symplectic structures that depend on the existence of a compatible Riemannian metric. The present notes deal mostly with the complex case, and are designed to introduce a selection of topics and examples in differential geometry, accessible to anyone with an acquaintance of the definitions of smooth manifolds and vector bundles.

One basic problem is, given a Riemannian manifold (M,g), to determine whether there exists an orthogonal complex structure J on M, and to classify all such J. A second problem is, given (M,g), to determine whether there exists an orthogonal almost-complex structure for which the corresponding 2-form ω is closed. The first problem is more tractible in the sense that there is an easily identifiable curvature obstruction to the existence of J, and this obstruction can be interpreted in terms of an auxiliary almost-complex structure on the twistor space of M. This leads to a reasonably complete resolution of the problem in the first non-trivial case, that of four real dimensions. The theory has both local and global aspects that are illustrated in Pontecorvo's classification [89] of bihermitian anti-self-dual 4-manifolds. Much less in known in higher dimensions, and some of the basic classification questions concerning orthogonal complex structures on Riemannian 6-manifolds remain unanswered.

The second problem encompasses the so-called Goldberg conjecture. If one believes this, there do not exist compact Einstein almost-Kähler manifolds for which the metric is not actually Kähler [50]. A thorough investigation of the associated geometry requires an exhaustive analysis of high order curvature jets, and we explain this briefly in the 4-dimensional case. Although we shall pursue the resulting theory elsewhere, readers interested in almost-Kähler manifolds may find some sections of these notes, such as those describing curvature, relevant to this subject.

Complex and almost-complex manifolds are introduced in the first section, and there is a discussion of integrability conditions in terms of exterior forms and their decomposition into types. As an application, we discuss left invariant complex structures on Lie groups, which provide a casual source of examples throughout the notes. The second section is devoted to a study of almost-Hermitian manifolds. It incorporates a brief description of Hodge theory for the $\bar{\partial}$ operator, in order to explain what is special in the Kähler case, though this topic is covered in detail in many standard texts. We also study the space of almost-complex structures compatible with the standard inner product on $\mathbb{R}^{2n}$, and explain how this leads to the definition of twistor space. Attention is given to the case of four dimensions, and examples arising from self-duality, in preparation for more detailed treatment.

The third section introduces the theory of connections on vector bundles by way of covariant differentiation. Basic results for connections compatible with either a Riemannian or symplectic structure exhibit a certain amount of duality involving symmetric versus skew-symmetric tensors. Analogous identities are derived for the complex case, and these lead to the obstruction mentioned above. The subsequent analysis of the Riemann curvature tensor of an almost-Hermitian manifold is partially lifted from [38], and leads straight on to a treatment of the four-dimensional case in the last section. This is followed by a discussion of special Kähler metrics that arise from the study of algebraically integrable systems. The notes conclude with the re-interpretation of a 4-dimensional hyperkähler example constructed earlier from a solvable Lie group.

One aim of the original courses was to introduce the audience to the material underlying some current research papers. The author had partly in mind Apostolov–Grantcharov–Gauduchon [7], Freed [42], Hitchin [59], Poon–Pedersen [87], Gelfand–Retakh–Shubin [47], and there are some direct references to these. The author had useful conversations with the majority of these authors, and is grateful for their help. Even when coverage of a particular topic is scant, we have tried to supply some of the basic references. In this way, readers will know where to look for a more comprehensive and effective treatment.

The author's first exposure to differential geometry took place during one of Brian Steer's undergraduate tutorials, in which some scheduled topic was replaced by a discussion of connections on manifolds. This was after ground had been prepared by the inclusion of books such as [81] in a vacation reading list. I am for ever grateful for Brian's guidance and the influence he had in my choice of research.

Contents

1 Complex manifolds

Holomorphic functions

Let U be an open set of $\mathbb{C}$, and $f : U \to \mathbb{C}$ a continuously differentiable mapping. We may write

$$f = u + iv = u(x,y) + iv(x,y), \tag{1.1}$$

where $x+iy$ and $u+iv$ are complex coordinates on the domain and target respectively. The function f is said to be *holomorphic* if the resulting differential $df = f_* : \mathbb{R}^2 \to \mathbb{R}^2$ is complex linear. This means that the Jacobian matrix of partial derivatives commutes with multiplication by i, so that

$$\begin{pmatrix} u_x & u_y \\ v_x & v_y \end{pmatrix}\begin{pmatrix} 0 & -1 \\ 1 & 0 \end{pmatrix} = \begin{pmatrix} 0 & -1 \\ 1 & 0 \end{pmatrix}\begin{pmatrix} u_x & u_y \\ v_x & v_y \end{pmatrix},$$

and one obtains the Cauchy–Riemann equations

$$\begin{cases} u_x &= v_y, \\ v_x &= -u_y. \end{cases}$$

The differential of a holomorphic function then has the form

$$df = u_x \begin{pmatrix} 1 & 0 \\ 0 & 1 \end{pmatrix} + v_x \begin{pmatrix} 0 & -1 \\ 1 & 0 \end{pmatrix}.$$

This corresponds to multiplication by the complex scalar $u_x + iv_x$, which coincides with the complex limit

$$f'(z) = \lim_{h \to 0} \frac{f(z+h) - f(z)}{h}.$$

Now let $z = x+iy$ and $\bar{z} = x - iy$, so that the mapping (1.1) may be regarded as a function $f(z,\bar{z})$ of the variables $z, \bar{z}$. Define complex-valued 1-forms

$$dz = dx + idy, \quad d\bar{z} = dx - idy,$$

which are dual to the tangent vectors

$$\frac{\partial}{\partial z} = \frac{1}{2}\left(\frac{\partial}{\partial x} - i\frac{\partial}{\partial y}\right), \quad \frac{\partial}{\partial \bar{z}} = \frac{1}{2}\left(\frac{\partial}{\partial x} + i\frac{\partial}{\partial y}\right).$$

Then f is holomorphic if and only if

$$\frac{\partial f}{\partial \bar{z}} = 0,$$

which is the case whenever f may be expressed explicitly in terms of z alone. Indeed, Cauchy's theorem implies that f is holomorphic if and only if it is complex analytic, i.e. if for every $z_0 \in U$ it has a convergent power series expansion in $z - z_0$ valid on some disk around z_0.

Now suppose that U is an open set of $\mathbb{C}^m$ and that $F : U \to \mathbb{C}^n$ is a continuously differentiable mapping. It is convenient to identify $\mathbb{C}^m$ with $\mathbb{R}^{2m}$ by means of the association

$$(z^1, \ldots, z^m) \leftrightarrow (x^1, \ldots, x^m, y^1, \ldots, y^m),$$

where $z^r = x^r + iy^r$. Then F is *holomorphic* if and only if

$$dF \circ \begin{pmatrix} 0 & -I \\ I & 0 \end{pmatrix} = \begin{pmatrix} 0 & -I \\ I & 0 \end{pmatrix} \circ dF,$$

where I denotes an identity matrix of appropriate size. Equivalently,

$$\frac{\partial F^j}{\partial \bar{z}^k} = 0, \quad j = 1, \ldots, n, \; k = 1, \ldots, m. \tag{1.2}$$

A complex manifold is a smooth manifold M, equipped with a smooth atlas with the additional property that its transition functions are holomorphic wherever defined. Thus, a complex manifold has an open covering $\{U_\alpha\}$ and local diffeomorphisms

$$\phi_\alpha : U_\alpha \to \phi(U_\alpha) \subseteq \mathbb{C}^m$$

with the property that $\phi_\alpha \circ \phi_\beta^{-1}$ is holomorphic whenever $U_\alpha \cap U_\beta \neq \emptyset$. If x is a point of U_α and $\phi_\alpha = (z^1, \ldots, z^m)$ then $z^1, \ldots, z^m$ are called *local holomorphic coordinates* near x. The complex structure is deemed to depend only on the atlas up to the usual notion of equivalence, whereby two atlases are equivalent if their mutual transition functions are holomorphic.

Example 1.3 Let Γ denote the additive subgroup of $\mathbb{C}^m$ generated by a set of $2m$ vectors, linearly independent over $\mathbb{R}$. Then $M = \mathbb{C}^m/\Gamma$ is diffeomorphic to the real torus $\mathbb{R}^{2m}/\mathbb{Z}^{2m} = (\mathbb{R}/\mathbb{Z})^{2m} \cong (S^1)^{2m}$. Let $\pi : \mathbb{C}^m \to M$ be the projection. An atlas is constructed from pairs (U_α, ϕ_α) where $\pi \circ \phi_\alpha$ is the identity on U_α, and the transition functions are translations by elements of $\mathbb{Z}^{2m}$ in $\mathbb{C}^m$.

Fix a point of M, and let T denote real the tangent space to M at that point, $T_\mathbb{C}$ its complexification $T \oplus iT$. The endomorphism J of T defined by

$$(d\phi_\alpha)^{-1} \circ \begin{pmatrix} 0 & -I \\ I & 0 \end{pmatrix} \circ d\phi_\alpha$$

is independent of the choice of chart, since $d(\phi_\alpha \circ \phi_\beta^{-1}) = d\phi_\alpha \circ d\phi_\beta^{-1}$ is complex-linear. The same is true of the subspace $T^{1,0}$ of $T_\mathbb{C}$ generated by the tangent vectors $\partial/\partial z^1, \ldots, \partial/\partial z^m$, since this coincides with the i eigenspace $\{X - iJX\colon X \in T\}$, and

$$\frac{\partial}{\partial z^r} = \sum_{s=1}^{m} \frac{\partial w^s}{\partial z^r} \frac{\partial}{\partial w^s}.$$

Setting

$$T^{0,1} = \overline{T^{1,0}} = \left\langle \frac{\partial}{\partial \bar{z}_1}, \ldots, \frac{\partial}{\partial \bar{z}_m} \right\rangle$$

gives a decomposition

$$T_{\mathbb{C}} = T^{1,0} \oplus T^{0,1}.$$

There is a dual decomposition

$$T^*_{\mathbb{C}} = \Lambda^{1,0} \oplus \Lambda^{0,1}, \tag{1.4}$$

where $\Lambda^{1,0}$ is the annihilator of $T^{0,1}$ spanned by $dz^1, \ldots, dz^m$. Varying from point to point, we may equally well regard $T^*_{\mathbb{C}}$ as the complexified cotangent *bundle*, and (1.4) as a decomposition of it into conjugate subbundles. The endomorphism J acts on T^* by the rule $(J\alpha)(v) = \alpha(Jv)$ and this implies that

$$J\,dz^r = i\,dz^r, \quad \text{or} \quad J\,dx^r = -dy^r.$$

Thus a function $x + iy$ with real and imaginary components x, y is holomorphic if and only if $dx - J dy = 0$.

A *holomorphic function* on a complex manifold M is a mapping $f\colon M \to \mathbb{C}$ such that $f \circ \phi^{-1}$ is holomorphic on U for any chart (U, ϕ). If $f \circ \phi^{-1}$ is allowed to have poles at isolated points then f is merely *meromorphic.* On a *compact* complex manifold M with charts $\phi_\alpha : U_\alpha \to \mathbb{C}^m$, any holomorphic function is constant, whereas the field of meromorphic functions can be used to define the *algebraic dimension* $a(M)$, which satisfies $a(M) \leqslant m$.

Holomorphic mappings between complex manifolds are defined by reference to charts and the resulting condition (1.2), with the corresponding generalization to the meromorphic case. Two complex manifolds M, N are called *biholomorphic* if there exists a bijective holomorphic mapping $f : M \to N$; in this case f^{-1} is automatically holomorphic. Biholomorphism is the natural equivalence relation between complex manifolds, although other notions are important in the realm of algebraic geometry. For example, if there exist open sets $U \subseteq M$ and $V \subseteq N$ and a biholomorphic map $f : U \to V$, then M and N are *birational.* For surfaces this implies $a(M) = a(N)$.

Exercises 1.5 (i) Prove that if M is a compact complex manifold, any holomorphic mapping from M to the complex numbers $\mathbb{C}$ is necessarily constant. Give an example of a complex manifold for which there is no holomorphic mapping $\mathbb{C} \to M$ (it may help to know that such a complex manifold is called *hyperbolic*).

(ii) Let Γ be a discrete group, and M a complex manifold. Say what is meant by a *holomorphic* action of Γ on M, and give sufficient conditions for the set M/Γ of cosets to be a complex manifold for which the projection $\pi : M \to M/\Gamma$ is holomorphic.

Examples and special classes

The basic compact example of a complex manifold is the complex projective space

$$\mathbb{CP}^m = \{1\text{-dimensional subspaces of } \mathbb{C}^{m+1}\} = \frac{\mathbb{C}^{m+1} \setminus \{0\}}{\mathbb{C}^*}.$$

A point of $\mathbb{CP}^m$ is denoted $[Z^0, Z^1, \ldots, Z^m]$, and represents the span of a non-zero vector in $\mathbb{C}^{m+1}$. An open set U_α is defined by the condition $Z^\alpha \neq 0$, for each $\alpha = 0, 1, \ldots, m$, and holomorphic coordinates are given by Z^r/Z^α on U_α. For example, fix $\alpha = 0$ and set $z^r = Z^r/Z^0$. Then transition functions on $U_0 \cap U_1$ are given by

$$(z^1, \ldots, z^m) \longmapsto \left(\frac{1}{z^1}, \frac{z^2}{z^1}, \ldots, \frac{z^m}{z^1}\right), \quad z^1 \neq 0,$$

and are clearly holomorphic.

Definition 1.6 *A compact complex manifold M is projective if there exists a holomorphic embedding of M into $\mathbb{CP}^n$ for some n.*

Projective manifolds are most easily defined as the zeros of homogeneous polynomials in $\mathbb{CP}^n$. For example, if F is a homogeneous polynomial of degree d in $Z^0 = X, Z^1 = Y, Z^2 = Z$ then

$$M_F = \{[X, Y, Z] \in \mathbb{CP}^2 \colon F(X, Y, Z) = 0\}$$

is a 1-dimensional complex submanifold of the projective plane $\mathbb{CP}^2$ provided $F_X = F_Y = F_Z = 0$ implies $X = Y = Z = 0$. Note that the vanishing of the partial derivatives of F at a point implies the vanishing of F at that point, by Euler's formula. A complex manifold of complex dimension 1 is called a *Riemann surface*, and is necessarily orientable.

Example 1.7 To be more specific, consider the hypersurface M_F of degree d defined by the equation $X^d + Y^d + Z^d = 0$. The projection

$$\begin{aligned} M_F &\to \mathbb{CP}^1 \\ (X, Y, Z) &\longmapsto (Y, Z) \end{aligned}$$

is a 'branched $d:1$ covering': each fibre has d points unless $Y^d + Z^d = 0$, an equation that has d solutions in $\mathbb{CP}^1$. The resulting topology can be derived from the basic set-theoretic properties of the Euler characteristic χ, that imply that

$$\chi(M_F) = d\,\chi(\mathbb{CP}^1 \setminus d \text{ points}) + d = d(2-d) + d.$$

Thus the genus of the resulting compact oriented surface satisfies $2 - 2g = -d^2 + 3d$, and

$$g = \frac{(d-1)(d-2)}{2}. \tag{1.8}$$

The degree-genus formula (1.8) is always valid for a smooth plane curve of degree d (see [66] for a thorough treatment). The cases $d=1,2$ are elementary. If $d=3$ then $g=1$, and a smooth cubic curve in $\mathbb{CP}^2$ is topologically a torus. It is known that any such cubic is projectively equivalent to

$$C_\lambda\colon\ Y^2Z = X(X-Z)(X-\lambda Z), \quad \lambda\notin\{0,1\},$$

and that C_λ, C_μ are equivalent iff

$$\mu\in\left\{\lambda, \frac{1}{\lambda}, 1-\lambda, \frac{1}{1-\lambda}, \frac{\lambda}{\lambda-1}, \frac{\lambda-1}{\lambda}\right\},$$

in which case

$$j = \frac{(\mu^2-\mu+1)^3}{\mu^2(\mu-1)^2}$$

has a unique value. The manifold C_μ is biholomorphic to the quotient group $\mathbb{C}^2/\langle 1,\tau\rangle$, where τ belongs to the upper half-plane. The special case $\tau=i$ corresponds to $j=27/4$, but in general the mapping $\tau\longmapsto j$ involves non-trivial function theory.

If $d=4$ then $g=3$, so no smooth curve in $\mathbb{CP}^2$ has genus 2. But that does not of course mean to say that a torus with two holes does not admit a complex structure. Indeed, any oriented surface can be made into a complex manifold by first embedding it in $\mathbb{R}^3$. The classical isothermal coordinates theorem asserts that, given a surface in $\mathbb{R}^3$, there exist local coordinates x,y such that the first fundamental form equals $f(dx^2+dy^2)$ for some positive function f. It then suffices to define $J(\partial/\partial x)=(\partial/\partial y)$; transition functions will be automatically holomorphic.

Any Riemann surface of genus at least 2 is biholomorphic to Δ/Γ, where Δ denotes the open unit ball in $\mathbb{C}$, and Γ is a discrete group acting holomorphically. Many important classes of complex manifolds may be defined by taking discrete quotients of an open set of $\mathbb{C}^m$.

Example 1.9 We have already mentioned complex tori $\mathbb{C}^m/\Gamma$, in which the lattice Γ is a subgroup of the abelian group $(\mathbb{C}^m,+)$. The *Kodaira surface S* is obtained in a not dissimilar manner by regarding $\mathbb{C}^2$ as the real nilpotent group of matrices

$$\begin{pmatrix} 1 & x & u & t\\ 0 & 1 & y & 0\\ 0 & 0 & 1 & 0\\ 0 & 0 & 0 & 1\end{pmatrix}$$

under multiplication. If Γ denotes the subgroup for which x,y,u,t are integers then S is the quotient $\Gamma\backslash\mathbb{C}^2$ of $\mathbb{C}^2$ by left-multiplication of Γ. This action leaves invariant the 1-forms

$$dx, \quad dy, \quad du-x\,dy, \quad dt,$$

and therefore the closed 2-form $\eta = (dx + idy) \wedge (du - xdy + idt)$ which defines a holomorphic symplectic structure on S [69]. Foundations of a general theory of such compact quotients of nilpotent Lie groups are established in [79]. To cite just two applications of Kodaira surfaces, refer to [46, 55].

Other complex surfaces defined as discrete quotients, include Hopf and Inoue surfaces [14]. A *Hopf surface* is a compact complex surface whose universal covering is biholomorphic to $\mathbb{C}^2 \setminus \{0\}$, and is called *primary* if $\pi_1 \cong \mathbb{Z}$. Any primary Hopf surface is homeomorphic to $S^1 \times S^3$.

Example 1.10 Let Γ denote the infinite cyclic group generated by a non-zero complex number λ with $|\lambda| \neq 1$. Then $M_\lambda = (\mathbb{C}^m \setminus \{0\})/\langle\lambda\rangle$ is a complex manifold diffeomorphic to $S^1 \times S^{2m-1}$. This is an example of a Calabi–Eckmann structure that exists on the product of any two odd-dimensional spheres [30]. When $m = 2$, the complex manifold M_λ is a primary Hopf surface.

An almost-complex structure on a real $2m$-dimensional vector space $T \cong \mathbb{R}^{2m}$ is a linear mapping $J : T \to T$ with $J^2 = -\mathbf{1}$. An almost-complex structure on a differentiable manifold is the assignment of an almost-complex structure J on each tangent space that varies smoothly. We have seen that any complex manifold is naturally equipped with such a tensor. Conversely, an almost-complex structure J is said to be *integrable* if M has the structure of a complex manifold with local coordinates $\{z^r\}$ for which $J\,dz^r = i\,dz^r$. This means that the almost-complex structure has locally the standard form

$$J = \sum_{r=1}^{m} \left(\frac{\partial}{\partial y^r} \otimes dx^r - \frac{\partial}{\partial x^r} \otimes dy^r \right). \tag{1.11}$$

In general this will not be true, and we shall refer to the pair (M, J) as an almost-complex manifold.

Definition 1.12 *A hypercomplex manifold is a smooth manifold equipped with a triple I_1, I_2, I_3 of integrable complex structures satisfying*

$$I_1 I_2 = I_3 = -I_2 I_1. \tag{1.13}$$

It is easy to see that such a manifold must have real dimension $4k$ for some $k \geqslant 1$. Less obvious is the fact that the integrability of just I_1 and I_2 implies (in the presence of (1.13)) that of I_3. Thus, a hypercomplex manifold can be defined by the existence of an anticommuting pair of complex structures. It follows that $aI_1 + bI_2 + cI_3$ is a complex structure for any $(a, b, c) \in S^2$.

Exercises 1.14 (i) Show that the set $\mathbb{F}_n$ of full flags in $\mathbb{C}^n$ (that is, sequences $\{0\} \subset V_1 \subset \cdots \subset V_n = \mathbb{C}^n$ of subspaces with $\dim V_i = i$) can be given the structure of a complex manifold. The choice of an appropriate line bundle determines a holomorphic embedding $\mathbb{F}_3 \to \mathbb{CP}^7$.

(ii) Show that if $m = 2k$ and $\lambda \in \mathbb{R}$ in Example 1.10 then M is hypercomplex. Explain why certain Hopf surfaces admit two distinct families of hypercomplex structures [45].

(iii) Let $m > 1$, and suppose that $M = \mathbb{C}^m/\Gamma$ is a complex torus as in Example 1.3. Find sufficient conditions on Γ so that M is projective. In this case M is called an *abelian variety.*

Use of differential forms

On any almost-complex manifold, we may extend (1.4) by defining

$$\textstyle\bigwedge^k T^*_{\mathbb{C}} = \bigoplus_{p+q=k} \Lambda^{p,q}, \tag{1.15}$$

where

$$\Lambda^{p,q} \cong \textstyle\bigwedge^p(\Lambda^{1,0}) \otimes \bigwedge^q(\Lambda^{0,1}).$$

These summands represent either vector bundles or the vector spaces corresponding to the fibres of these bundles at a given (possibly unspecified) point, depending on the context. On a complex manifold, $\Lambda^{p,q}$ is spanned by elements $dz^{i_1} \wedge \cdots \wedge dz^{i_p} \wedge d\bar{z}^{j_1} \wedge \cdots \wedge d\bar{z}^{j_q}$, but it is still well defined in the presence of an almost-complex structure. To avoid complexifying spaces unnecessarily, we shall make occasional use of the notation

$$[\![\Lambda^{p,q}]\!] \quad (p \neq q), \qquad [\Lambda^{p,p}] \tag{1.16}$$

of [92] for the real subspaces of $\Lambda^k T^*$ of dimension $2pq$ and p^2, with complexifications $\Lambda^{p,q} \oplus \Lambda^{q,p}$ and $\Lambda^{p,p}$ respectively.

The *Nijenhuis tensor* N of J is defined by

$$N(X,Y) = [JX,JY] - [X,Y] - J[JX,Y] - J[X,JY], \tag{1.17}$$

and

$$\begin{aligned} -N(X,Y) &= \Re\mathrm{e}\Big([X - iJX, Y - iJY] + iJ[X - iJX, Y - iJY]\Big) \\ &= 8\,\Re\mathrm{e}\Big([X^{1,0}, Y^{1,0}]^{0,1}\Big). \end{aligned} \tag{1.18}$$

Lemma 1.19 *The following are equivalent:*

(i) $d(\Gamma(\Lambda^{1,0})) \subseteq \Gamma(\Lambda^{2,0} \oplus \Lambda^{1,1})$;
(ii) $\Gamma(T^{1,0})$ *is closed under Lie bracket of vector fields;*
(iii) $N(X,Y) = 0$ *for all vector fields* X, Y.

To see the equivalence of (i) and (ii), let $\alpha \in \Gamma(\Lambda^{1,0})$, and use the formula

$$2d\bar{\alpha}(A,B) = A(\bar{\alpha}B) - B(\bar{\alpha}A) - \bar{\alpha}[A,B] = -\bar{\alpha}[A,B], \quad A,B \in \Gamma(T^{1,0}).$$

The equivalence of (ii) and (iii) follows from (1.18). □

The composition

$$\Gamma(\Lambda^{1,0}) \xrightarrow{d} \Gamma(\bigwedge^2 T^*_{\mathbb{C}}) \to \Gamma(\Lambda^{0,2})$$

is an element of

$$\mathrm{Hom}(\Lambda^{1,0}, \Lambda^{0,2}) \cong \Lambda^{0,2} \otimes (\Lambda^{1,0})^* \subset \bigwedge^2 T^*_{\mathbb{C}} \otimes T_{\mathbb{C}}, \tag{1.20}$$

since any differentiation is cancelled out by the projection. It follows that the value of $N(X,Y)$ at a point p depends only on the values of X, Y at p, and is essentially the real part of the element in (1.20). It is clear that the Nijenhius tensor N vanishes on a complex manifold. The converse is a deep result of [83] that took some years to prove after it was first conjectured.

Theorem 1.21 *An almost-complex structure J is integrable if N is identically zero.*

Example 1.22 An almost-complex structure J can equally well be defined by its action on each *co*tangent space T^*. Define one on $\mathbb{R}^4$ with coordinates (x, y, u, t) by setting

$$J(dx) = -dy, \quad J(dt) = -du + xdy,$$

so that the space of $(1,0)$-forms is generated by $\alpha^1 = dx + idy$ and $\alpha^2 = dt + i(du - xdy)$. Since $d\alpha^2$ is a multiple of $\alpha^1 \wedge \bar{\alpha}^1$, Theorem 1.21 implies that there must exist local holomorphic coordinates z^1, z^2. In fact, we may take

$$z^1 = x + iy, \quad z^2 = t + iu - ixy + \tfrac{1}{2}y^2,$$

since $dz^2 = \alpha^2 - iy\alpha^1$. This example is relevant to Example 1.9, since J passes to the compact quotient S.

From now on, we shall denote the space $\Gamma(\Lambda^{p,q})$ of smooth forms of type (p,q) (that is, smooth sections of the vector bundle with fibre $\Lambda^{p,q}$) by $\Omega^{p,q}$. Now suppose that M is a complex manifold. Then we may write $d = \partial + \bar{\partial}$, where

$$\partial : \Omega^{p,q} \to \Omega^{p+1,q}, \quad \bar{\partial} : \Omega^{p,q} \to \Omega^{p,q+1}.$$

Since $d^2 = 0$ we get

$$\partial^2 = 0, \quad \partial\bar{\partial} + \bar{\partial}\partial = 0, \quad \bar{\partial}^2 = 0,$$

and this gives $\bigoplus_{p,q} \Omega^{p,q}$ the structure of a bigraded bidifferential algebra [82].

It is convenient to use the language of spectral sequences, and we set $E_0^{p,q} = \Omega^{p,q}$ and $d_0 = \bar{\partial} : E_0^{p,q} \to E_0^{p,q+1}$. The Dolbeault cohomology groups of M are given by

$$H^{p,q}_{\bar{\partial}} = E_1^{p,q} = \frac{\ker(\bar{\partial}|\Omega^{p,q})}{\bar{\partial}(\Omega^{p,q-1})}, \tag{1.23}$$

and there is a linear mapping $d_1 : E_1^{p,q} \to E_1^{p+1,q}$ induced from ∂.

The rules of a spectral sequence decree that, given

$$d_r : E_r^{p,q} \to E_r^{p+r,q-r+1}, \quad {d_r}^2 = 0,$$

spaces $E_{r+1}^{p,q}$ are defined as the cohomology groups of the complex $(E_r^{p,q}, d_r)$. In the present case, successive differentials are obtained by diagram chasing, and d_r will vanish for $r > n+1$, so we may write $E_\infty^{p,q} = E_{n+2}^{p,q}$. Frölicher's theorem asserts that the spectral sequence converges to the deRham cohomology of M, which means that

$$H^k(M, \mathbb{R}) = \bigoplus_{p+q=k} E_\infty^{p,q},$$

in a way compatible with a natural filtration on $H^k(M, \mathbb{R})$.

To illustrate this, suppose that M has complex dimension $m = 3$. We obtain six non-zero instances of d_2. For example, to define $d_2 \colon E_2^{1,2} \to E_2^{3,1}$, suppose that $[[x]] \in E_2^{1,2}$ with $x \in E_0^{1,2}$. Then $\partial x = \bar\partial y$ for some $y \in E_0^{2,1}$, and we set $\alpha([[x]]) = [[\partial y]]$. There are only two non-zero instances of d_3, mapping $E_3^{0,2} \to E_3^{3,0}$ and $E_3^{0,3} \to E_3^{3,1}$.

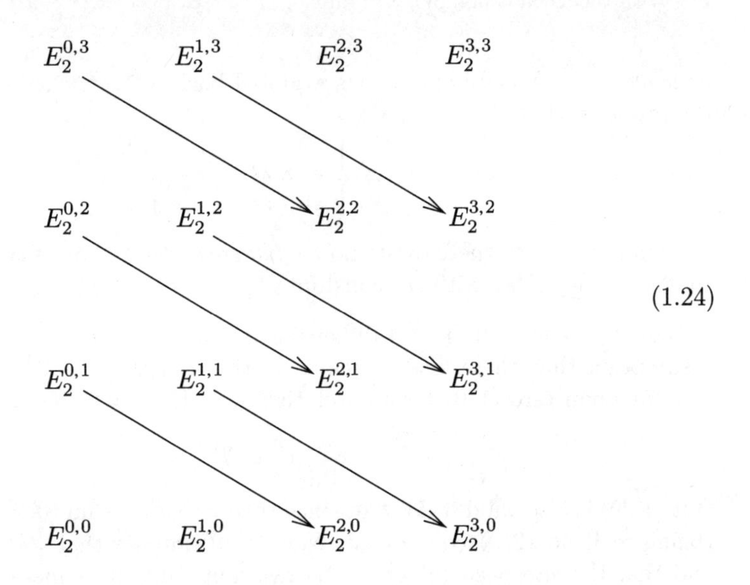

(1.24)

The conclusion is that there is an isomorphism

$$\begin{aligned} H^3(M, \mathbb{R}) \cong (\ker d_3 \text{ in } E_3^{0,3}) &\oplus (\ker d_2 \text{ in } E_2^{1,2}) \\ &\oplus (\operatorname{coker} d_2 \text{ in } E_2^{2,1}) \oplus (\operatorname{coker} d_3 \text{ in } E_3^{3,0}). \end{aligned}$$

On a compact manifold for which (1.23) are finite-dimensional, one deduces that

$$b_k \leqslant \sum_{p+q=k} h^{p,q}, \tag{1.25}$$

since $\dim E_{r+1}^{p,q} \leqslant \dim E_r^{p,q}$ for all r. It is known that if M is a compact complex surface (i.e., $n = 2$) then all the differentials d_2 vanish so there is equality [14]. The wealth of homogeneous spaces in [108] provide test cases for $n = 3$, though explicit examples for which d_r does not vanish for $r \geqslant 3$ are rare (in the literature, but probably not in real life). Instances arising from invariant complex structures on compact Lie groups are described in [88].

Exercises 1.26 (i) Discover the definition of the *sheaf* $\mathcal{O}$ of germs of local holomorphic functions on a complex manifold M, and that of the Čech cohomology groups of $\mathcal{O}$.

(ii) Find out how to prove the following, at least for $q = 1$. Let $m \in M$, and suppose that $\alpha \in \Omega^{0,q}$ satisfies $\bar\partial\alpha = 0$ with $q \geqslant 1$. Then there exists a $(0, q-1)$-form β on a neighbourhood of m such that $\bar\partial\beta = \alpha$.

(iii) Explain why this $\bar\partial$-Poincaré lemma implies that the complex

$$0 \to \Omega^{0,0} \to \Omega^{0,1} \to \Omega^{0,2} \to \cdots \to \Omega^{0,n} \to 0$$

is a *resolution* of $\mathcal{O}$, and that the Dolbeault cohomology groups $E_1^{0,q}$ coincide with the Čech cohomology groups of $\mathcal{O}$.

Example 1.27 A manifold X has a global basis $\{e^1, e^2, e^3, e^4\}$ of 1-forms with the property that

$$de^i = \begin{cases} 0, & i = 1, 2, \\ e^1 \wedge e^2, & i = 3, \\ e^1 \wedge e^3, & i = 4. \end{cases} \tag{1.28}$$

We shall show that there exists no *complex* structure J on X with the property that $Je_i = \sum_j J_i^j e_j$ with J_i^j constant.

Let T^* denote the real 4-dimensional space spanned by the e^i. Under the assumption that there exists such a J, the subspace $\langle e^1, e^2, e^3\rangle_{\mathbb{C}}$ of $T^*_{\mathbb{C}}$ must contain a non-zero $(1,0)$ form α relative to J. If $d\alpha = 0$ then $\alpha \in \langle e^1, e^2\rangle_{\mathbb{C}}$, and

$$e^1 \wedge e^2 \in \Lambda^{1,1}. \tag{1.29}$$

But (1.29) is also valid if $d\alpha \neq 0$, since then $e^1 \wedge e^2$ has no $(0,2)$-component and (being real) no $(2,0)$ component. Now (1.29) implies that $Je^1 \wedge Je^2 = e^1 \wedge e^2$ and that the subspace $\langle e^1, e^2\rangle$ is J-invariant. But the same argument with e^3 replaced by e^4 would give that $\langle e^1, e^3\rangle$ is J-invariant, which is impossible.

Structures on Lie groups

The last exercise is one about a Lie group in disguise. The equations (1.28) determine a Lie algebra structure on any tangent space T to X, and it follows that we may take X to be a corresponding (nilpotent) group. The conclusion is that this Lie group admits no left-invariant complex structure. By contrast, left-invariant complex structures exist on any compact Lie group of even dimension (references are given below). We shall illustrate the case of $SU(2)\times SU(2)$ in this subsection, that was motivated by papers such as [87] describing the deformation of invariant complex structures.

A left-invariant complex structure on a Lie group G is determined by an almost-complex structure J on its Lie algebra $\mathfrak{g}$ satisfying $N=0$ in which the brackets of (1.20) are now interpreted in a purely algebraic sense. The i eigenspace $\mathfrak{g}^{1,0}$ of J is always a *complex Lie algebra*, but the Lie group G itself is not in general complex. The integrability condition is satisfied in the following special cases:

(i) $[JX,Y]=J[X,Y]$. In this case we may write $J=i$ so $(\mathfrak{g},i)$ is a complex Lie algebra isomorphic to $\mathfrak{g}^{1,0}$, and G *is* a complex Lie group.

(ii) $[JX,JY]=[X,Y]$. This condition is equivalent to asserting that d maps the subspace Λ of $(1,0)$-forms $\mathfrak{g}_{\mathbb{C}}$ into $\Lambda^{1,1}$, or that $\mathfrak{g}^{1,0}$ is an *abelian* Lie algebra.

From now on, let $G=SU(2)\times SU(2)$; as a manifold G is the same thing as $S^3\times S^3$. The space of left-invariant 1-forms on the first $SU(2)$ factor is modelled on the Lie algebra $\mathfrak{su}(2)$, and it follows that there is a global basis $\{e^1,e^2,e^3\}$ of 1-forms with the property that

$$de^1=e^2\wedge e^3,\quad de^2=e^3\wedge e^1,\quad de^3=e^1\wedge e^2.$$

Fix a similar basis $\{e^4,e^5,e^6\}$ of left-invariant 1-forms for the second $SU(2)$ factor. Left-invariant almost-complex structures on G are determined by almost-complex structures on the vector space $\mathbb{R}^6$ spanned by the e^i. As an example, consider the almost-complex structure J_0 whose space Λ of $(1,0)$-forms is spanned by

$$\sigma^1=e^1+ie^2,\quad \sigma^2=e^3+ie^4,\quad \sigma^3=e^5+ie^6.$$

It is easy to check that $d\sigma^i$ has no $(0,2)$-component for each i, so that J_0 is integrable.

There is nothing terribly special about J_0. Fix the inner product for which $\{e^i\}$ is an orthonormal basis, and set $U=\langle e^1,e^2,e^3\rangle$, $V=\langle e^4,e^5,e^6\rangle$. Then J_0 belongs to a family of complex structures parametrized by

$$(\mathrm{u},\mathrm{v})\in S^2\times S^2\subset U\times V, \tag{1.30}$$

which is the orbit of J_0 under right translation by $SU(2)\times SU(2)$. The ordered pair (u,v) determines the structure with $(1,0)$-forms $\{\mathrm{u}',\mathrm{u}+i\mathrm{v},\mathrm{v}'\}$ where

$$\begin{aligned}&\mathrm{u}' \text{ is an isotropic vector in } \langle e^1,e^2,e^3\rangle_{\mathbb{C}} \text{ orthogonal to u, and}\\ &\mathrm{v}' \text{ is an isotropic vector in } \langle e^4,e^5,e^6\rangle_{\mathbb{C}} \text{ orthogonal to v.}\end{aligned} \tag{1.31}$$

In this way, J_0 corresponds to $\mathrm{u}=e^3$ and $\mathrm{v}=e^4$.

Consider now deformations of the complex structure J_0 without reference to an inner product. Suppose that J is an almost-complex structure whose space

Λ of $(1,0)$-forms is spanned by $\alpha^1, \alpha^2, \alpha^3$, and set

$$\eta = \alpha^{123} \wedge \bar{\alpha}^{123}, \qquad \xi = \alpha^{123} \wedge \bar{\sigma}^{123}$$

($\bar{\alpha}^{123}$ is short for $\bar{\alpha}^1 \wedge \bar{\alpha}^2 \wedge \bar{\alpha}^3$, etc). The necessary condition $\Lambda \cap \overline{\Lambda} = \{0\}$ is equivalent to $\eta \neq 0$. On the other hand, we may regard the set of J satisfying $\xi \neq 0$ as a large affine neighbourhood of J_0, on which row echelon reduction allows us to take

$$\begin{aligned}
\alpha^1 &= \sigma^1 + r\bar{\sigma}^1 + s\bar{\sigma}^2 + t\bar{\sigma}^3, \\
\alpha^2 &= \sigma^2 + u\bar{\sigma}^1 + v\bar{\sigma}^2 + w\bar{\sigma}^3, \\
\alpha^3 &= \sigma^3 + x\bar{\sigma}^1 + y\bar{\sigma}^2 + z\bar{\sigma}^3,
\end{aligned}$$

with $r, s, t, u, v, w, x, y, z \in \mathbb{C}$. It is convenient to let X denote the 3×3 matrix formed by these nine coefficients as displayed. The correspondence $J \leftrightarrow X$ is then one-to-one, and $J_0 \leftrightarrow 0$. Structures (such as $-J_0$) for which $\xi = 0$ can only be obtained by letting some of the entries of X become infinite.

Proposition 1.32 *Any left-invariant complex structure on G admits a basis $\{\mathrm{u}', \mathrm{u} + \tau\mathrm{v}, \mathrm{v}'\}$ of $(1,0)$-forms satisfying* (1.30) *and* (1.31)*, with $\tau \in \mathbb{C} \setminus \mathbb{R}$.*

Proof By Lemma 1.19, we require that each $d\alpha^i$ have no $(0,2)$-component relative to J. This is equivalent to the equations

$$d\alpha^i \wedge \alpha^{123} = 0, \quad i = 1, 2, 3, \tag{1.33}$$

since wedging with the $(3,0)$-form α^{123} annihilates everything but the $(0,2)$-component of $d\alpha^i$. Computations show that this implies that $v \neq 1$. Furthermore, setting $a = v - 1$,

$$\text{(i)} \qquad a^2 X = \begin{pmatrix} s^2(a+1) & sa^2 & swa \\ sa(a+2) & a^2(a+1) & wa^2 \\ sw(a+2) & wa(a+2) & w^2(a+1) \end{pmatrix},$$

$$\text{(ii)} \qquad \eta = \frac{-8i(1 - |v|^2)(|a|^2 + |s|^2)^2(|a|^2 + |w|^2)^2}{|a|^4} e^{12\cdots 6}.$$

We may now define

$$\begin{aligned}
\mathrm{u}' &= a(a\alpha^1 - s\alpha^2) = (a^2 - s^2)e^1 + i(a^2 + s^2)e^2 - 2as\, e^3, \\
\mathrm{v}' &= a(w\alpha^2 - a\alpha^3) = 2iaw\, e^4 - (a^2 + w^2)e^5 - i(a^2 - w^2)e^6,
\end{aligned}$$

and τ can be expressed as a function of v. □

The sign of η is determined by $|v|$, and the set $\mathcal{C}$ of invariant complex structures inducing the same orientation as J_0 has $|v| < 1$. The above proposition allows us to identify $\mathcal{C}$ with $\Delta \times (S^2 \times S^2)$, where Δ is the unit disk in $\mathbb{C}$.

If $v = 0$ then X is a skew-symmetric matrix and we recover the subset $S^2 \times S^2$ described above.

The fact that any even-dimensional compact Lie group admits a complex structure is due to Samelson [97] and Wang [104]. This theory was generalized by Snow [99], who described the moduli space of so-called 'regular' invariant complex structures on any reductive Lie group such as $GL(2m, \mathbb{R})$. The nilpotent case is covered by [32] and the author's paper [95].

Exercises 1.34 (i) The manifold $S^3 \times S^1$ can be identified with the Lie group $SU(2) \times U(1)$. As such, it has a global basis $\{e^1, e^2, e^3, e^4\}$ of left-invariant 1-forms such that

$$de^1 = e^{23}, \quad de^2 = e^{31}, \quad de^3 = e^{12}, \quad de^4 = 0.$$

Deduce that $S^3 \times S^1$ has a left-invariant hypercomplex structure, but no abelian complex structures.

(ii) Let $\mathfrak{g}$ be the Lie algebra of an even-dimensional compact Lie group, and let $\mathfrak{g} = \mathfrak{t} \oplus \bigoplus [\![\mathfrak{g}_\alpha]\!]$ be a root space decomposition in which the big sum is taken over a choice of positive roots. Noting that each real 2-dimensional space $[\![\mathfrak{g}_\alpha]\!]$ carries a natural almost-complex structure, show that any almost-complex structure on $\mathfrak{t}$ can be extended to an almost-complex structure J on $\mathfrak{g}$ with $N = 0$.

(iii) Compute the spaces and maps in the array (1.24) for the complex manifold $(S^3 \times S^3, J_0)$, and determine $\sum_{p=0}^{3} h^{p,3-p}$.

2 Almost-Hermitian metrics

Let M be a smooth manifold. A Riemannian metric on M is the assignment of a smoothly varying inner product (that is, a positive-definite symmetric bilinear form) on each tangent space. We write $g(X, Y)$ (rather than $\langle X, Y \rangle$) to denote the inner product of two tangent vectors or fields X, Y. Thus, g is a smooth section over M of $T^* \otimes T^*$, where T^* is the cotangent bundle, and indeed $g \in \Gamma(\bigodot^2 T^*)$, where $\bigodot^2 T^*$ is the symmetrized tensor product.

Suppose that M is a Riemannian manifold of dimension $2n$. An almost-complex structure J on M is said to be *orthogonal* if $g(JX, JY) = g(X, Y)$. Note that this condition depends only on the conformal class of g. Indeed, when $n = 1$ such a J is identical to the oriented conformal structure determined by g; given this, J is essentially 'rotation by 90°'. An almost-complex structure J is orthogonal if and only if its space Λ of $(1,0)$ forms is totally isotropic at each point. The triple (M, g, J) with J orthogonal is called an *almost-Hermitian manifold.*

We shall first define the fundamental 2-form of an almost-Hermitian manifold, and then study the special case in which the manifold is Kähler. After a summary of some relevant Hodge theory, we consider the parametrization of compatible complex structures on a given even-dimensional Riemannian manifold.

The Kähler condition

Given an almost-Hermitian manifold (M, g, J), the tensor

$$\omega(X, Y) = g(JX, Y) \tag{2.1}$$

is a 2-form on M. This 2-form is non-degenerate in the sense that it defines an isomorphism $T \to T^*$ at each point. Equivalently, the volume form

$$v = \frac{\omega^n}{n!} \tag{2.2}$$

of M is nowhere zero. To check the constants, suppose that $n = 3$ and that $g = \sum_{i=1}^{3} e^i \otimes e^i$ at a fixed point $m \in M$. A standard almost-complex structure is determined by setting $\omega = e^{12} + e^{34} + e^{56}$, with the convention $e^{ij} = e^i \wedge e^j = e^i \otimes e^j - e^j \otimes e^i$. Then $\omega^3 = 6e^{123456} = 6v$.

A standard almost-complex structure J on $\mathbb{R}^{2n}$ is the *linear transformation* described by the matrix

$$\mathbb{J} = \begin{pmatrix} 0 & -I \\ I & 0 \end{pmatrix} \tag{2.3}$$

where I is the $n \times n$ identity matrix. The stabilizer of J in $GL(2n, \mathbb{R})$ is isomorphic to $GL(n, \mathbb{C})$; it is the set of matrices A for which $A^{-1}\mathbb{J}A = \mathbb{J}$ or $A\mathbb{J} = \mathbb{J}A$. The matrix $\mathbb{J}$ may also be regarded as that of a standard non-degenerate skew bilinear form ω_0 on $\mathbb{R}^{2n}$. It follows that the choice of ω_0 determines a reduction to the subgroup $Sp(2n, \mathbb{R})$, determined by the set of matrices such that $A^T\mathbb{J}A = \mathbb{J}$.

The metric g determines a reduction to the orthogonal group $O(2n)$ and the intersection of any two of the subgroups $GL(n, \mathbb{C})$, $Sp(2n, \mathbb{R})$, $O(2n)$ is isomorphic to the unitary group $U(n)$. An almost-Hermitian structure on a manifold is therefore determined by a reduction of its principal bundle of frames to $U(n)$.

The 2-form ω defines an 'almost-symplectic' structure on M, which is called 'symplectic' if $d\omega = 0$. The latter condition is analogous to the integrability of J, since the Darboux theorem asserts that $d\omega = 0$ if and only if there exist real coordinates $x^1, \ldots, x^n, y^1, \ldots, y^n$ such that

$$\omega = \sum_{r=1}^{n} dx^r \wedge dy^r \tag{2.4}$$

(compare (1.11)) in a neighbourhood U of any given point.

Definition 2.5 *An almost-Hermitian manifold M is said to be Hermitian if J is integrable, and Kähler if in addition $d\omega = 0$.*

Since

$$\omega(X - iJX, Y - iJY) = ig(X - iJX, Y - iJY) = 0, \quad X, Y \in T,$$

we can assert that ω is a $(1,1)$-form. Relative to local holomorphic coordinates on a Hermitian manifold, we may write

$$\omega = \tfrac{1}{2} i \sum_{k,l} \omega_{kl} dz^k \wedge d\bar{z}^l. \tag{2.6}$$

Applying (2.1) backwards, we see that this gives rise to a corresponding expression

$$g = \tfrac{1}{2} \sum_{k,l} \omega_{kl} dz^k \odot d\bar{z}^l$$

of the Riemannian metric, with the convention

$$dz \odot d\bar{z} = dz \otimes d\bar{z} + d\bar{z} \otimes dz = 2(dx \otimes dx + dy \otimes dy).$$

Lemma 2.7 *Let M be a Kähler manifold and $m \in M$. Then there exists a real-valued function ϕ on a neighbourhood of m such that $\omega = \frac{1}{2} i\partial\bar{\partial}\phi$, so that $\omega_{kl} = \partial^2\phi/\partial z^k \partial \bar{z}^l$.*

Proof Since $d\omega = 0$, there exists a $(1,0)$-form α on a neighbourhood of m such that $\omega = d(\alpha + \bar{\alpha})$. But then $\bar{\partial}\bar{\alpha} = 0$ and, by Exercise 1.26(ii), there exists a complex-valued function f on a possibly smaller neighbourhood such that $\bar{\partial} f = \bar{\alpha}$. Setting $\phi = \frac{1}{2} i(\bar{f} - f)$ gives

$$2\omega = d(\partial\bar{f} + \bar{\partial} f) = \bar{\partial}\partial\bar{f} + \partial\bar{\partial} f = i\partial\bar{\partial}\phi,$$

as required. □

Example 2.8 The most obvious example of such a 'Kähler potential' is $\phi = \sum_{r=1}^n |Z^r|^2$ on $\mathbb{R}^{2n} = \mathbb{C}^n$. This gives rise to the flat Kähler metric

$$\omega = \frac{1}{2} i \sum_{r=1}^n dZ^r \wedge d\bar{Z}^r.$$

The standard Kähler metric on $\mathbb{CP}^n$ is constructed most invariantly by starting from the function $\psi = \log \sum_{r=0}^n |Z^r|^2$ on $\mathbb{C}^{n+1} \setminus \{0\}$. Then

$$\partial\bar{\partial}\psi = \partial\bar{\partial} \log\left(1 + \sum_{r=1}^n |z^r|^2\right) = \pi^*\omega,$$

where ω has the form (2.1) in local coordinates on U_0. However, the construction shows that ω is well defined globally on $\mathbb{CP}^n$. One recovers, for $n = 1$, the form and metric

$$\omega = \frac{i\, dz \wedge d\bar{z}}{2(1 + |z|^2)^2}, \qquad g = \frac{dx^2 + dy^2}{(1 + x^2 + y^2)^2}$$

with constant positive Gaussian curvature.

On a compact Kähler manifold M of real dimension $2n$, the powers ω^k of ω represent non-zero cohomology classes in $H^{2k}(M,\mathbb{R})$ for all $k \leqslant n$. This is because a global equation $\omega^n = d\sigma$ would imply (by Stokes' theorem) that $\int_M \omega^n = 0$, contradicting (2.2). Since $H^2(\mathbb{CP}^n,\mathbb{R})$ is well known to be 1-dimensional, the 2-form ω constructed above can be normalized so as to belong to $H^2(\mathbb{CP}^n,\mathbb{Z})$. The power of the Kähler condition is that any complex submanifold M of $\mathbb{CP}^n$ is automatically Kähler, since ω on $\mathbb{CP}^n$ pulls back to a closed form on M compatible with the induced metric and complex structure. On the other hand, it is known that a necessary and sufficient condition for a compact complex manifold M to satisfy Definition 1.6 is that it possess a Kähler metric ω with $[\omega] \in H^2(M,\mathbb{Z})$ (see for example [54]).

The restriction of the standard dot product on $\mathbb{R}^n$ to any subspace is an inner product, so any submanifold of $\mathbb{R}^n$ has an induced Riemannian metric. A wider class of metrics can be constructed by considering certain types of submanifolds of a real vector space endowed with a bilinear form of mixed signature. A classical example of this is the pseudosphere $-x^2 - y^2 + z^2 = 1$ of negative Gaussian curvature isometrically embedded in the 'Lorentzian' space $\mathbb{R}^{1,2}$. Additional structures on the ambient vector space $\mathbb{R}^n$ can sometimes be used to induce an almost-Hermitian structure on a hypersurface or submanifold.

Example 2.9 Identify $\mathbb{R}^7$ with the space of imaginary Cayley numbers, which is endowed with a (non-associative) product $\times$. Then any hypersurface M of $\mathbb{R}^7$ has an almost-complex structure J, compatible with the induced metric, defined by $JX = \mathbf{n} \times X$ where $X \in T_m M$ and $\mathbf{n}$ is a consistently-defined unit normal vector at m. The Nijenhuis tensor of J can be related to the second fundamental form of M, and Calabi proved that J is integrable if and only if M is minimal [28, 91]. A generalization of this phenomenon for submanifolds of $\mathbb{R}^8$ is described in [21].

There is also a Kähler version of Definition 1.12:

Definition 2.10 *A manifold is hyperkähler if it admits a Riemannian metric g which is Kähler relative to complex structures I_1, I_2, I_3 satisfying* (1.13).

The significance of the existence of such complex structures (at least in four dimensions) will become clearer after Proposition 2.25. The integrability requires that the 2-form ω_i associated to I_i is closed for each i. If we fix I_1 then $\eta = \omega_2 + i\omega_3$ is a $(2,0)$-form, which is closed and so holomorphic since the $(2,1)$-form $\bar\partial\eta$ vanishes. In fact, a hyperkähler manifold M has dimension $4k$, and (choosing an appropriate basis $\{\alpha^1,\ldots,\alpha^n\}$ of $(1,0)$-forms at each point) $\eta^k = \alpha^{12\cdots n}$ trivializes $\Lambda^{n,0}$, so M is holomorphic symplectic. Conversely, with the same algebraic set-up, the closure of η implies that I_1 is integrable; this is because

$$0 = d(\eta^k)^{n-1,2} = (d\alpha^1)^{0,2} \wedge \alpha^{23\cdots n} + (d\alpha^2)^{0,2} \wedge \alpha^{3\cdots n1} + \cdots$$

and $(d\alpha^i)^{0,2} = 0$ for all i.

For future reference, we describe the standard 2-forms associated to a flat hyperkähler structure on $\mathbb{R}^{4k}$. Consider coordinates (x^r, y^r, u^r, v^r) with $1 \leqslant i \leqslant k$. Then

$$\begin{cases} \omega_1 = \sum_{r=1}^{k} (dx^r \wedge dy^r + du^r \wedge dv^r), \\ \omega_2 = \sum_{r=1}^{k} (dx^r \wedge du^r + dv^r \wedge dy^r), \\ \omega_3 = \sum_{r=1}^{k} (dx^r \wedge dv^r + dy^r \wedge du^r). \end{cases} \tag{2.11}$$

Exercises 2.12 (i) Show that the mapping $A \longmapsto A\mathbb{J}$ identifies the Lie algebra $\mathfrak{sp}(2n, \mathbb{R})$ with the space of symmetric matrices. Show that $\mathfrak{sp}(2, \mathbb{R})$ is isomorphic to the Lie algebra $\mathfrak{sl}(2, \mathbb{R})$ of real 2×2 matrices of trace zero.

(ii) Further to Example 2.8, show that the function $\phi + \log \phi$, where $\phi = \sum_{r=0}^{n} |Z^r|^2$, is the potential for a Kähler metric that extends to the blow-up of $\mathbb{C}^{n+1}$ at the origin.

(iii) Let $n = 2$. The subgroup of $GL(8, \mathbb{R})$ preserving all three 2-forms (2.11) is exactly $Sp(2)$. Determine the stabilizer of each of the two 2-forms

$$\Phi_{\pm} = \omega_1 \wedge \omega_1 + \omega_2 \wedge \omega_2 \pm \omega_2 \wedge \omega_3,$$

referring to [24, 92] if necessary.

Summary of Hodge theory

It turns out that a global version of Lemma 2.7 is valid on a compact Kähler manifold. In order to explain this, we sketch the essentials of Hodge theory in the context of Dolbeault cohomology. More details can be found in many standard texts [54, 57, 105, 106, 107].

Let M be a complex manifold of real dimension $2n$. An inner product on the complex space $\Omega^{p,q}$ is defined by

$$\langle\langle \alpha, \beta \rangle\rangle = \int_M g(\alpha, \bar{\beta}) v = \int \alpha \wedge *\bar{\beta} = \int \bar{\beta} \wedge *\alpha,$$

where g is the Riemannian metric (extended as a complex *bilinear* form), and

$$* : \Lambda^{p,q} \to \Lambda^{n-q,n-p}$$

is the complexification of a *real* isometry satisfying $*^2\alpha = (-1)^{p+q}\alpha$. The double brackets emphasize the global nature of the pairing. If $\alpha \in \Omega^{p,q}$ and $\beta \in \Omega^{p,q-1}$ then

$$d(\beta \wedge *\bar{\alpha}) = \bar{\partial}(\beta \wedge *\bar{\alpha}) = \bar{\partial}\beta \wedge *\bar{\alpha} - (-1)^{p+q} \beta \wedge \bar{\partial}(*\bar{\alpha}),$$

and it follows that
$$\langle\langle\bar\partial\beta,\alpha\rangle\rangle = -\langle\langle\beta,*\partial*\alpha\rangle\rangle.$$

Although the space $\Omega^{p,q}$ of smooth forms is not complete for the norm defined above, we may regard $\bar\partial^* = -*\partial*$ as the adjoint of the mapping (2.13) below.

Fix a Dolbeault cohomology class $c \in H^{p,q}_{\bar\partial}$. To represent c uniquely, one seeks a $\bar\partial$-closed (p,q)-form α of least norm, thus *orthogonal* to the image of
$$\bar\partial : \Omega^{p,q-1} \to \Omega^{p,q}. \tag{2.13}$$

Since $\langle\langle\alpha,\bar\partial\gamma\rangle\rangle = \langle\langle\bar\partial^*\alpha,\gamma\rangle\rangle$ for all γ, we therefore need $\bar\partial^*\alpha = 0$. Now $\langle\langle\alpha,\Delta_{\bar\partial}\alpha\rangle\rangle = \|\bar\partial\alpha\|^2 + \|\bar\partial^*\alpha\|^2$, where
$$\Delta_{\bar\partial} = \bar\partial\bar\partial^* + \bar\partial^*\bar\partial \tag{2.14}$$
is the $\bar\partial$-*Laplacian*, so we require α to belong to the space
$$\mathcal{H}^{p,q} = \{\alpha \in \Omega^{p,q} \colon \Delta_{\bar\partial}\alpha = 0\}$$
of *harmonic* (p,q)-forms. It also follows from above that α is orthogonal to the image $\Delta_{\bar\partial}(\Omega^{p,q})$ if and only if α is harmonic.

Suppose for a moment that $\Omega^{p,q}$ were a finite-dimensional vector space. This is not as unrealistic as it seems, because examples we have considered earlier based on Lie groups and or their quotients possess subcomplexes consisting of *invariant* differential forms with constant coefficients, so one could restrict to these to obtain a more restricted type of cohomology. In this situation, $\Omega^{p,q}$ is the direct sum of $\Delta_{\bar\partial}\Omega^{p,q}$ and its orthogonal complement $\mathcal{H}^{p,q}$. The Hodge theorem asserts that this applies in the general case:

Theorem 2.15 *Let M be a compact complex manifold. Then $\mathcal{H}^{p,q}$ is finite-dimensional, and there is a direct sum decomposition $\Omega^{p,q} = \mathcal{H}^{p,q} \oplus \Delta_{\bar\partial}\Omega^{p,q}$.*

It follows that, modulo a finite-dimensional space, the Laplacian is invertible. For given $\alpha \in \Omega^{p,q}$, there exists $G\alpha \in \Omega^{p,q}$ such that $(\Delta_{\bar\partial}G\alpha - \mathbf{1})\alpha$ is harmonic. The mapping G is called the *Green's operator*. One has additional orthogonal direct sums
$$\begin{aligned}\Omega^{p,q} &= \mathcal{H}^{p,q} \oplus \bar\partial(\bar\partial^*\Omega^{p,q}) \oplus \bar\partial^*(\bar\partial\Omega^{p,q})\\ &= \mathcal{H}^{p,q} \oplus \bar\partial\Omega^{p,q-1} \oplus \bar\partial^*\Omega^{p,q+1}.\end{aligned}$$

To see the second equality, observe that if for example $\beta \in \Omega^{p,q-1}$ then $\beta - \Delta_{\bar\partial}G\beta$ is harmonic, and so $\bar\partial\beta = \bar\partial\bar\partial^*\gamma$ where $\gamma = \bar\partial G\beta$.

Corollary 2.16 $H^{p,q}_{\bar\partial} \cong \mathcal{H}^{p,q}$.

Proof If $\gamma \in \Omega^{p,q}$ is $\bar\partial$-closed then γ is orthogonal to $\bar\partial^*\Omega^{p,q+1}$. Thus, the kernel of $\bar\partial$ on $\Omega^{p,q}$ equals $\mathcal{H}^{p,q} \oplus \bar\partial\Omega^{p,q-1}$. It follows that each cohomology class has a unique harmonic representative. □

Corollary 2.17 $H^{p,q}_{\bar\partial} \cong H^{n-p,n-q}_{\bar\partial}$.

Proof Since $\Delta_{\bar\partial} * \bar\alpha = \overline{*\Delta_{\bar\partial}\alpha}$, the composition of $*$ with conjugation determines an isomorphism from $\mathcal{H}^{p,q}$ to $\mathcal{H}^{n-p,n-q}$. □

We shall now return to the Kähler condition. From (2.2) we may deduce that $*\omega = \omega^{n-1}/(n-1)!$ Moreover, if e^1 is a unit 1-form with $Je^1 = -e^2$, then $e^1 \wedge e^2 \wedge \omega^{n-1} = (n-1)!v$, so $*e^1 = e^2 \wedge \omega^{n-1}/(n-1)!$ It follows that, if f is a function,

$$\begin{aligned}\bar\partial^*(f\omega) = - * \partial * (f\omega) &= -\frac{1}{(n-1)!} * \partial(f\omega^{n-1})\\ &= -\frac{1}{(n-1)!} * (\partial f \wedge \omega^{n-1})\\ &= J(\partial f) = i\partial f.\end{aligned}$$

This is a special case of

Lemma 2.18 *Let* $\alpha \in \Omega^{p,q}$. *Then* $\bar\partial^*(\omega\wedge\alpha) - \omega\wedge\bar\partial^*\alpha = i\partial\alpha$.

We omit the general proof, which is most easily carried out with connections and theory developed in Section 3. The lemma is illustrated by the diagram

$$\begin{array}{ccc} & & \Omega^{p+1,q+1}\\ & \nearrow & \downarrow\\ \Omega^{p,q} & \to & \Omega^{p+1,q}\\ \downarrow & \nearrow & \\ \Omega^{p,q-1} & & \end{array}$$

Let $L : \bigwedge^k T^* \to \bigwedge^{k+2} T^*$ denote the operation of wedging with ω, and

$$\Lambda = L^* = (-1)^k * L *$$

its adjoint relative to $\langle\langle .,. \rangle\rangle$ (equivalently g). Then

$$\begin{aligned} [\bar\partial^*, L] &= i\partial, \quad [\partial^*, L] = -i\bar\partial,\\ [\bar\partial, \Lambda] &= i\partial^*, \quad [\partial, \Lambda] = -i\bar\partial^*;\end{aligned}$$

the first equation is a restatement of the lemma, and the others immediate consequences. Substituting the last two lines into an expansion of

$$\Delta_d = dd^* + d^*d = (\partial+\bar\partial)(\partial+\bar\partial)^* + (\partial+\bar\partial)^*(\partial+\bar\partial)$$

to eliminate the mixed terms on the right-hand side reveals that

$$\Delta_d = \partial\partial^* + \partial^*\partial + \bar\partial\bar\partial^* + \bar\partial^*\bar\partial = \Delta_\partial + \Delta_{\bar\partial},$$

and that $\Delta_\partial = \Delta_{\bar\partial}$. Hodge theory for the exterior derivative now yields

Theorem 2.19 *On a compact Kähler manifold* $H^k(M,\mathbb{R})$ *is isomorphic to*

$$\{\alpha \in \Gamma(\textstyle\bigwedge^k T^*M) \colon \Delta_d \alpha = 0\} = \bigoplus_{p+q=k} \mathcal{H}^{p,q},$$

and there is equality in (1.25).

Complex conjugation gives an isomorphism $\mathcal{H}^{p,q} \cong \overline{\mathcal{H}^{q,p}}$, and so $h^{p,q} = h^{q,p}$. In particular, the Betti numbers b_{2i+1} are all even. Moreover the mapping

$$L^{n-k} : H^k(M,\mathbb{R}) \to H^{2n-k}(M,\mathbb{R}) \tag{2.20}$$

is an isomorphism; this is the so-called *hard Lefschetz property.* The failure of (2.20) underlies almost all known examples of manifolds that do not admit a Kähler metric.

The goal of this subsection is achieved by

Lemma 2.21 *Let M be a compact Kähler manifold, and let $\alpha \in \Omega^{p,q}$ be closed* $(p, q \geqslant 1)$. *Then $\alpha = \partial\bar\partial\beta$ for some $\beta \in \Omega^{p-1,q-1}$*.

Proof In the notation (2.14), we know that $\alpha - \Delta_{\bar\partial} G\alpha$ is annihilated by $\Delta_{\bar\partial} = \frac{1}{2}\Delta_d$. Thus

$$0 = d\alpha - d\Delta_{\bar\partial} G\alpha = -(\partial + \bar\partial)\bar\partial\bar\partial^* G\alpha = -\partial\bar\partial(\bar\partial^* G\alpha),$$

using the fact that $\bar\partial G = G\bar\partial$. □

Known obstructions to the existence of Kähler metrics on compact manifolds stem from (2.20) or this $\partial\bar\partial$ lemma. Either can be used to establish the *formality* of the de Rham cohomology of Kähler manifolds [33]. More recent references to this topic can be found in [78, 101].

Exercises 2.22 (i) Let M be a compact complex manifold of dimension n, and let α be a $(k, 0)$-form satisfying $\bar\partial\alpha = 0$. Prove that if M is Kähler *or* if $k = n - 1$ then one may conclude that $d\alpha = 0$.

(ii) Describe the set $\mathcal{S}$ of all symplectic forms on the manifold in Example 1.22 of the type $\omega = \sum \omega_{ij} e^i \wedge e^j$ with ω_{ij} constant. Is $\mathcal{S}$ connected? What is its dimension?

(iii) Compute the action of d on the basis $\{e^i \wedge e^j \colon i < j\}$ of 2-forms of X in Example 1.27. What can you say about the *Betti numbers* of Y? Is the mapping $H^1(Y,\mathbb{R}) \to H^3(Y,\mathbb{R})$ induced by wedging with the symplectic form $\omega = e^{14} + e^{23}$ an isomorphism?

Orthogonal complex structures

If we are given a complex manifold (M, J), there is no difficulty in choosing a Hermitian metric. Namely, pick any metric h and then define

$$g(X,Y) = h(X,Y) + h(JX, JY); \tag{2.23}$$

this renders J orthogonal. One can do the same sort of thing if M is hypercomplex by defining

$$g(X,Y) = \sum_{k=0}^{3} h(I_k X, I_k Y),$$

where $I_0 = \mathbf{1}$. For then g will be 'hyperhermitian' in the sense that $g(JX,JY) = g(X,Y)$ whenever $J = \sum a_r I_r$ with $\sum {a_r}^2 = 1$.

The reverse problem has a very different character. Given an oriented Riemannian manifold (M,g) of even dimension $2n$, it is not in general a straightforward job to find a complex structure J for which (M,g,J) is Hermitian. In the presence of an assigned metric, any such J is referred to as an *orthogonal complex structure*, or OCS for short.

Let Z_n denote the set of orthogonal almost-complex structures on $\mathbb{R}^{2n}$ compatible with a standard orientation. The group $SO(2n)$ acts transitively on Z_n, and using the notation (2.3),

$$Z_n = \{A^{-1}\mathbb{J}A\colon A \in SO(2n)\}.$$

If

$$A = \begin{pmatrix} A_1 & A_2 \\ A_3 & A_4 \end{pmatrix}$$

where each A_i is $n \times n$ then $A\mathbb{J} = \mathbb{J}A$ iff $A_1 = A_4$ and $A_2 = -A_3$, and the orthogonality implies that $A_1 + iA_2$ is a unitary matrix. Thus the stabilizer of $\mathbb{J}$ is isomorphic to $U(n)$, and Z_n can be identified with the homogeneous space $SO(2n)/U(n)$.

Any almost-complex structure on $\mathbb{R}^{2n}$ is completely determined by the corresponding space $\Lambda = \Lambda^{1,0}$ of $(1,0)$-forms in (1.4). Given Λ, we may obviously define J by $J(v + \overline{w}) = iv - i\overline{w} = iv + \overline{iw}$. The mapping

$$J \mapsto \Lambda$$

identifies Z_n with one component of the subset of maximal isotropic subspaces of $\mathbb{C}^{2n}$ in the Grassmannian $\mathbb{G}\mathrm{r}_n(\mathbb{C}^{2n})$. This gives Z_n a natural complex structure. In fact, $SO(2n)/U(n)$ is a Hermitian symmetric space and admits a Kähler metric compatible with this complex structure.

Let $\{e_1, \ldots, e_{2n+2}\}$ be an oriented orthonormal basis of $\mathbb{R}^{2n+2}$. Define a mapping

$$\begin{array}{c} Z_{n+1} \\ \Big\downarrow \pi \\ S^{2n} \subset \langle e_2, \ldots, e_{2n+2} \rangle, \end{array} \tag{2.24}$$

where S^{2n} is the sphere, by setting $\pi(J) = Je_1$. Then $\pi^{-1}(e_2)$ is the set of oriented orthogonal complex structures on $\langle e_1, e_2 \rangle^{\perp} = T_{e_2}S^{2n}$. Thus (2.24) is a bundle with fibre Z_n. Of course, Z_1 is a point. The next result shows that Z_2 is a 2-sphere.

Proposition 2.25 *π is an isomorphism for $n=1$.*

Proof Let J be an almost-complex structure compatible with the metric and orientation of $\mathbb{R}^4$. From above we may write

$$-Je_1 = ae_2 + be_3 + ce_4, \quad a^2+b^2+c^2=1.$$

Passing to the dual basis, the 2-form associated to J is

$$\omega = e^1 \wedge (-Je^1) + f \wedge (-Jf),$$

where f is any unit 1-form orthogonal to both e^1 and Je^1. Since $be^2 - ae^3$ and $ce^2 - ae^4$ are both orthogonal to Je^1,

$$f \wedge (-Jf) = \frac{1}{a}(be^2 - ae^3) \wedge (ce^2 - ae^4) = ae^{34} + be^{42} + ce^{23}$$

is completely determined. □

This proof shows that the 2-form of any compatible almost-complex structure on $\mathbb{R}^4$ can be written $a\omega_1 + b\omega_2 + c\omega_3$, where $a^2+b^2+c^2=1$, and

$$\begin{cases} \omega_1 = e^{12} + e^{34}, \\ \omega_2 = e^{13} + e^{42}, \\ \omega_3 = e^{14} + e^{23}. \end{cases} \tag{2.26}$$

Definition 2.27 *The span of $\omega_1, \omega_2, \omega_3$ is the space Λ^+ of self-dual 2-forms on $\mathbb{R}^4$.*

The above argument shows that Λ^+ depends only on the choice of metric and orientation. A fixed almost-complex structure J gives a splitting

$$\Lambda^2 T^* = [\![\Lambda^{2,0}]\!] \oplus [\Lambda^{1,1}], \tag{2.28}$$

corresponding to a reduction to $U(1) \times SU(2)$. Moreover, $[\![\Lambda^{2,0}]\!]$ is the real tangent space to $S^2 \subset \Lambda^+$ at the point ω associated to J.

If (M, g, J) is an almost-Hermitian 4-manifold then

$$d\omega = \theta \wedge \omega$$

for some 1-form θ (to see this, write $\theta = \sum a_i e^i$). The object θ is (modulo a universal constant) the so-called *Lee form* [76]. Suppose that M is a 4-manifold with a global basis $\{e^1, e^2, e^3, e^4\}$ of 1-forms, so that ω_i is defined by (2.26). Let I_i denote the corresponding almost-complex structure, and θ_i the Lee form, for $i = 1, 2, 3$.

Lemma 2.29 *I_1 is integrable if and only if $\theta_2 = \theta_3$.*

Proof The space of $(1,0)$-forms relative to I_1 is spanned by

$$\alpha^1 = e^1 + ie^2, \quad \alpha^2 = e^3 + ie^4.$$

As in (1.33), I_1 is integrable if and only if

$$\begin{aligned} 0 \ = \ d\alpha^i \wedge (\omega_2 + i\omega_3) &= \alpha^i \wedge (d\omega_2 + id\omega_3) \\ &= \alpha^i \wedge (\omega_2 \wedge \theta_2 + i\omega_3 \wedge \theta_3) \\ &= \tfrac{1}{2}\alpha^i \wedge (\omega_2 - i\omega_3) \wedge (\theta_2 - \theta_3), \end{aligned}$$

since $\alpha^i \wedge (\omega_2 + i\omega_3) = 0$. The result follows. □

This innocent looking lemma has the following important consequences:

(i) If I_1 and I_2 are anticommuting complex structures then $I_3 = I_1 I_2$ is integrable and M is hypercomplex. In this case, the θ_i coincide and represent a common Lee form.
(ii) If $\theta_2 = 0 = \theta_3$ then $(M, I_1, \omega_2 + i\omega_3)$ is holomorphic symplectic. Moreover, M is hyperkähler if and only if $\theta_i = 0$ for all i.

Similar statements hold in higher dimensions. Refer to [40] for examples of hypercomplex structures in eight real dimensions for which the above theory can easily be applied.

Example 2.30 The following theory was investigated by Barberis [12, 13]. Define real 1-forms by

$$E^1 = e^{-t}\,dx, \quad E^2 = e^{-t}\,dv, \quad E^3 = e^{-2t}(du + cx\,dv), \quad E^4 = dt,$$

where $c \in \mathbb{R}$. Then

$$\begin{cases} dE^1 = E^1 \wedge E^4, \\ dE^2 = E^2 \wedge E^4, \\ dE^3 = cE^1 \wedge E^2 + 2E^3 \wedge E^4, \\ dE^4 = 0. \end{cases}$$

Consider the metric

$$g_c = \sum_{i=1}^{4} E^i \otimes E^i = e^{-2t}(dx^2 + dv^2) + e^{-4t}(du + cx\,dv)^2 + dt^2. \tag{2.31}$$

The Lee forms defined by the basis $\{E^i\}$ are

$$\theta_1 = (c-2)e^4, \qquad \theta_2 = \theta_3 = -3e^4,$$

so I_1 is integrable. There are two special cases:

(i) $c = 2$. Then $d\omega_1 = 0$ and (g_2, I_1, ω_1) is Kähler.
(ii) $c = -1$. Then I_2 and I_3 are also integrable, and g_{-1} is hypercomplex.

It is known that g_2 is isometric to the symmetric metric on complex hyperbolic space $\mathbb{C}H^2$. Indeed, x, t, u, v are coordinates on a solvable Lie group that acts simply transitively on $\mathbb{C}H^2$, and the E^i generate the dual of its Lie algebra $\mathfrak{g}$. Being Einstein, g_2 is one of the metrics that crops up in Jensen's classification [61]. Here though, we shall focus on g_{-1}, and show that it is *conformally* hyperkähler.

Setting $c = -1$ and $s = e^t$ in (2.31) gives

$$s^3 g_{-1} = s(dx^2 + dv^2 + ds^2) + \frac{1}{s}(du - x\,dv)^2.$$

This is a very special case of the Gibbons–Hawking ansatz, that classifies all hyperkähler metrics with a triholomorphic S^1 action [48]. It is easy to see directly that $s^3 g_{-1}$ is hyperkähler. Guided by (2.11), define

$$\begin{cases} \omega_1 = s\,dx \wedge dv + (du - x\,dv) \wedge ds, \\ \omega_2 = dx \wedge (du - x\,dv) + s\,ds \wedge dv, \\ \omega_3 = s\,dx \wedge ds + dv \wedge (du - x\,dv). \end{cases} \tag{2.32}$$

Relative to $s^3 g_{-1}$, these are associated to a triple of almost-Hermitian structures I_1, I_2, I_3 for which clearly $d\omega_i = 0$ for all i.

In the above description of $s^3 g_{-1}$, one regards the 4-dimensional space as an S^1-bundle over the half-space $\mathbb{R}^2 \times \mathbb{R}^+$. Then u is a fibre coordinate and $du - x\,dv$ a connection 1-form. The projection to (x, v, s)-space is the so-called *hyperkähler moment mapping* defined by the S^1 group of isometries. The metrics g_2, g_{-1} can also be characterized (amongst all invariant ones on 4-dimensional Lie groups) by the conditions that $W_+ = 0$ and $W_- \neq 0$ [96]. (See Section 4 for a description of the Weyl tensor $W_+ + W_-$.)

Passing to higher dimensions, it can be shown that the space Z_3 of almost-complex structures on $\mathbb{R}^6$ is isomorphic to the projective space $\mathbb{CP}^3$. To visualize this equivalence, one needs a scheme whereby the four coordinates of $\mathbb{R}^4$ are linked combinatorially to the six coordinates of $\mathbb{R}^6$. Such a 'tetrahedral' isomorphism is described in [2], and exploited in [11], to describe structures on 6-manifolds. Using homogeneous coordinates on $\mathbb{CP}^3$, the four points $(1,0,0,0)$, $(0,1,0,0)$, $(0,0,1,0)$, $(0,0,0,1)$ can be identified with those almost-complex structures that have 2-forms

$$\pm e^{12} \pm e^{34} \pm e^{56}$$

(with an even number of minus signs to fix the orientation) relative to an orthonormal basis of $\mathbb{R}^6$.

In general, one can show that the image of a section $s : S^{2n} \to Z_{n+1}$ is a complex submanifold if and only if the almost-complex structure determined by s is integrable. Since Z_{n+1} is Kähler, the submanifold (and so S^{2n}) would itself have to be Kähler. Since $b_2(S^{2n}) = 0$ for $n > 1$, the sphere S^{2n} cannot admit an OCS unless $n = 1$. Actually, it is well known that S^{2n} does not admit any *almost*-complex structure unless $n = 1$ or 3, so the force of this statement is restricted to S^6. Properties of the Lie group $SO(8)$ enable the total space Z_4 over S^6 to be identified with $SO(8)/(SO(2)\times SO(6))$, itself a complex quadric in $\mathbb{CP}^7$. Note that, as soon as a point is removed, $S^{2n}\setminus\{x\}$ (being conformally equivalent to $\mathbb{R}^{2n}$) does admit OCS's for any n.

The fibration (2.24) can be generalized by replacing the base by any even-dimensional Riemannian manifold M, and taking the associated bundle whose fibre $\pi^{-1}(m)$ (again Z_n) parametrizes compatible almost-complex structures on T_mM [17, 85]. The total space ZM admits two natural almost-complex structures, denoted J_1 and J_2 in [91], which are best defined with the aid of the Levi-Civita connection on M (see Corollary 3.17). Whilst an example of J_1 is provided by the complex structure on $Z_{n+1}=ZS^{2n}$ described above, J_2 is never integrable. On the other hand, if $\psi:\Sigma\to ZM$ is a J_2-holomorphic mapping from a Riemann surface Σ, its projection $\pi\circ\psi:\Sigma\to M$ is a harmonic mapping. This universal property characterizes many other types of 'twistor bundles' that have also proved useful in classifying OCS's on symmetric spaces [26, 25].

When M is 4-manifold, Proposition 2.25 tells us that its twistor space can be identified with the 2-sphere bundle $S(\Lambda^+T^*M)$. In this case, (2.24) is the celebrated Penrose fibration, and Atiyah–Hitchin–Singer [10] showed that J_1 is integrable if and only if half of the Weyl curvature tensor of M vanishes (we shall return to this fact in §4). Such twistor spaces provide important examples of complex 3-dimensional manifolds of various algebraic dimensions [90]. Other applications were developed by Bryant [22].

Exercises 2.33 (i) Let $\{e^1,\dots,e^6\}$ be a basis of 1-forms on $\mathbb{R}^6$, and set

$$\begin{aligned}\Phi &= (e^1+ie^2)\wedge(e^3+ie^4)\wedge(e^5+ie^6)\\ &= e^{135}-e^{146}-e^{236}-e^{245}+i(e^{136}+e^{145}+e^{235}-e^{246}).\end{aligned}$$

Show that the almost-complex structure for which Φ is a $(3,0)$-form at each point is integrable if $d\Phi=0$. Give an example to show that the condition $d(\Re\mathrm{e}\Phi)=0$ is not sufficient to guarantee integrability.

(ii) By associating to each point of Z_3 its fundamental form, we may regard $Z_3\cong\mathbb{CP}^3$ as a submanifold of $\bigwedge^2(\mathbb{R}^6)^*$. Describe the intersection of this submanifold with the orthogonal complement of each of the following 2-forms: $e^{12}+e^{34}+e^{56}$, $e^{12}+e^{34}$, and e^{12}.

(iii) Explain the complex quadric interpretation of Z_4 mentioned above. Find out how to define a subbundle (with fibre $\mathbb{CP}^2$) of Z_4 over S^6 that was exploited in [21] to classify pseudo-holomorphic curves in S^6.

3 Compatible connections

In this section, we consider in turn connections preserving g, ω and J, with emphasis on the torsion-free condition. Since g and ω are both bilinear forms, the corresponding theories can be developed in parallel. The study of connections preserving the linear transformation J has more complicated aspects relating to the Nijenhuis tensor of J.

Preliminaries

Let $V \to M$ be a vector bundle. Typically, this will be one of $TM = T, T^*$, $\mathrm{End}\, T = T^* \otimes T$ (which contains the tensors $\mathbf{1}$ and J) or $T^* \otimes T^*$ (which contains g and ω). Let $\Gamma(V)$ denote the space of smooth sections of V over M. Thus, $\Gamma(T^*) = \Omega^1$ is the space of 1-forms over M, and $\Gamma(T) = \mathfrak{X}$ the space of vector fields.

A *connection* on V is an $\mathbb{R}$-linear mapping $\nabla : \Gamma(V) \longrightarrow \Gamma(T^* \otimes V)$ such that

$$\nabla(fv) = df \otimes v + f\nabla v, \tag{3.1}$$

whenever v is a section of V and f is a smooth function on M. The section ∇v is called the 'covariant derivative' of v. If $X \in \mathfrak{X}$ and $C_X : T^* \otimes V \to V$ is the corresponding contraction then $\nabla_X = C_X \circ \nabla$ satisfies

$$\nabla_X(fv) = (Xf)v + f\nabla_X v.$$

The operator ∇_X is tensorial in X, in the sense that for fixed $v \in \Gamma(V)$ the value $(\nabla_X v)_m$ at a point $m \in M$ depends only on the value X_m.

Let $\nabla, \widetilde{\nabla}$ be two connections on V. To measure their difference, set

$$\widetilde{\nabla}_X = \nabla_X + A_X. \tag{3.2}$$

Since $A_X(fY) = fA_X Y$, the element A_X is an endomorphism of V that depends linearly on X. Conversely, given a tensor A with values in $T^* \otimes \mathrm{End}\, V$ at each point, and a connection ∇, then (3.2) satisfies (3.1) and is itself a connection. The space of connections on a vector bundle is therefore an affine space modelled on the vector space $\Gamma(T^* \otimes \mathrm{End}\, V)$.

A connection on the tangent bundle T determines one on T^* by the rule

$$X(\alpha Y) = (\nabla_X \alpha)(Y) + \alpha(\nabla_X Y), \quad \alpha \in \Gamma(T^*). \tag{3.3}$$

Given a connection ∇ on T, and so on T^*, its torsion $\tau = \tau(\nabla)$ may be defined as $d - \wedge \circ \nabla$, where $\wedge$ denotes the mapping $\alpha \otimes \beta \longmapsto \alpha \wedge \beta$. It follows from (3.1) that τ is tensorial, and thus a section of

$$\mathrm{Hom}(T^*, \textstyle\bigwedge^2 T^*) \cong \bigwedge^2 T^* \otimes T \cong \mathrm{Hom}(\bigwedge^2 T, T).$$

By regarding τ as a linear mapping $\bigwedge^2 T \to T$, it is easy to prove the

Lemma 3.4 $\tau(X,Y) = \nabla_X Y - \nabla_Y X - [X,Y]$.

A connection on T is determined locally by its Christoffel symbols:

$$\nabla_{\partial/\partial x^i}\frac{\partial}{\partial x^j}=\sum_k \Gamma^k_{ij}\frac{\partial}{\partial x^k}$$

or

$$\nabla\frac{\partial}{\partial x^j}=\sum \Gamma^k_{ij}\,dx^i\otimes\frac{\partial}{\partial x^k}.$$

In dual language, this becomes

$$\nabla dx^i=-\sum \Gamma^i_{jk}\,dx^j\otimes dx^k.$$

Thus,

$$\tau(dx^i)=-\sum \Gamma^i_{jk}\,dx^j\wedge dx^k,$$

and ∇ is torsion-free or 'symmetric' iff $\Gamma^i_{jk}=\Gamma^i_{kj}$.

Example 3.5 Let M be a real 2-dimensional submanifold of $\mathbb{R}^3$, parametrized locally by a smooth vector-valued function $\mathbf{r}(p,q)$. The classical first fundamental form $E\,dp^2+2F\,dp\,dq+G\,dq^2$ defined by Gauss is the same thing as the induced Riemannian metric g with components $g_{11}=E$, $g_{12}=F=g_{21}$, $g_{22}=G$ relative to the local coordinates $p=x^1$, $q=x^2$. Let $\mathbf{n}$ be a unit normal vector to M, and $\sum h_{ij}dx^i dx^j=L dp^2+2M dp dq+N dq^2$ the second fundamental form. The formula

$$\frac{\partial^2\mathbf{r}}{\partial x^i\partial x^j}=\Gamma^1_{ij}\frac{\partial\mathbf{r}}{\partial x^1}+\Gamma^2_{ij}\frac{\partial\mathbf{r}}{\partial x^2}+h_{ij}\mathbf{n}$$

then determines a torsion-free connection on TM that is independent of the choice of coordinates.

Given an arbitrary connection on T, its is easy to check that the one defined by (3.2) with $A_XY=-\frac{1}{2}\tau(X,Y)$ is torsion-free. In terms of Christoffel symbols, this amounts to setting

$$\tilde{\Gamma}^i_{jk}=\tfrac{1}{2}(\Gamma^i_{jk}+\Gamma^i_{kj}).$$

In this way, a connection on T can be regarded as composed of two pieces, namely its torsion τ and the torsion-free connection $\nabla-\frac{1}{2}\tau$. If $\widetilde{\nabla}$ and $\widetilde{\nabla}+A'$ are *both* torsion-free then $A'_XY=A'_YX$, so that (at each point) we may write

$$A'\in\bigodot^2T^*\otimes T\subset T^*\otimes T^*\otimes T=T^*\otimes\operatorname{End}T.$$

It follows that mapping $\nabla\longmapsto(\tau,\nabla-\frac{1}{2}\tau)$ is an affine isomorphism mimicking the isomorphism

$$T^*\otimes\operatorname{End}T\cong(\bigwedge^2T^*\otimes T)\oplus(\bigodot^2T^*\otimes T)$$

of vector spaces.

We next recall the well-known way in which a connection on a vector bundle V extends to operators on differential forms of all degree with values in V. Let ∇ be a connection on V. Define $\mathbb{R}$-linear operators

$$\nabla_k : \bigwedge^k T^* \otimes V \longrightarrow \bigwedge^{k+1} T^* \otimes V \tag{3.6}$$

for each $k \geqslant 1$ by setting

$$\nabla_k(\alpha \otimes v) = d\alpha \otimes v + (-1)^k \alpha \wedge \nabla v.$$

This gives a sequence

$$\Gamma(T) \xrightarrow{\nabla} \Gamma(T^* \otimes V) \xrightarrow{\nabla_1} \Gamma(\bigwedge^2 T^* \otimes V) \to \cdots \tag{3.7}$$

The *curvature* ρ of ∇ is defined to be the composition

$$\nabla_1 \circ \nabla : \Gamma(V) \to \Gamma(T^* \otimes V) \to \Gamma(\bigwedge^2 T^* \otimes V).$$

The fact that $d^2 = 0$ implies that $\rho(fv) = f\rho(v)$ for $v \in \Gamma(V)$ and f a function. It follows that ρ determines a section of

$$V^* \otimes \bigwedge^2 T^* \otimes V \cong \bigwedge^2 T^* \otimes \operatorname{End} V, \tag{3.8}$$

that we denote by R to avoid confusion with the operator ρ acting on V. The associated curvature operator $R_{XY} \in \operatorname{End} V$ is also defined for any vector fields X, Y, and it may be calculated in the following alternative manner.

Lemma 3.9 $R_{XY} = \nabla_X \nabla_Y - \nabla_Y \nabla_X - \nabla_{[X,Y]}$.

Proof Follows by writing $\nabla v = \sum \alpha_i \otimes v_i$ for 1-forms α^i. □

One of the most obvious features of the curvature is that if $\varphi \in \Gamma(V)$ satisfies $\nabla\varphi = 0$, then $\rho(\varphi) = 0$, or equivalently $R_{XY}\varphi = 0$ for all X, Y. Typically, V will be some auxiliary vector bundle associated to the manifold (such as $\bigwedge^k T^*$), and ∇ and ρ denote connection and curvature induced from that of the tangent bundle T.

It is easy to show that for any k, the composition

$$\nabla_{k+1} \circ \nabla_k : \bigwedge^k T^* \otimes V \to \bigwedge^{k+2} T^* \otimes V \tag{3.10}$$

is given by $\alpha \otimes v \longmapsto \alpha \wedge \rho(v)$. A connection ∇ is said to be *flat* if its curvature ρ vanishes; in this case, the operators (3.6) give rise to a complex. The vanishing of (3.10) is also the integrability condition for the 'horizontal' distribution D on the total space of V, defined as follows. At a point $v \in \pi^{-1}(m) \in V$, the subspace D_v of the tangent space to V equals $s_*(T_m M)$, where s is any section of V satisfying $s(m) = v$ and $\nabla s|_m = 0$.

Exercises 3.11 (i) Check that (3.3) does define a connection, and define a connection ∇ on $\operatorname{End} T$ with the property that $\nabla\mathbf{1} = 0$.

(ii) Let ∇ be torsion-free, and σ any 2-form. Prove a formula that expresses $d\sigma(X, Y, Z)$ as a universal constant times

$$(\nabla_X\sigma)(Y, Z) + (\nabla_Y\sigma)(Z, X) + (\nabla_Z\sigma)(X, Y).$$

(iii) Let ∇ be a torsion-free connection on the cotangent bundle T^*. Prove the first Bianchi identity, namely that $\wedge \circ \rho = 0$ where $\wedge : \bigwedge^2 T^* \otimes T^* \to \bigwedge^3 T^*$.

(iv) A vector field on the total space of V is called *horizontal* if it lies in the distribution D defined directly above. Verify that $\rho = 0$ is equivalent to the condition that the Lie bracket of any two horizontal vector fields is itself horizontal. Deduce the following result.

Theorem 3.12 *Suppose that $\rho = 0$ and $m \in M$. Then on some neighbourhood of m, there exists a basis $\{v_1, \ldots, v_k\}$ of sections of V satisfying $\nabla v_i = 0$ for all i.*

Riemannian and symplectic connections

Now suppose that the metric g is 'covariant constant' relative to a connection ∇ on the tangent bundle T, so that $\nabla g = 0$. This condition involves the natural extension of ∇ to the vector bundle $T^* \otimes T^*$, and means that

$$X(g(Y, Z)) = g(\nabla_X Y, Z) + g(Y, \nabla_X Z) \tag{3.13}$$

for all vector fields X, Y, Z. In local coordinates, with $g = \sum g_{jk} dx^j \otimes dx^k$, the condition becomes

$$\partial_i g_{jk} = \Gamma_{ij}^{\ell} g_{\ell k} + g_{j\ell}\Gamma_{ik}^{\ell} = \Gamma_{ijk} + \Gamma_{ikj}, \tag{3.14}$$

where $\partial_i = \partial/\partial x^i$. From now on, starting with (3.14), we adopt the Einstein convention whereby summation is understood whenever there are repeated indices, one up one down.

Given another connection $\widetilde{\nabla} = \nabla + A$ on T for which $\widetilde{\nabla} g = 0$, set

$$\phi(X, Y, Z) = g(A_X Y, Z).$$

In classical language, we have simply lowered an index of A by setting $\phi_{ijk} = g_{rk} A_{ij}^r$. Then (3.13) implies that $\phi(X, Y, Z) = -\phi(X, Z, Y)$ or $\phi_{ijk} = -\phi_{ikj}$, and

$$\phi \in T^* \otimes \textstyle\bigwedge^2 T^* \subset T^* \otimes T^* \otimes T^*.$$

If ∇ and $\widetilde{\nabla}$ are, in addition, both torsion-free then

$$\phi \in (\textstyle\bigodot^2 T^* \otimes T^*) \cap (T^* \otimes \bigwedge^2 T^*) = \ker f, \tag{3.15}$$

where

$$f\colon T^*\otimes\Lambda^2T^*\subset T^*\otimes T^*\otimes T^*\to\Lambda^2T^*\otimes T^* \tag{3.16}$$

is the obvious composition. The symmetric group $\mathfrak{S}_3$ acts by permutations on $T^*\otimes T^*\otimes T^*$ and $e=(1,2,3)^3=((1,2)(2,3))^3$ acts as both $+1$ and -1 on $\ker f$. Thus f is an isomorphism and $\phi=0$.

Combined with the remarks following (3.2), the above discussion establishes the existence of the *Riemannian* or *Levi-Civita connection:*

Corollary 3.17 *There exists a unique connection ∇ on T for which $\nabla g=0$ and $\tau=0$.*

Let us repeat the above procedure with a non-degenerate 2-form ω in place of g. If ∇ is a torsion-free connection preserving ω then

$$d\omega=\wedge\nabla\omega=0.$$

Conversely, suppose that $d\omega=0$. In the notation of (2.4), the connection on U characterized by $\nabla dx^r=0=\nabla dy^r$ is torsion-free, and these locally defined connections can be combined with a partition of unity. It follows that (M,ω) is a symplectic manifold if and only if there exists a torsion-free connection with $\nabla\omega=0$.

Now suppose that ∇ and $\widetilde{\nabla}=\nabla+A$ are two (not necessarily torsion-free) connections satisfying $\nabla\omega=0=\widetilde{\nabla}\omega$, and set

$$\psi(X,Y,Z)=\omega(A_XY,Z). \tag{3.18}$$

Then $\psi\in T^*\otimes\bigodot^2T^*$, and the analogue of (3.16) assigns ψ to the difference $\tau(\widetilde{\nabla})-\tau(\nabla)$ of the torsions. This assignment can be identified with the middle mapping in the naturally defined exact sequence

$$0\to\textstyle\bigodot^3T^*\to T^*\otimes\bigodot^2T^*\to\Lambda^2T^*\otimes T^*\to\Lambda^3T^*\to 0.$$

If $\widetilde{\nabla},\nabla$ are both torsion-free, then ψ is a section of $\bigodot^3T^*$.

In the terminology of [47], a *Fedosov manifold* is a symplectic manifold (M,ω) endowed with a choice of torsion-free connection satisfying $\nabla\omega=0$. In local coordinates, the compatibility condition $\nabla\omega=0$ is equivalent to

$$\partial_i\omega_{jk}=\Gamma_{ij}^{\ell}\omega_{\ell k}-\omega_{j\ell}\Gamma_{ik}^{\ell}.$$

If we modify the Christoffel symbols by setting $\Gamma_{ijk}=\Gamma_{ij}^{r}\omega_{rk}$ (in contrast to the more conventional meaning (3.14)), then Γ_{ijk} is totally symmetric. Note that in the coordinates of (2.4), the only non-zero values of ω_{rk} are ±1.

The above situation can be summarized by the following diagram, in which $\sim$ denotes affine isomorphism. It is valid on any manifold equipped with a metric g and a non-degenerate 2-form ω, not necessarily compatible. The less familiar lower mapping $\sim$ is described in (iii) below, and [47] raises the question of computing the circular composition.

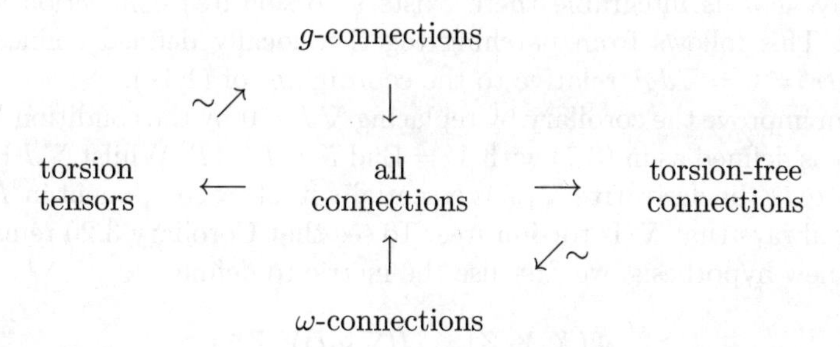

Exercises 3.19 (i) Prove that the connection defined in Example 3.5 coincides with the Levi-Civita connection of the induced metric g on the surface M.

(ii) If ∇^g, ∇^h denote the Levi-Civita connections for the metrics related by (2.23), try to find an expression for the difference tensor $A_X = \nabla^g_X - \nabla^h_X$.

(iii) Let $\Delta_{ijk} = \Delta_{ij}^m \omega_{mk}$ be the Christoffel symbols of a torsion-free connection on a manifold equipped with a non-degenerate 2-form ω. Show that

$$\Gamma_{ijk} = (\Delta_{ijk} + \Delta_{ikj} - \Delta_{jki}) + \tfrac{1}{2}(\partial_k \omega_{ij} - \partial_i \omega_{jk} - \partial_j w_{ki})$$

are the Christoffel symbols of a connection preserving ω. Simplify the last bracket under the symplectic assumption $dw = 0$.

Derivatives of J

Suppose that (M, J) is an almost-complex manifold. Choose any torsion-free connection ∇ on (the tangent bundle of) M. Using (1.17) and Lemma 3.4, this allows us to write

$$N(X,Y) = \sigma_X Y - \sigma_Y X,$$

where

$$\begin{aligned} \sigma_X Y &= \nabla_{JX} JY - \nabla_X Y - J\nabla_{JX} Y - J\nabla_X JY \\ &= (\nabla_{JX} J)Y - J(\nabla_X J)Y, \end{aligned}$$

and given that $\nabla_X (J \circ J) = 0$,

$$\sigma_X = \nabla_{JX} J + (\nabla_X J)J.$$

From the analogous property of N (1.18), we see that σ is the tensor characterized by

$$\sigma_X Y = -8\,\Re\mathrm{e}\,(\nabla_{X^{1,0}} Y^{1,0})^{0,1}.$$

Thus, $\sigma = 0$ if and only if $\nabla_A B \in T^{1,0}$ for all vector fields A, B of type $(1,0)$.

Corollary 3.20 *If ∇ is a torsion-free connection such that $\nabla J = 0$ then J is integrable.*

Conversely, if J is integrable there exists a torsion-free connection such that $\nabla J = 0$. This follows from patching together locally defined connections for which $\nabla dx^i = 0 = \nabla dy^j$ relative to the coordinates of (1.11).

We can improve the corollary by replacing $\nabla J = 0$ by the condition $\nabla_1 J = 0$, where ∇_1 is defined as in (3.7) with $V = \operatorname{End} T = T^* \otimes T$. Whilst ∇J has values in $T^* \otimes T^* \otimes T$, the derivative $\nabla_1 J$ is essentially its skew component in $\bigwedge^2 T^* \otimes T$, assuming always that ∇ is torsion-free. To see that Corollary 3.20 remains valid with the new hypothesis, we first use the metric to define

$$\Phi(X, Y, Z) = g((\nabla_X J)Y, Z). \tag{3.21}$$

Observe that

$$g(\sigma_X Y, Z) = \Phi(JX, Y, Z) + \Phi(X, JY, Z),$$

and that to obtain $g(N(X,Y), Z)$ we need to skew the right-hand side over X and Y. The result of doing this is however the same as first skewing in the first two positions of Φ. Thus,

$$g(N(X,Y), Z) = (\wedge \circ \Phi)(JX, Y, Z) + (\wedge \circ \Phi)(X, JY, Z), \tag{3.22}$$

where $\wedge : T^* \otimes T^* \otimes T^* \to \bigwedge^2 T^* \otimes T^*$ is the antisymmetrization. The condition $\nabla_1 J = 0$ implies that $\wedge \circ \Phi = 0$, so that $N = 0$.

Lemma 3.23 *If ∇ is the Levi-Civita connection then*

$$\Phi(X, Y, Z) = -\Phi(X, Z, Y) = -\Phi(X, JY, JZ).$$

Proof The first inequality follows because J, and so $\nabla_X J$, is skew-adjoint relative to g. The second since $\nabla_X J \circ J = -J \circ \nabla_X J$. □

For the remainder of Section 3, ∇ denotes the Levi-Civita connection. Lemma 3.23 implies that

$$\Phi \in T^* \otimes [\![\Lambda^{2,0}]\!], \tag{3.24}$$

reflecting the fact that $\nabla_X \omega \in [\![\Lambda^{2,0}]\!]$ for all X. There are injective mappings

$$\alpha : \Lambda^{3,0} \to \Lambda^{1,0} \otimes \Lambda^{2,0},$$
$$\beta : \Lambda^{1,0} \to \Lambda^{2,1},$$

given by $\alpha(A \wedge B \wedge C) = A \otimes (B \wedge C) + B \otimes (C \wedge A) + C \otimes (A \wedge B)$ and $\beta(A) = A \wedge \omega$, where A, B, C are (1,0)-forms and ω is the fundamental 2-form. Let $V = \ker \alpha^*$, and $\Lambda_0^{2,1} = \ker \beta^*$, where the asterisk denotes adjoint with respect to the natural Hermitian metric. Then V and $\Lambda_0^{2,1}$ are known to be irreducible under $U(n)$. It follows that the space in (3.24) has four components under the action of $U(n)$ for $n \geqslant 3$, and leads to the characterization in [53] of $2^4 = 16$ classes of almost-Hermitian manifolds. For example, the $(1,2)$-component of Φ can be identified with $(d\omega)^{1,2}$, and the vanishing of this is exactly complementary to the condition that $N = 0$.

Corollary 3.25 *M is Kähler if and only if $\nabla J = 0$.*

The condition $\nabla J = 0$ is equivalent to asserting that the holonomy group is contained in $U(n)$.

Suppose that M is Hermitian so that $N = 0$. The remark after (3.16), together with (3.22), implies that $(A, B, C) \longmapsto \Phi(A, B, C)$ is zero for all $A, B, C \in T^{1,0}$. Also $\Phi(A, B, \overline{C})$, so $\sigma = 0$.

Corollary 3.26 *The following are equivalent:*

(i) *J is integrable,*

(ii) $\nabla_{JX} J = J(\nabla_X J)$,

(iii) $A, B \in \Gamma(T^{1,0}) \Rightarrow \nabla_A B \in \Gamma(T^{1,0})$.

Condition (iii) has important consequences for the curvature tensor.

Exercises 3.27 (i) Let $\Lambda_0^{1,1}$ denote the orthogonal complement of ω in $\Lambda^{1,1}$. Show that as a $U(n)$-module it can be identified with the Lie algebra $\mathfrak{su}(n)$ (and adjoint representation), and that $\Lambda_0^{1,1}$ is irreducible.

(ii) Let M be an almost-Hermitian manifold of real dimension $2n \geqslant 6$. Deduce that Φ lies in a real vector space that is the direct sum of four $U(n)$-invariant real subspaces of dimensions $2d$, $n^3 - n^2 - 2d$, $2n$, $n^3 - n^2 - 2n$, where $d = \binom{n}{3}$. What happens when $n = 2$?

(iii) Let M be a 6-dimensional almost-Hermitian manifold. A sequence of differential operators

$$0 \to \Gamma(\Lambda^{0,0}) \to \Gamma(\Lambda^{1,0} \oplus \Lambda^{0,1}) \to \Gamma(\Lambda_0^{1,1} \oplus \Lambda^{2,0}) \to \Gamma(\Lambda_0^{2,1}) \to 0$$

is defined by composing d with the appropriate projection. Find necessary and sufficient conditions on Φ that guarantee that this is a complex [92].

The Riemann curvature tensor

Let M be a Riemannian manifold of dimension $2n$, and ∇ its Levi-Civita connection. It follows from the fact that $\nabla_X g = 0$ that the operator R_{XY} is a skew-adjoint transformation of the inner product space (T, g). The tensor

$$R(X, Y, Z, W) = g(R_{XY} Z, W)$$

is therefore skew not just in X, Y, but also in Z, W. Let $\{e_i\}$ be an orthonormal basis of vector fields, and set $R_{ijkl} = R(e_i, e_j, e_k, e_l)$, so that

$$R_{ijkl} = R_{jilk}. \tag{3.28}$$

The fact that ∇ is torsion-free gives the Bianchi identity

$$R_{ijkl} + R_{jkil} + R_{kijl} = 0 \tag{3.29}$$

(see Exercises 3.11), with the well-known consequence.

Lemma 3.30 $R_{ijkl} = R_{klij}$.

Proof This can be reformulated in terms of the alternating group $\mathfrak{U}_4$ that acts on $\bigotimes^4 T^*$ by permuting the factors. Let R be an element of $\bigotimes^4 T^*$ satisfying (3.28) and (3.29), so that it makes sense to consider the span $\mathbb{R}\mathfrak{U}_4 \cdot R$ of the orbit of R. Let B be the subgroup of $\mathfrak{U}_4$ generated by the double transposition $(12)(34)$, C that generated by the cycle (123), and $e \in \mathfrak{U}_4$ the identity. Then the two elements in any 'coset' $\sigma B \cdot R$ are equal, and the three elements in $\tau C \cdot R$ add to zero. Since $|\sigma B \cap \tau C| \leqslant 1$ and $|BC \cap CB| = 4$, it follows that

$$e + (12)(34) - (13)(42) - (14)(23) \in \mathbb{R}\mathfrak{U}_4$$

annihilates R. Thus, $(13)(42) \cdot R = R$, as required. □

The lemma implies that $\bigwedge^2(\bigwedge^2 T^*)$ injects into $T^* \otimes \bigwedge^3 T^*$, and R belongs to the space

$$\mathcal{R} = \ker\left(\bigodot^2(\bigwedge^2 T^*) \longrightarrow \bigwedge^4 T^*\right).$$

It is well known that $\mathcal{R}$ consists of three irreducible components under the action of $O(2n)$ for $n \geqslant 2$ (see Exercises 3.11), and these are described classically as follows. The Ricci tensor is defined by

$$R_{il} = R_{ijkl} g^{jk},$$

and represents a contraction $\mathcal{R} \to \bigodot^2 T^*$. Its 'trace' $s = R_{il} g^{il}$, obtained by further contraction $\bigodot^2 T^* \to \mathbb{R}$, is by definition the scalar curvature. The manifold is *Einstein* if the Ricci tensor is proportional to the metric, and this forces $2nR_{il} = s g_{il}$. The second Bianchi identity can be used to prove that in this case s is constant, still assuming that $n \geqslant 2$ [51].

Let

$$\begin{aligned} A_{ijkl} &= R_{jk} g_{il} - R_{jl} g_{ik} - R_{ik} g_{jl} + R_{il} g_{jk}, \\ B_{ijkl} &= s(g_{jk} g_{il} - g_{jl} g_{ik}). \end{aligned}$$

Recalling that the real dimension equals $2n$, we seek functions $a(n), b(n)$ such that

$$R_{ijkl} = W_{ijkl} + a(n) A_{ijkl} + b(n) B_{ijkl}, \quad W_{ijkl} g^{jk} = 0.$$

This will ensure that W represents the Weyl tensor, by definition that part of the curvature with zero Ricci contraction. The second equation implies that $1/a(n) = 2(n-1)$ and $1/b(n) = -2(n-1)(2n-1)$. Since

$$A_{ijkl} g^{jk} g^{il} = 2(2n-1)s, \quad B_{ijkl} g^{jk} g^{il} = 2n(2n-1)s,$$

it also follows that

$$R_{ijkl} = W_{ijkl} + C_{ijkl} + \frac{1}{2n(2n-1)} B_{ijkl},$$

where $C_{ijkl} = (A_{ijkl} - (1/n) B_{ijkl})/(2n-2)$ represents the trace-free Ricci tensor.

Exercises 3.31 (i) Let T be a real inner product space of dimension $2n \geqslant 4$. Relative to $O(2n)$, there are equivariant isomorphisms

$$\begin{aligned} \bigodot^2(\bigwedge^2 T^*) &\cong \mathbb{R} \oplus \bigodot^2_0 T^* \oplus \bigwedge^4 T^* \oplus \mathcal{W}, \\ \bigwedge^2(\bigwedge^2 T^*) &\cong \bigwedge^2 T^* \oplus \mathcal{X}, \\ T^* \otimes \bigwedge^3 T^* &\cong \bigwedge^4 T^* \oplus \bigwedge^2 T^* \oplus \mathcal{Y}, \end{aligned}$$

where all the summands are irreducible. Assuming this, compute the dimensions of $\mathcal{X}, \mathcal{Y}, \mathcal{W}$, and show that they all vanish if $n = 3$. Using Schur's lemma (the elementary fact that any G-homomorphism between irreducible spaces is either zero or an isomorphism), show that $\mathcal{X}$ and $\mathcal{Y}$ are isomorphic, and $\mathcal{W}$ lies in the kernel of the skewing map $\bigodot^2(\bigwedge^2 T^*) \to \bigwedge^4 T^*$.

(ii) Let M be an Einstein manifold. Its curvature tensor R may be regarded as an element of $\mathbb{R} \oplus \mathcal{W}$ at each point. Show that the tensor $S_{lm} = R^{ij}{}_{kl} R^k{}_{ijm}$ is determined by a linear mapping $\bigodot^2 \mathcal{W} \to T^* \otimes T^*$. Deduce from Section 4 that if $\dim M = 4$ then S_{lm} is a scalar multiple of g_{lm} (this is called the 'super-Einstein' condition).

Now let (M, g, J) be an almost-Hermitian manifold. In arbitrary dimension, the type decomposition (2.28) induces a decomposition of real vector spaces

$$\bigodot^2(\bigwedge^2 T^*) = \mathcal{S}_1 \oplus \mathcal{S}_2 \oplus \mathcal{S}_3 \oplus \mathcal{S}_4,$$

where

$$\begin{aligned} \mathcal{S}_1 &= [\Lambda^{1,1} \odot \Lambda^{1,1}], \\ \mathcal{S}_2 &= [\Lambda^{2,0} \odot \Lambda^{0,2}], \\ \mathcal{S}_3 &= [\![\Lambda^{2,0} \odot \Lambda^{2,0}]\!], \\ \mathcal{S}_4 &= [\![\Lambda^{2,0} \odot \Lambda^{1,1}]\!]. \end{aligned}$$

(If U_1, U_2 are subspaces of V then $U_1 \odot U_2$ denotes the image of $U_1 \otimes U_2$ under the symmetrization $V \otimes V \to \bigodot^2 V$. Addition of complex conjugates is understood when there are double brackets, in accordance with (1.16).) For example, $R \in \bigodot^2(\bigwedge^2 T^*)$ belongs to the subspace $\mathcal{S}_1$ if and only if

$$R(X, Y, Z, W) = R(X, Y, JZ, JW) \tag{3.32}$$

for all tangent vectors X, Y, Z, W.

Lemma 3.33 *If M is Kähler then its Riemann tensor R belongs to the space $\mathcal{R}_1 = \mathcal{R} \cap \mathcal{S}_1$.*

Proof The Levi-Civita connection ∇ induces a connection on $\operatorname{End} T$, whose curvature is induced from the natural Lie algebra action of $\operatorname{End} T$ on $\operatorname{End} T$. Thus, if R_{XY} is the operator of Lemma 3.9 then

$$0 = (\nabla_1 \nabla J)_{XY} = R_{XY}(J) = R_{XY} \circ J - J \circ R_{XY}. \tag{3.34}$$

When we convert R into a tensor with all lower indices, the relation $R_{XY} \circ J = J \circ R_{XY}$ translates into the equation (3.32). □

Define $\mathcal{R}_2$ so that the right-hand side of

$$\mathcal{R} \cap (\mathcal{S}_1 \oplus \mathcal{S}_2) = \mathcal{R}_1 \oplus \mathcal{R}_2$$

is an orthogonal sum, and set

$$\mathcal{R}_3 = \mathcal{R} \cap \mathcal{S}_3, \qquad \mathcal{R}_4 = \mathcal{R} \cap \mathcal{S}_4.$$

Since the images of $\mathcal{S}_1 \oplus \mathcal{S}_2$, $\mathcal{S}_3$, $\mathcal{S}_4$ in $\bigwedge^4 T^*$ are mutually orthogonal, it follows that

$$\mathcal{R} = \bigoplus_{i=1}^{4} \mathcal{R}_i.$$

This notation is consistent with the spaces $\mathcal{L}_j = \bigoplus_{i=1}^{j} \mathcal{R}_i$, $j = 1, 2, 3$, defined by Gray [52], although a full analysis of the $U(n)$-components of $\mathcal{R}$ was subsequently carried out by Tricerri and Vanhecke [102, 38].

Lemma 3.35 *If M is Hermitian then $R \in \mathcal{R}_1 \oplus \mathcal{R}_2 \oplus \mathcal{R}_4$.*

Proof This follows from Corollary 3.26. Suppose that J is integrable, and let A, B, C be vector fields of type $(1, 0)$. Then

$$R_{AB}C = \nabla_A \nabla_B C - \nabla_B \nabla_A C - \nabla_{[A,B]} C$$

also has type $(1, 0)$, and so $g(R_{AB}C, D) = 0$ for all $A, B, C, D \in T^{1,0}$. Put another way, R has no component in $\Lambda^{2,0} \otimes \Lambda^{2,0}$ nor (taking complex conjugates) in $\Lambda^{0,2} \otimes \Lambda^{0,2}$. The result follows from the definition of $\mathcal{R}_3$. □

If $R \in \mathcal{R}_3$ then $R = \sigma + \bar{\sigma}$ where $\sigma \in \Lambda^{2,0} \otimes \Lambda^{2,0}$. Since $\Lambda^{2,0}$ is isotropic, the Ricci contraction annihilates σ (and similarly $\bar{\sigma}$). Thus, $\mathcal{R}_3$ is a component of the subspace $\mathcal{W}$ of $\mathcal{R}$ generated by Weyl tensors. A dimension count shows that $\dim \mathcal{R}_3 > \frac{1}{8} \dim \mathcal{W}$ for $n \geqslant 2$ [93], and this gives some idea of how the existence of a single OCS conditions the Weyl tensor.

Let R_i denote the component of the Riemann tensor R in $\mathcal{R}_i$. The following result, taken from [38], highlights the fundamental nature of the space $\mathcal{R}_1$ of Kähler curvature tensors.

Proposition 3.36 *The tensors R_2, R_3, R_4 are linear functions of $\nabla^2 J$.*

Proof The kernel of the mapping $R \longmapsto R(J)$ of (3.34) equals $\mathcal{R}_1$, and $R(J)$ can be identified with $R_2 \oplus R_3 \oplus R_4$. □

Let M be a Kähler manifold of real dimension $2n$, and let $\kappa = \Lambda^{n,0}$ denote its *canonical line bundle.* Let ξ be a local section of κ, so that the Levi-Civita connection satisfies

$$\nabla \xi = i\alpha \otimes \xi \tag{3.37}$$

for some *real* 1-form α. The curvature of κ is given by

$$\rho(\xi) = \nabla_1 \nabla \xi = i\, d\alpha \wedge \xi,$$

and is, to all intents and purposes, the same as the closed 2-form $\Omega = id\alpha$. The latter will not in general be globally exact, since α is only defined locally.

Algebraically, Ω can be identified with the image of R under the contraction

$$\textstyle\bigwedge^2 T^* \otimes \operatorname{End} T \to \bigwedge^2 T^* \otimes (\operatorname{End} \kappa) \cong \bigwedge^2 T^*, \tag{3.38}$$

given in lowered index notation by

$$R_{ijkl} \longmapsto R_{ijkl}\omega^{kl} = -(R_{iklj} + R_{iljk})\omega^{kl} = 2R_{kilj}\omega^{kl},$$

where ω is the fundamental 2-form. It follows that

$$\Omega(X, Y) = 2S(JX, Y)$$

is (twice) the *Ricci form*, a $(1,1)$-form manufactured in the natural way from the Ricci tensor, just as ω is from g.

Example 3.39 Further to Example 2.30, the canonical bundle is generated by $\eta = \omega_2 + i\omega_3$. To determine α in (3.37), note that $i\alpha \wedge \eta = d\eta = -3e^4 \wedge \eta$. Thus $(i\alpha + 3e^4) \wedge \eta = 0$, so $i\alpha + 3e^4$ must be a $(1,0)$-form and $\alpha = -3e^3$. Hence, $\Omega = -6i\omega_1$, showing that g_2 is Einstein.

We conclude this section by discussing very briefly properties of the curvature of a torsion-free symplectic connection. Let M be a manifold of dimension $2n \geqslant 4$ with a symplectic form ω, and let ∇ be a connection on M satisfying $\tau = 0$ and $\nabla\omega = 0$. The curvature of the induced connection on T^* is a linear mapping $\rho : T^* \to \bigwedge^2 T^* \otimes T^*$, and we define

$$R_{ijkl} = \omega_{ir}\rho(e^r)(e_k, e_l, e_j),$$

with summation over r. It follows easily that $R_{ijkl} = R_{jikl}$, whence

$$R \in \ker\left(\textstyle\bigodot^2 T^* \otimes \bigwedge^2 T^* \longrightarrow T^* \otimes \bigwedge^3 T^*\right). \tag{3.40}$$

Exterior powers of T^* are not irreducible for $Sp(2n, \mathbb{R})$. For example, $\bigwedge^2 T^*$ contains the trivial 1-dimensional space $\langle\omega\rangle$ generated by the symplectic form ω; more generally wedging with ω determines an equivariant mapping $\bigwedge^k T^* \to \bigwedge^{k+2} T^*$. In this sense, the situation is dual to that of the orthogonal group, and symmetric powers of T^* *are* irreducible for $Sp(2n, \mathbb{R})$. It follows that the kernel in (3.40) contains only two $Sp(2n, \mathbb{R})$-irreducible summands. The conclusion is that there is a unique 'Ricci tensor' in the symplectic situation, but no way of defining scalar curvature [103].

Recent problems associated to the theory of symplectic connections can be found in [27] and references therein.

Exercises 3.41 (i) Consider the manifold X of Example 1.27. Verify that $\omega = e^1 \wedge e^4 + e^2 \wedge e^3$ is closed. Determine possible constants τ^i_j for which $\nabla e^i = \sum_j \tau^i_j \otimes e^j$ defines a torsion-free connection on TX for which $\nabla\omega = 0$. Compute the curvature of this connection.

(ii) Let M be a hyperkähler manifold (Definition 2.10). Its holomorphic cotangent space $E = \Lambda^{1,0}$ has a quaternionic structure, and an even tensor product such as $\bigodot^4 E$ is the complexification of a real space $[\bigodot^4 E]$. By referring to [92] or [60], explain why the curvature tensor of M may be regarded as an element of $[\bigodot^4 E]$ at each point.

4 Further topics

This section is devoted to applications of the preceding ones, and is divided into three subsections. The first two are concerned with four-dimensional Riemannian geometry, and the third with the theory of special Kähler manifolds.

By way of preliminaries, we specialize the discussion of the curvature of the Levi-Civita connection on an almost-Hermitian manifold in Section 3 to the case of four real dimensions. Orientation plays an important role, and attention is focussed on the semi-Weyl tensor W_+. This leads to a summary of well-known relations with topology. In the second subsection, results are applied to give a crude classification of 4-dimensional Riemannian manifolds, distinguished according to the number of orthogonal complex structures that exist locally. This is rounded off by a description of constraints arising from the decomposition of W_+ in the context of almost-Kähler manifolds.

In highlighting the role of curvature, it seems only right to include a situation in which the curvature is zero. Many interesting integrability conditions can be interpreted by the vanishing of the curvature of some connection or family of connections. Such situations are playing an increasingly important role in differential geometry, and we illustrate one involving a class of flat symplectic connections. This is used in the definition of special Kähler metrics and associated hyperkähler metrics.

Curvature in four dimensions

Let M be an almost-Hermitian manifold of real dimension 4, and let T denote the tangent space at an arbitrary point. Forgetting about J for the moment, the curvature operator is an endomorphism $\hat{R}$ of

$$\bigwedge^2 T^* = \Lambda^+ \oplus \Lambda^-. \tag{4.1}$$

It therefore decomposes under the action of $SO(4)$ as a block matrix

$$\hat{R} = \begin{pmatrix} A_+ & B \\ B^T & A_- \end{pmatrix},$$

where $A_\pm$ is a symmetric matrix, corresponding to a self-adjoint endomorphism of $\Lambda^\pm$. For example, regarding $\hat{\omega}_1^\pm = (e^{12} \pm e^{34})/\sqrt{2}$ as unit 2-forms, we see that

$$\begin{aligned}(A_+)_{11} &= R(\hat{\omega}_1^+, \hat{\omega}_1^+) = \tfrac{1}{2}(R_{1212} + 2R_{1234} + R_{3434}),\\ B_{11} &= R(\hat{\omega}_1^+, \hat{\omega}_1^-) = \tfrac{1}{2}(R_{1212} - 2R_{1234} + R_{3434}).\end{aligned} \tag{4.2}$$

Hence, using (3.29),

$$\operatorname{tr} A_+ = \operatorname{tr} A_- = \tfrac{1}{4}\sum_{i,j} R_{ijij} = \tfrac{1}{4}\, s,$$

where s is the scalar curvature.

The traceless endomorphisms $W_\pm = A_\pm - \frac{1}{12} s\mathbf{1}$ represent two halves of the Weyl tensor, whereas $B \in \operatorname{Hom}(\Lambda^-, \Lambda^+) \cong \bigodot_0^2 T^*$ represents the trace-free part of the Ricci tensor. The last isomorphism is obtained by contracting on the middle two indices as in the example

$$\begin{aligned} 2\hat{\omega}_1^+ \otimes \hat{\omega}_2^- &= (e^{12} + e^{34}) \otimes (e^{13} - e^{42})\\ &= (e^1e^2 - e^2e^1 + e^3e^4 - e^4e^3)(e^1e^3 - e^3e^1 - e^4e^2 + e^2e^4)\\ &\longmapsto (e^1e^4 - e^2e^3 - e^3e^2 + e^4e^1) = (e^1 \odot e^4 - e^2 \odot e^3).\end{aligned}$$

Symbolically, we may now write

$$R = (W_+, W_-, B, s) \qquad [5, 5, 9, 1], \tag{4.3}$$

where the numbers in square brackets indicate the dimension of the corresponding space of tensors. It follows that M is Einstein if and only if B is identically zero, and M is conformally flat if and only if $W_+ \equiv 0 \equiv W_-$.

Definition 4.4 *M is half conformally flat if either W_+ or W_- vanishes identically. In the latter case, M is called self-dual, and in the former case anti-self-dual or ASD.*

Now suppose that M is almost-Hermitian, and recall that J reduces Λ^+ into the sum $\langle\omega\rangle \oplus [\![\Lambda^{2,0}]\!]$ of real subspaces of dimension 1 and 2. Relative to this, we have

Lemma 4.5 *$\hat{R}$ decomposes as a symmetric matrix*

$$\left(\begin{array}{c|c||c} \frac{1}{4}s^* & W' & B' \\ \hline \bullet & W'' + \frac{1}{8}(s - s^*) & B'' \\ \hline\hline \bullet & \bullet & W_- + \frac{1}{12}s \end{array}\right) \tag{4.6}$$

with $\operatorname{tr}(W'') = 0$.

The effect of J is therefore to refine (4.3) by the additional splittings

$$B = (B', B'') \qquad [3, 6],$$

$$W_+ = (s - 3s^*, W', W'') \qquad [1, 2, 2],$$

with the indicated dimensions. The metric is said to be $*$*Einstein* if $B' = 0$ and $W' = 0$, and *strongly* $*$*Einstein* if, in addition, s^* is constant.

Lemma 4.7 (i) *If M is Kähler, then $W', W'', s - s^*, B''$ all vanish.*
(ii) *If M is Hermitian then W'' vanishes.*

Proof This is a re-interpretation of Lemmas 3.33 and 3.35. The Kähler condition implies that R is effectively an endomorphism of $\Lambda^{1,1}$. It follows that all tensors above involving $[\![\Lambda^{2,0}]\!]$ vanish. Put another way, one eliminates all the components corresponding to a big cross $+$ in the matrix (4.6), and (i) follows. If J is integrable, we have seen that R has no component in

$$\Lambda^{2,0} \otimes \Lambda^{2,0} \cong \mathrm{Hom}(\Lambda^{0,2} \otimes \Lambda^{2,0}).$$

The latter corresponds to the trace-free part of the real space $\mathrm{Hom}([\![\Lambda^{2,0}]\!], [\![\Lambda^{2,0}]\!])$, and is represented by W''. □

In the Kähler case, we may therefore write

$$R = (W_-, B', s) \qquad [5, 3, 1],$$

and M is ASD if and only if $s = 0$. Kähler metrics satisfying $s \equiv 0$ are called 'scalar-flat Kähler' or SFK, and their study forms part of the more general theory of extremal Kähler metrics introduced in [29]. The tensor W_- is the so-called Bochner tensor that in higher dimensions is generalized by a corresponding component of the space $\mathcal{R}_1$ of Lemma 3.33 [109]. Self-dual Kähler surfaces are studied in [34, 4, 23]. Finally, we remark that an almost-Hermitian 4-manifold satisfying $W'' = 0$ is said to have 'Hermitian Weyl tensor'.

We have already seen that on a Kähler manifold, the Ricci tensor can be extracted as the curvature of the canonical bundle. In the case of an arbitrary 4-dimensional almost-Hermitian manifold, we can easily distinguish the components of R that contribute to the curvature 2-form Ω of the canonical bundle $\kappa = \Lambda^{2,0}$. The contraction (3.38) amounts to selecting the first row of (4.6) and, identifying W', B' with 2-forms,

$$\Omega = \tfrac{1}{4} s^* \omega + W' + B', \qquad W' \in [\![\Lambda^{2,0}]\!] \subset \Lambda^+,\ B' \in \Lambda^-. \tag{4.8}$$

Corollary 4.9 *The induced connection on the canonical line bundle is self-dual if M is Einstein, and ASD if M is SFK.*

The forms in (4.8) may be used to represent the first Chern class

$$c_1(T^{1,0}) = c^+ + c^- \in H^+ \oplus H^- \tag{4.10}$$

of M, where $H^\pm = \{\alpha \in \Gamma(\Lambda^\pm)\colon d\alpha = 0\}$. The corollary is in theory relevant to constructions of self-dual metrics by Joyce and others, described by [62, 31] and references therein.

Let M be a connected compact oriented 4-manifold M. Its Betti numbers are defined by $b_i = \dim H^i(M, \mathbb{R})$, and satisfy $b_0 = 1 = b_4$ and $b_1 = b_3$ by

Poincaré duality. The Euler characteristic is then

$$\chi = \sum_{i=0}^{4}(-1)^i b_i = 2 - 2b_1 + b_2,$$

and is positive if M is simply connected. Applying Hodge theory to the decomposition (4.1) yields $b_2 = b^+ + b^-$, where $b^\pm = \dim H^\pm$. The quantity

$$\sigma = b^+ - b^-$$

is the signature of the real quadratic form associated to the intersection form

$$H_2(M,\mathbb{Z}) \times H_2(M,\mathbb{Z}) \to H_4(M,\mathbb{Z}) \cong \mathbb{Z} \tag{4.11}$$

on homology. We now explain how both quantities χ, σ can be computed from a knowledge of the curvature tensor R of the Levi-Civita connection.

Recall that R is a matrix of 2-forms ($\rho_{ij} = \sum R_{ijkl} e^k \wedge e^l$). Invariant polynomials on the Lie algebra $\mathfrak{so}(4)$, isomorphic to the space (4.1), provide the following antisymmetric matrices of 2-forms:

$$\rho = \begin{pmatrix} 0 & \rho_{12} & \rho_{13} & \rho_{14} \\ \bullet & 0 & \rho_{23} & \rho_{24} \\ \bullet & \bullet & 0 & \rho_{34} \\ \bullet & \bullet & \bullet & 0 \end{pmatrix}, \quad *\rho = \begin{pmatrix} 0 & \rho_{34} & -\rho_{24} & \rho_{23} \\ \bullet & 0 & \rho_{14} & -\rho_{13} \\ \bullet & \bullet & 0 & \rho_{12} \\ \bullet & \bullet & \bullet & 0 \end{pmatrix}.$$

Here, $*$ can be regarded as the involution acting as ± 1 on $\Lambda^\pm$. In terms of these matrices,

$$\begin{aligned} -\operatorname{tr}(\rho^2) &= \sum_{i,j} {\rho_{ij}}^2, \\ \tfrac{1}{4}\operatorname{tr}(\rho(*\rho)) &= \rho_{12}\rho_{34} + \rho_{13}\rho_{42} + \rho_{14}\rho_{23}. \end{aligned}$$

The latter is the so-called Pfaffian, formally the square root of $\det \rho$.

The Hirzebruch signature theorem implies that σ equals $\frac{1}{3}p_1$, where p_1 is the first Pontrjagin number. The latter can be computed by Chern–Weil theory as

$$\begin{aligned} p_1 &= -\frac{1}{8\pi^2}\int \operatorname{tr}(\rho \wedge \rho) \\ &= \frac{1}{4\pi^2}\int \Big((|A_+|^2 + |B|^2) - (|B^T|^2 + |A_-|^2)\Big)\upsilon \\ &= \frac{1}{4\pi^2}\int \Big(|W_+|^2 - |W_-|^2\Big)\upsilon. \end{aligned}$$

Thus if M is ASD then $\sigma \leqslant 0$, with equality if M is conformally flat. On the other hand, by Chern's theorem,

$$\begin{aligned} \chi &= \frac{1}{32\pi^2}\int \operatorname{tr}(\rho \wedge *\rho) \\ &= \frac{1}{8\pi^2}\int \Big((|A_+|^2 - |B|^2) - (|B^T|^2 - |A_-|^2)\Big)\upsilon \\ &= \frac{1}{8\pi^2}\int \Big(|W_+|^2 + |W_-|^2 + 6\Big(\frac{1}{12}s\Big)^2 - 2|B|^2\Big)\upsilon. \end{aligned}$$

Thus, if $B = 0$,

$$\chi = \frac{2}{3}|\sigma| + \frac{1}{8\pi^2}\int \left(2|W_\pm|^2 + \frac{1}{24}s^2\right)v.$$

This provides the celebrated Hitchin–Thorpe inequality:

Corollary 4.12 *An Einstein 4-manifold satisfies $\chi \geqslant \frac{3}{2}|\sigma|$, with equality iff $s = 0$ and $W_\pm = 0$.*

If M is almost-complex then there are corresponding formulae for the indices of the Dolbeault complexes discussed in Section 2. In particular, the *arithmetic index* $1 - h^{0,1} + h^{0,2}$ equals

$$\tfrac{1}{12}(c_1^2 + c_2) = \tfrac{1}{12}(p_1 + 3\chi) = \tfrac{1}{4}(\chi + \sigma).$$

So a compact 4-manifold can only have an almost-Hermitian structure if

$$\chi + \sigma \equiv 0 \bmod 4.$$

Conversely, if the second Stiefel–Whitney class $w_2(M)$ is the reduction mod 2 of $c \in H^2(M,\mathbb{Z})$ and $c^2 = 2\chi + 3\sigma$ then M admits an almost-complex structure J for which $c = c_1$. If M is simply connected, such a c exists in H^+ if and only if b^+ is odd.

Example 4.13 The complex projective plane $\mathbb{CP}^2$ has

$$p_1 = c_1{}^2 - 2c_2 = (3x)^2 - 2(3x^2) = 3x^2,$$

where $x = c_1(L)$ is the positive generator of $H^2(\mathbb{CP}^2,\mathbb{Z})$ (and L denotes the standard holomorphic line bundle), which is consistent with the fact that $\chi = 3$ and $\sigma = 1$. More generally, one may consider the connected sum

$$m\mathbb{CP}^2 \# n\overline{\mathbb{CP}}^2 \tag{4.14}$$

of m copies of $\mathbb{CP}^2$ and n copies of the same smooth manifold with reversed orientation. The result has signature $\sigma = m - n$ and Euler characteristic $\chi = 2 + m + n$. If $m \geqslant n$, one requires $m \leqslant 4 + 5n$ for the possibility of an Einstein metric. Equality is not possible for the following reason. If

$$R = \begin{pmatrix} 0 & 0 \\ 0 & W_- \end{pmatrix} \tag{4.15}$$

then R has no component in $\bigwedge^2 T^* \otimes \operatorname{End}\Lambda^+$ and the bundle $\Lambda^+ T^*M$ is flat. It follows that $b^+ = 3$ and $\sigma \leqslant 3$.

The equality $b^+ = 3$ occurs for a *K3 surface*, by definition a simply-connected compact complex surface with $c_1 = 0$. Any two are known to be diffeomorphic, and an example is a quartic hypersurface K in $\mathbb{CP}^3$. The Chern classes of K

are easily computed from the formula $T\mathbb{CP}^3|_K = TK \oplus L^4$; we obtain and $c_2 = 6x^2$ where x now denotes the pull-back of the generator of $H^2(\mathbb{CP}^3, \mathbb{Z})$. Thus, $\sigma = -16$ and $\chi = 24$, so $b^+ = 3$, $b^- = 19$ and $b_2 = 22$.

Let M be a compact, oriented, simply-connected, 4-manifold, with (as we always assume) a smooth structure. The choice of a Riemannian metric reduces the structure group of the tangent bundle to $SO(4)$, and this lifts to $Spin(4) \cong SU(2) \times SU(2)$ (equivalently, the Stiefel–Whitney class $w_2 \in H^2(M, \mathbb{Z}_2)$ vanishes) if and only if the intersection form (4.11) takes only even values. In this spin case it is known that $\sigma \equiv 0 \bmod 16$, and in the light of the work of Donaldson and Freedman it is conjectured that

$$\frac{b_2}{|\sigma|} \geqslant \frac{11}{8}$$

[35, 43, 44]. If this is the case, M would necessarily be homeomorphic to a connected sum $mK\#n(S^2 \times S^2)$. By contrast, the topological classes of non-spin 4-manifolds are exhausted by the connected sums (4.14).

The condition (4.15) implies that the manifold M is locally Ricci-flat Kähler. If it is simply connected then, by Theorem 3.12, there exists an orthonormal basis $\{\omega_1, \omega_2, \omega_3\}$ of parallel sections and M is hyperkähler. In this case, M is necessarily diffeomorphic to T^4 or a K3 surface [58]. Of course, the torus has a flat metric and compatible hyperkähler structure, whereas any K3 surface admits a hyperkähler metric by Yau's deep theorem [110], a new account of which can be found in Joyce's book [63].

The question of exactly which compact smooth 4-manifolds can admit Einstein metrics has been pursued by LeBrun and collaborators, by using Seiberg–Witten theory to refine Corollary 4.12. For example, it can be shown that the topological manifold underlying a K3 surface has infinitely many smooth structures, but the above argument shows that only the standard one admits an Einstein metric. This situation is not untypical [73].

Structures on 4-manifolds

Recall that, at each point, W_+ is a self-adjoint linear transformation of the 3-dimensional space Λ^+. Let its eigenvalues be $\lambda_1 \leqslant \lambda_2 \leqslant \lambda_3$ with $\sum \lambda_i = 0$, and let $\{\sigma_1, \sigma_2, \sigma_3\}$ be a corresponding basis of orthonormal eigenvectors. The following result can be found in [80, 93]:

Lemma 4.16 *If J is an OCS on M, oriented so that $\omega \in \Gamma(M, \Lambda^+T^*M)$. Then*

$$\pm\omega = \sqrt{\frac{\lambda_1 - \lambda_2}{\lambda_1 - \lambda_3}}\sigma_1 \pm \sqrt{\frac{\lambda_2 - \lambda_3}{\lambda_1 - \lambda_3}}\sigma_3. \tag{4.17}$$

Proof We shall show that the fundamental 2-form ω lies in the span of σ_1, σ_3, leaving determination of the coefficients as an exercise. The condition $W'' = 0$

implies that W_+ is represented by the matrix

$$\begin{pmatrix} 2\lambda & x & y \\ x & -\lambda & 0 \\ y & 0 & -\lambda \end{pmatrix}.$$

Computation of its characteristic polynomial reveals that $\lambda = \lambda_2$ is the middle eigenvalue. The corresponding eigenvector is the column vector

$$\sigma_2 = \pm\left(0, \frac{-y}{\sqrt{x^2+y^2}}, \frac{x}{\sqrt{x^2+y^2}}\right)^T,$$

and the result follows from the fact that ω is represented by $(1,0,0)^T$. □

We shall call the four elements (4.17) the *roots* of the tensor W_+. This terminology is justified by spinor language in which W_+ is represented by a quartic polynomial, and (4.17) are (projectivizations of) the roots of this polynomial. The lemma allows us to perform a classification in terms of the existence of compatible complex structures. There are two cases according to whether W_+ is identically zero or not, and we proceed to consider each in turn.

Hermitian manifolds with $W_+ \neq 0$

A generic Riemannian metric will not admit any compatible complex structure locally, since the roots of W_+ will determine non-integrable almost-complex structures. Incidentally, this situation shows that the converse of Lemma 4.7(ii) is false, although Hermitian Weyl tensor together with an additional curvature condition is known to imply that $N = 0$ (see Theorem 4.22 below).

A generic Hermitian metric compatible with a given complex structure J will admit only $\pm J$ as a compatible complex structure, as the other root will be non-integrable. A more special situation is that in which two of the eigenvalues $\lambda_1, \lambda_2, \lambda_3$ coincide (as do the roots of W_+ in pairs). This occurs if and only if $W' = 0 = W''$, or equivalently

$$W_+ = \begin{pmatrix} \frac{1}{6}s & 0 & 0 \\ 0 & -\frac{1}{12}s & 0 \\ 0 & 0 & -\frac{1}{12}s \end{pmatrix}. \tag{4.18}$$

We have seen that any Kähler metric has this property, but (4.18) can also hold in other cases.

A Riemannian version of the so-called *Goldberg–Sachs theorem* implies that an Einstein metric satisfies (4.18) (and so is $*$Einstein) if and only if it is Hermitian relative to the eigenform σ_1 [6]. The only compact example known of an Einstein–Hermitian manifold that is not Kähler is $\mathbb{CP}^2$ blown up at at one point, with Page's metric that has a non-trivial group of isometries. If there is another, the underlying complex surface must be biholomorphic to $\mathbb{CP}^2$ blown up at either two or three points [72].

Example 4.19 A simple instance of a $*$Einstein non-Einstein metric satisfying (4.18) is the one

$$g = dx^2 + dy^2 + dt^2 + (dz - xdy)^2$$

naturally defined in Example (1.9) [1]. The formal similarity with (2.31) is part of a more general construction [49, 56].

Non-ASD *bihermitian metrics* have the property that the eigenvalues of W_+ are distinct on a dense open set, but that all the almost-complex structures defined by Lemma 4.16 are integrable. In this case, the twistor space of M has four sections that constitute the zero set of the Nijenhuis tensor of the almost-complex structure J_1. Following the initial examples of [67], a general construction of such metrics has been given in [7], and we describe this in the next paragraph. A contrasting situation described in [70] is that in which a 4-manifold has two complex structures determining *opposite* orientations.

If $T = \mathbb{R}^4$ and v is a non-zero element of $\bigwedge^4 T^*$, then the formula $\alpha \wedge \beta = B(\alpha, \beta)v$ defines a bilinear form B on $\bigwedge^2 T^*$. This bilinear form is known to have signature $(3, 3)$ and defines a double covering from $SL(4, \mathbb{R})$ to a connected component of $O(3, 3)$ [95]. Moreover, there is a bijective correspondence between 3-dimensional subspaces Λ^+ of $\bigwedge^2 T^*$ on which B is positive-definite, and oriented conformal structures on T. Now let M be an oriented 4-manifold, and suppose that Φ_r, $r = 1, 2, 3$, is a triple of real symplectic forms on M for which $v = \Phi_r \wedge \Phi_r$ is independent of r, and relative to which the matrix of B is

$$\begin{pmatrix} 1 & 0 & 0 \\ 0 & 1 & p \\ 0 & p & 1 \end{pmatrix}, \quad |p| < 1.$$

It follows that there exists a Riemannian metric g for which $|\Phi_r| = 2$ (in accordance with the convention used in (4.2)), and $\Lambda^+ = \langle \Phi_1, \Phi_2, \Phi_3 \rangle$. Setting $\Lambda^{2,0} = \langle \Phi_0 + i\Phi_r \rangle$ determines an almost-complex structure J_r for $r = 1$ and $r = 2$, with the property that $J_1 J_2 + J_2 J_1 = -2p\mathbf{1}$. However, Lemma 2.29 shows that J_1 and J_2 are both *integrable*, so (M, g, J_1, J_2) is bihermitian.

This method can be used to prove the existence of non-ASD bihermitian structures on any 4-manifold with a hyperhermitian metric. The latter fits into the next category.

Hermitian manifolds with $W_+ = 0$

Recall that M is called *anti-self-dual* (ASD) if W_+ is identically zero. It follows from Lemma 4.16 that this will hold whenever M admits at least three independent OCS's compatible with the orientation, around each point. Conversely, $W_+ \equiv 0$ implies that the twistor space is a complex manifold, and any almost-complex structure on $T_m M$ extends to a complex structure on a neighbourhood of $m \in M$. Thus, there are actually infinitely many OCS's locally.

The elementary compact examples are S^4 (with its conformally flat metric) and $\overline{\mathbb{CP}}^2$ (with its symmetric metric of constant holomorphic sectional curvature).

Following results of Poon [90], LeBrun [71], and others, general constructions of ASD metrics were found by Floer [41], Donaldson and Friedman [36], and Taubes [100] who proved

Theorem 4.20 *If M is a compact oriented smooth 4-manifold then $M\#n\overline{\mathbb{CP}}^2$ admits an ASD metric for all sufficiently large n.*

Estimating the minimal n for a given M is non-trivial; for example, $n = 14$ suffices for $M = \mathbb{CP}^2$.

One can divide the class of ASD metrics into subcases according to the number of OCS's that exist *globally* on M, which we now suppose to be compact.

The classification of ASD Hermitian surfaces depends crucially on the parity of b_1 [19]. If b_1 is even then the metric is conformally equivalent to a Kähler one, which is therefore SFK. Such metrics were investigated in [65], which established analogues of Theorem 4.20 for SFK under blowing up. If b_1 is odd then M cannot carry a Kähler metric. However, the Lee form is closed and it follows that the metric is *locally* conformally Kähler. The surfaces in question are the so-called Type VII_0 ones.

A classification of bihermitian ASD metrics has been given in [89]; these are metrics with $W_+ = 0$ but admitting two OCS's J, J' compatible with the orientation, for which the set A of points where $J' = \pm J$ is a proper subset of M. In fact, if $A \neq \emptyset$ then A is a union of complex curves whose existence restricts the possibilities in the type VII_0 case. If $A = \emptyset$ the structure is called *strongly bihermitian* and M is necessarily hyperhermitian, so that there are then infinitely many OCS's on M. It follows that M is either a Hopf surface or hyperkähler [20]. In the latter case, M is a torus, or a K3 surface with a Calabi–Yau metric.

We remark that in each dimension $4k \geqslant 8$, there exist flat hyperkähler metrics on finite quotients of a torus T^{4k} and at least two compact irreducible hyperkähler manifolds that are not diffeomorphic [16, 86]. Moreover, the Euler characteristic of any compact hyperkähler manifold of dimension $4k$ satisfies $k\chi \equiv 0 \bmod 24$ [94].

Exercises 4.21 (i) Let ∇ denote its Levi-Civita connection of the metric g in Example 4.19, and set $e^i = \sum_j \sigma^i_j \otimes e^j$. Determine the 1-forms σ^i_j, using the formulae $de^i = \sum_j \sigma^i_j \wedge e^j$ and $\sigma^i_j + \sigma^j_i = 0$. Compute the curvature of g using the formula $R^i{}_{jk\ell} e^k \wedge e^\ell = d\sigma^i_j - \sum_k \sigma^i_k \wedge \sigma^k_j$ (find $R^i{}_{jk\ell}$ for as many (i,j,k,ℓ) as are necessary to determine R from its known symmetries). Hence, determine the eigenvalues of W_+.

(ii) Complete the proof of Lemma 4.16, and show that σ_2 is proportional to the component of $d\theta$ (the exterior derivative of the Lee form) in $\bigwedge^+$.

Almost-Kähler 4-manifolds

Let (M, g, J) be an almost-Hermitian 4-manifold. In terms of the decomposition of ∇J, the condition that ω be closed is exactly complementary to the integrability of J. The two conditions together imply that $\nabla J = 0$ and the structure is Kähler. We have mentioned above the little that is known about Einstein–Hermitian metrics that are not Kähler, and the situation is analogous for Einstein almost-Kähler (EaK) metrics that are not Kähler. Such metrics admit a compatible non-integrable almost-complex structure J for which $d\omega = 0$, and have been studied in connection with the Goldberg conjecture mentioned in the Introduction.

The following compilation of results is indicative of progress in this area.

Theorem 4.22 *Let (M, g, J) be a compact EaK 4-manifold. Then J is necessarily integrable (and so M is Kähler) if any one of the following conditions applies:*

(i) *s is non-negative* [98];
(ii) *s^* is constant* [8];
(iii) $W' = 0$ [9];
(iv) $W'' = 0$ [4].

Turning to the local problem, Example 2.30 provides a simple Einstein strictly almost-Kähler structure. We showed there that the metric $s^3 g$ is hyperkähler, and therefore Einstein. Reverse the orientation of this example by considering the 2-forms ω_i^- formed from (2.32) by changing signs. In particular,

$$\omega_2^- = dx \wedge (du - x\,dv) - s\,ds \wedge dv$$

is a closed 2-form for which the corresponding almost-complex structure I_2^- is *not* integrable. We could have also chosen ω_3^-; the non-integrability follows because $d\omega_1^- \neq 0$. Such examples were first discovered in [84], and are completely characterized locally by the condition $W' = 0$ [9]. Different examples appear in [5], though share the property that the metric becomes hyperkähler relative to the opposite orientation.

The metrics described in the previous paragraph are necessarily non-complete. An attempt to construct EaK metrics compatible with the standard symplectic form on $\mathbb{R}^4$ reveals that there is a non-trivial obstruction to extending the 3-jet of such a metric g to a 4-jet. This obstruction derives from the formula

$$\Delta s^* = -\tfrac{1}{4}(3s^* - s)(s^* - s) + 12|W'|^2 - 8|W''|^2 + 4\langle \nabla\Phi, \nabla\omega \rangle, \qquad (4.23)$$

where $\Phi \in [\![\Lambda^{2,0}]\!]$ satisfies $-4W' = \Phi \odot \omega$, that appears in [37]. (Observe that the left-hand side depends on $j_4(g)$ whereas the right-hand side is determined by $j_3(g)$.) Integrating (4.23) leads to (i) above.

When $W'' = 0$, the right-hand side of (4.23) is non-negative, and a maximum principle implies that $s^* - s$ is identically zero. This yields (iv), which remains valid when the Einstein condition $B = 0$ is relaxed to the J-invariant Ricci condition $B'' = 0$, provided $5\chi + 6\sigma \neq 0$ [4].

Almost-Kähler manifolds occur naturally in work relating Seiberg–Witten theory to curvature. The main estimate [74] implies that on a compact almost-Hermitian 4-manifold (M, g, J) for which the SW equations have a solution for every metric conformally related to g then

$$\int_M \left(|W^+| - \frac{1}{\sqrt{6}} s \right)^2 \geqslant 72\pi^2 (c^+)^2 \tag{4.24}$$

(see (4.10)). Equality in (4.24) implies that M is almost-Kähler with $W' = 0 = W''$ (see (4.18)); in this case we have seen that if g is Einstein then it is also Kähler.

Special Kähler manifolds

We use the notation (2.1), and follow closely [42].

Definition 4.25 *A special Kähler manifold is an almost-Hermitian manifold (M, g, J) admitting a flat torsion-free connection ∇ for which (i) $\nabla\omega = 0$ and (ii) $\nabla_1 J = 0$.*

Given that ∇ is torsion-free, (i) implies that $d\omega = 0$ and (ii) that $N = 0$. Thus (M, g, J) is indeed Kähler, justifying the terminology.

We shall denote the Levi-Civita connection of g by $\widetilde{\nabla}$ in this final subsection. Whilst $\widetilde{\nabla}$ is also a torsion-free symplectic connection, it is of course not in general flat and so not equal to ∇. For example, we have seen that $\widetilde{\nabla}_1 J = 0$ if and only if $\widetilde{\nabla} J = 0$ and condition (ii) is only useful for a non-metric connection. The tensor ψ defined in (3.18) is given by

$$\begin{aligned} \psi(X, Y, Z) &= g(J\widetilde{\nabla}_X Y - J\nabla_X Y, Z) \\ &= g(\widetilde{\nabla}_X(JY) - \nabla_X(JY) + (\nabla_X J)Y,\ Z). \end{aligned}$$

It follows that

$$\psi(X, Y, Z) - \psi(X, JY, JZ) = \Phi(X, Y, Z),$$

where Φ is defined in terms of the connection ∇ by (3.21).

Theorem 3.12 implies that there exist 1-forms $\{\alpha^i\}$ such that $\nabla\alpha^i = 0$ (so that $d\alpha^i = 0$) and $\omega = \sum_{r=1}^n \alpha^r \wedge \alpha^{n+r}$. Since $\tau(\nabla) = 0$ there exist charts $(x^1, \ldots, y^n)$ such that

$$\omega = \sum_{r=1}^n dx^r \wedge dy^{n+r}, \tag{4.26}$$

and

$$\nabla dx^r = 0 = \nabla dy^r.$$

This leads to an alternative definition of a special Kähler manifold, as an affine manifold with transition functions of the form $(x,y) \longmapsto P(x,y)+(a,b)$ with $P \in Sp(n,\mathbb{R})$. The associated metric $-w(J\cdot,\cdot)$ need only be pseudo-Riemannian for the theory to work.

If we trivialize the tangent bundle T by the parallel fields $\{(\partial/\partial x^1),\dots,((\partial/\partial y^n)\}$ then the operators in the sequence (3.7) reduce to ordinary d. Thus, locally $J=\nabla\zeta$ where

$$\zeta=\sum_{r=1}^{n}\left(V^r\frac{\partial}{\partial x^r}+U^r\frac{\partial}{\partial y^r}\right).$$

Hence

$$J=\sum\left(dV^r\otimes\frac{\partial}{\partial x^r}+dU^r\otimes\frac{\partial}{\partial y^r}\right),$$

and $Jdx^r=dV^r$, $Jdy^r=dU^r$. This means that

$$z^r=x^r-iV^r,\quad w^r=y^r-iU^r$$

are holomorphic functions. In this way, one obtains 'conjugate' holomorphic charts $(z^1,\dots,z^n)$, $(w^1,\dots,w^n)$.

Define holomorphic functions τ_{rs} by

$$dw^r=\sum_s\tau_{rs}dz^s,\quad\text{or}\quad\tau_{rs}=\frac{\partial w^r}{\partial z^s}.$$

Being a $(1,1)$-form,

$$\omega=\sum_r dx^r\wedge dy^r=J\omega=-\sum_r dU^r\wedge dV^r,$$

whence

$$\sum_r dz^r\wedge dw^r=-d\Big(\sum_r w^r\,dz^r\Big)$$

has zero real part, and so (being of type $(2,0)$) vanishes completely. Then

$$\sum w^r dz^r=d\mathcal{F}$$

for some locally-defined holomorphic function $\mathcal{F}$ that satisfies

$$w^r=\frac{\partial\mathcal{F}}{\partial z^r},\qquad\tau_{rs}=\frac{\partial^2\mathcal{F}}{\partial z^r\partial z^s}=\tau_{sr}.$$

Moreover,

$$2\omega=\Re\mathrm{e}\Big(\sum_r dz^r\wedge d\bar{w}^r\Big)=\Re\mathrm{e}\Big(\sum_{r,s}\bar{\tau}_{rs}dz^r\wedge d\bar{z}^s\Big)=-i\sum_{r,s}(\operatorname{Im}\tau_{rs})dz^r\wedge d\bar{z}^s,$$

so

$$\omega=\frac{1}{2}i\sum_{r,s}\omega_{rs}dz^r\wedge d\bar{z}^s,\qquad\omega_{rs}=-\operatorname{Im}\left(\frac{\partial^2\mathcal{F}}{\partial z^r\partial z^s}\right).$$

The following is also immediate:

Lemma 4.27 $\omega = -i\partial\bar\partial\big(\operatorname{Im}\sum_r \bar z^r(\partial\mathfrak{F}/\partial z^r)\big)$.

This means that the *real* function $\operatorname{Im}(\sum \bar z^r w^r)$ is a Kähler potential for the metric. By contrast, the function $\mathfrak{F}$ is called a holomorphic 'prepotential'. The flat Kähler metric on $\mathbb{C}^n$ is of course special Kähler with $\nabla = \widetilde\nabla$, and has $\mathfrak{F} = \frac{1}{2}\sum(z^r)^2$.

Let $P\colon T^*_{\mathbb{C}} \to \Lambda^{1,0}$ be the projection (of a 1-form to its component of type $(1,0)$). Thus

$$P = \frac{1}{2}(1-iJ) = \frac{1}{2}\sum_r\Big(dz^r\otimes\frac{\partial}{\partial x^r} + dw^r\otimes\frac{\partial}{\partial y^r}\Big).$$

But $P = \sum_r dz^r\otimes\partial_r$ where

$$\partial_r = \frac{\partial}{\partial z^r} = \frac{1}{2}\Big(\frac{\partial}{\partial x^r} + i\frac{\partial}{\partial U^r}\Big) = \frac{1}{2}\Big(\frac{\partial}{\partial x^r} + \tau_{rs}\frac{\partial}{\partial y^s}\Big)$$

(summation now understood over repeated indices). It follows that

$$\nabla\partial_r = \frac{1}{2}\frac{\partial\tau_{rs}}{\partial z^t}dz^t\otimes\frac{\partial}{\partial y^s} \in \Gamma(\Lambda^{1,0}\otimes T_{\mathbb{C}}).$$

Exercises 4.28 (i) Use the last formula to show that $\nabla_{\overline{\partial_r}}\partial_s = 0$ for all r,s. Since $\widetilde\nabla J = 0$, the Levi-Civita connection preserves types and $\omega(\widetilde\nabla_X\partial_s,\partial_t) = 0$ for all X. Deduce that $\psi(\bar\partial_r,\partial_s,\partial_t) = 0$.

(ii) By recalling that ω has the standard form $\sum dx^r\wedge dy^r$, show that

$$\psi(\partial_r,\partial_s,\partial_t) = -\frac{1}{4}\frac{\partial\tau_{st}}{\partial z^r}.$$

The exercises tell us that $\psi = \Xi + \overline\Xi$, where

$$\Xi = -\frac{1}{4}\sum_{r,s,t}\frac{\partial^3\mathfrak{F}}{\partial z^r\partial z^s\partial z^t}dz^r\otimes dz^s\otimes dz^t \in \Gamma(S^{3,0})$$

is a *holomorphic* cubic differential. This leads to an expression for the Riemann curvature (i.e. $\widetilde\nabla_1\circ\widetilde\nabla$) in terms of $\Xi\otimes\overline\Xi$, and the formula for the scalar curvature

$$s = R^i_{jik}g^{jk} = 4|\Xi|^2 \geqslant 0$$

of Lu [77], who deduced that if (M,g) is complete then $\Xi = 0$, so $R = 0$ and g is flat.

Example 4.29 A simple non-trivial special Kähler metric g is the one on the upper half-plane $H = \{z\colon \operatorname{Im} z > 0\}$ given by $\mathfrak{F} = z^3/6$, so that Ξ is a constant multiple of dz^3. For consistency with the above notation, we set $z = x - iV$ and

$$w = y - iU = \mathcal{F}'(z) = \tfrac{1}{2}z^2,$$

so that $\tau = z$. The real part of w is given by

$$y = \tfrac{1}{2}(x^2 - V^2), \tag{4.30}$$

and $VdV = xdx - dy$. If we set

$$\phi = \tfrac{1}{3}V^3 = \tfrac{1}{3}(x^2 - 2y)^{3/2} \tag{4.31}$$

then

$$\begin{aligned} g &= V(dx^2 + dV^2) \\ &= \frac{1}{V}\Big((x^2 + V^2)dx^2 - 2x\,dx\,dy + dy^2\Big) \\ &= \phi_{xx}dx^2 + 2\phi_{xy}dxdy + \phi_{yy}dy^2. \end{aligned}$$

Any special Kähler metric can in fact be expressed as

$$g = \sum \frac{\partial^2 \phi}{\partial x^i \partial x^j} dx^i dy^j$$

for a suitable function ϕ of Darboux coordinates [59].

Let M be a special Kähler manifold, and consider its *real* cotangent bundle $\pi : T^*M \to M$. Select Darboux coordinates $\{x^1, \ldots, x^n, y^1, \ldots, y^n\}$ on an open set $\mathcal{U}$ of M, so that (4.26) holds.

We choose to express a point p of $\pi^{-1}\mathcal{U}$ as $\sum(-u^r dx^r + v^r dy^r)$. This gives rise to a tautological 1-form on T^*M:

$$\tau = \sum_{r=1}^{n}(-u^r \pi^* dx^r + v^r \pi^* dy^r),$$

though we shall omit the pull-back symbol π^* by thinking of x^r, y^r as functions on T^*M. The u^r, v^r are 'fibre coordinates', and (for the moment) are unrelated to the functions U^r, V^r we defined on M.

Consider the coordinates (x^r, y^r, u^r, v^r) on $\pi^{-1}\mathcal{U}$, and the 2-forms given by (2.11). In the present context, Ω_2 is none other than $d\tau$ and equals the canonical real symplectic form (as defined on the cotangent bundle of any smooth manifold). Thus, Ω_2 is independent of the choice of coordinates, and extends to a 2-form globally on T^*M.

The 2-form Ω_1 may be written as $\pi^*\omega - \omega^*$, where ω^* is the 'dual' of ω under the identification ω itself provides between T_mM and the tangent space $V = T_p(T_m^*M) \cong T_m^*M$ to the fibre of π at a point p. Moreover, ω^* will be independent of the coordinates in the presence of a flat symplectic connection ∇. For in this case, we may write

$$T_p(T^*M) = V \oplus H,$$

where H is the 'horizontal' space determined by ∇ and (by definition) everywhere tangent to sections which are constant linear combinations of dx^r, dy^r.

So far so good, but in order to make the above construction more invariant, we shall replace Ω_3 by

$$\begin{aligned}\Omega_3' = d(J\tau) &= \sum(-dU^r \wedge dv^r + dV^r \wedge du^r)\\ &= \operatorname{Im}\sum(dw^r \wedge dv^r - dz^r \wedge du^r),\end{aligned}$$

where J is (the pull-back of) the complex structure on M. This will ensure that

$$\eta = \Omega_2 + i\Omega_3' = d(\tau + iJ\tau)$$

is (twice) the *holomorphic* symplectic form that exists on T^*M, when the latter is endowed with its natural complex structure I_1 extending J.

The triple $(\Omega_1, \Omega_2, \Omega_3')$ will then define a hyperkähler structure if H is a *complex* subspace of $(T_p(T^*M), I_1)$. This is true if M is special Kähler, since constant linear combinations of dx^r, dy^r (being the real parts of the holomorphic 1-forms dz^r, dw^r) are themselves I_1-holomorphic sections.

We may define a local section

$$s\colon \mathfrak{U} \to T^*M$$

by setting $u^r = U^r$ and $v^r = V^r$, so that the notation is amalgamated with what we did earlier. Once we do this,

$$\begin{aligned}s^*\Omega_1 &= \omega - \omega = 0,\\ s^*\Omega_2 &= -\operatorname{Im}\Big(\sum dz^r \wedge dw^r\Big) = 0,\\ s^*\Omega_3' &= -2\sum dU^r \wedge dV^r = 2\omega.\end{aligned}$$

This shows that s is *bi-Lagrangian* as a submanifold of T^*M. Thus any special Kähler manifold arises locally as a bi-Lagrangian submanifold of $(\mathbb{R}^{4n}, \Omega_1, \Omega_2)$, where Ω_1, Ω_2 are two standard real symplectic forms. This fact was established independently by Cortés and Hitchin [15, 59].

Since $0 = s^*\Omega_2 = s^*(d\tau)$, we may (on a possibly smaller open set $\mathfrak{U}'$) express $s^*\tau$ as

$$d\phi = \sum_{r=1}^{n}\left(\frac{\partial\phi}{\partial x^r}dx^r + \frac{\partial\phi}{\partial y^r}dy^r\right)$$

for some real-valued function ϕ so that $U^r = -\partial\phi/\partial x^r$ and $V^r = \partial\phi/\partial y^r$ on M (by analogy to the holomorphic equation $w^r = \partial\mathfrak{F}/\partial z^r$). Returning to T^*M, we see that Ω_3' equals

$$\sum_{r,s=1}^{n}\left(\frac{\partial^2\phi}{\partial x^r\partial x^s}dx^r \wedge dv^s + \frac{\partial^2\phi}{\partial y^r\partial x^s}dy^r \wedge dv^s\right.$$
$$\left. + \frac{\partial^2\phi}{\partial x^r\partial y^s}dx^r \wedge du^s + \frac{\partial^2\phi}{\partial y^r\partial y^s}dy^r \wedge du^s\right) = \sum_{r,s=1}^{2n}\frac{\partial^2\phi}{\partial X^r\partial X^s}dX^r \wedge dW^s,$$

in terms of coordinates $X^1, \ldots, X^{2n}$ on the base $\mathcal{U}'$ and $W^1, \ldots, W^{2n}$ for the fibres. Combined with the more standard expressions for Ω_1 and Ω_2 above this gives an explicit construction of hyperkähler metrics.

Example 4.32 This unites Examples 2.30 and 4.29. The 2-forms (2.32) fall within the above description. To see this, replace s by V, and define y by (4.30). Then

$$\begin{aligned} V\omega_1 &= 2(x^2 - y)dx \wedge dv - x(dx \wedge du + dy \wedge dv) + dy \wedge du \\ \omega_2 &= dx \wedge du + dv \wedge dy, \\ -\omega_3 &= dx \wedge dy + du \wedge dv. \end{aligned}$$

If we now take ϕ as in (4.31) then

$$\omega_1 = \phi_{xx} dx \wedge dv - \phi_{xy}(dx \wedge du + dy \wedge dv) + \phi_{yy} dy \wedge du.$$

The minus signs can be eliminated by changing the sign of x and v. Observe that the function ϕ appeared as the conformal factor converting a left-invariant hypercomplex metric into a hyperkähler one.

Remarks 4.33 (i) Independent proofs [39, 64] exist of the fact that there exists a hyperkähler metric on an open set of the cotangent bundle of *any* real analytic Kähler manifold.
(ii) An analogous theory of 'special complex manifolds' is developed in [3].

References

1. E. Abbena: An example of an almost Kähler manifold which is not Kählerian, *Boll. Unione Mat. Ital.* **3** (1984), 383–392.
2. E. Abbena, S. Garbiero and S. Salamon: Hermitian geometry on the Iwasawa manifold, *Boll. U.M.I.* **11**-B (1997), 231–249.
3. D.V. Alekseevsky, V. Cortés and C. Devchand: Special complex manifolds, *J. Geom. Phys.*, **42** (2002), 85–105.
4. V. Apostolov and J. Armstrong: Symplectic 4-manifolds with Hermitian Weyl tensor, *Trans. Amer. Math. Soc.* **352** (2000), 4501–4513.
5. V. Apostolov, D.M.J. Calderbank and P. Gauduchon: The geometry of weakly selfdual Kähler surfaces, `math.DG/0104233`.
6. V. Apostolov and P. Gauduchon: The Riemannian Goldberg–Sachs theorem, *Intern. J. Math.* **8** (1997), 421–439.
7. V. Apostolov, P. Gauduchon and G.Grantcharov: Bihermitian structures on complex surfaces, *Proc. London. Math. Soc.* **79** (1999) 414–428.
8. J. Armstrong: On four-dimensional almost Kähler manifolds, *Quart. J. Math.* **48** (1997), 405–415.
9. J. Armstrong: An ansatz for almost-Kähler, Einstein 4-manifolds, *J. Reine Angew. Math.* **542** (2002), 53–84.
10. M.F. Atiyah, N.J. Hitchin and I.M. Singer: Self-duality in four-dimensional Riemannian geometry, *Proc. Roy. Soc. Lond.* **A362** (1978), 425–461.

11. V. Apostolov, G. Grantcharov and S. Ivanov: Hermitian structures on twistor spaces, *Ann. Global Anal. Geom.* **16** (1998) 291–308.
12. M.L. Barberis: Homogeneous hyper-Hermitian metrics which are conformally hyper-Kähler, `math.DG/0009035`.
13. M.L. Barberis and I.G. Dotti Miatello: Hypercomplex structures on a class of solvable Lie groups, *Quart. J. Math.* **47** (1996), 389–404.
14. W. Barth, C. Peters and A. Van de Ven: 'Compact Complex Surfaces', Springer (1984).
15. O. Baues and V. Cortés: Realization of special Kähler manifolds as parabolic spheres, *Proc. Amer. Math. Soc.* **129** (2001), 2403–2407.
16. A. Beauville: Variétés Kählériennes dont la première class de Chern est nulle, *J. Differ. Geom.* **18** (1983), 755–782.
17. L. Bérard Bergery and T. Ochiai: On some generalizations of the construction of twistor space, 'Global Riemannian Geometry', Ellis Horwood (1984).
18. A. Besse: 'Einstein Manifolds', Springer (1987).
19. C. Boyer: Conformal duality and compact complex surfaces, *Math. Ann.* **274** (1986), 517–526.
20. C.P. Boyer, A note on hyperhermitian four-manifolds, *Proc. Amer. Math. Soc.* **102** (1988), 157–164.
21. R.L. Bryant: Submanifolds and special structures on the octonians, *J. Differ. Geom.* **17** (1982), 185–232.
22. R.L. Bryant: Conformal and minimal immersions into the 4-sphere, *J. Differ. Geom.* **17** (1982), 455–473.
23. R.L. Bryant: Bochner-flat metrics, *J. Amer. Math. Soc.* **14** (2001), 623–715.
24. R.L. Bryant and R. Harvey: Submanifolds in hyper-Kähler geometry, *J. Amer. Math. Soc.* **2** (1989), 1–31.
25. F. Burstall, O. Muškarov, G. Grantcharov and J. Rawnsley: Hermitian structures on Hermitian symmetric spaces, *J. Geom. Phys.* **10** (1993), 245–249.
26. F.E. Burstall and J.H. Rawnsley: 'Twistor theory for Riemannian symmetric spaces', LNM 1424, Springer (1990).
27. M. Cahen, S. Gutt and J. Rawnsley: Homogeneous symplectic manifolds with Ricci-type curvature, *J. Geom. Phys.* **38** (2001), 140–151.
28. E. Calabi: Métriques kählériennes et fibrés holomorphes, *Ann. Ec. Norm. Sup.* **12** (1979), 269–294.
29. E. Calabi: Extremal Kähler metrics, 'Seminar on Differential Geometry', Ann. of Math. Stud. 102, Princeton University Press, 1982.
30. E. Calabi and B. Eckmann: A class of compact, complex manifolds which are not algebraic, *Ann. of Math.* **58** (1953), 494–500.
31. D.M.J. Calderbank and H. Pedersen: Selfdual Einstein metrics with torus symmetry, `math.DG/0105263`.
32. L.A. Cordero, M. Fernández, A. Gray and L. Ugarte, Nilpotent complex structures on compact nilmanifolds, *Rend. Circolo Mat. Palermo* **49** suppl. (1997), 83–100.
33. P. Deligne, P. Griffiths, J. Morgan and D. Sullivan: Real homotopy theory of Kähler manifolds, *Invent. Math.* **29** (1975), 245–274.
34. A. Derzinski: Self-dual Kähler manifolds and Einstein manifolds of dimension four, *Compos. Math.* **49** (1983), 405–433.
35. S.K. Donaldson and P.B. Kronheimer: 'The Geometry of Four-Manifolds', Oxford University Press (1990).
36. S.K. Donaldson and R. Friedman: Connected sums of self-dual manifolds and deformations of singular spaces, *Nonlinearity* **2** (1989), 197–239.

37. T. Dragici: Almost Kähler 4-manifolds with J-invariant Ricci tensor, *Houston J. Math.* **25** (1999), 133–145.
38. A. Farinola, M. Falcitelli and S. Salamon: Almost-Hermitian geometry, *Differ. Geom. Appl.* **4** (1994), 259–282.
39. B. Feix: Hyperkähler metrics on the cotangent bundle, *J. Reine Angew. Math.* **532** (2001).
40. A. Fino and I.G. Dotti: Abelian hypercomplex 8-dimensional nilmanifolds, *Ann. Global Anal. Geom.* **18** (2000), 47–59.
41. A. Floer: Self-dual conformal structures on $\ell\mathbb{CP}^2$, *J. Differ. Geom.* **33** (1991), 551–573.
42. D. Freed: Special Kähler manifolds, *Commun. Math. Phys.* **203** (1999), 31–52.
43. M.H. Freedman: The topology of four-dimensional manifolds, *J. Differ. Geom.* **17** (1982), 357–453.
44. M. Furuta: Monopole equation and the 11/8 conjecture, *Math. Res. Letters* **8** (2001), 279–291.
45. P. Gauduchon: Structures de Weyl-Einstein, espaces de twisteurs et variétés de type $S^1 \times S^3$, *J. Reine Angew. Math.* **469** (1995), 1–50.
46. H. Geiges and J. Gonzalo: Contact geometry and complex surfaces, *Invent. Math.* **121** (1995), 147–209.
47. I. Gelfand, V. Retakh and M. Shubin: Fedosov manifolds, *Adv. in Math.* **136** (1998), 104–140.
48. G.W. Gibbons and S.W. Hawking: *Gravitational multi-instantons*, Phys. Lett. **B78** (1978), 430–442.
49. G.W. Gibbons and P. Rychenkova: Single-domain walls in M-theory, *J. Geom. Phys.* **32** (2000), 311–340.
50. S.I. Goldberg, Integrability of almost Kähler manifolds, *Proc. Amer. Math. Soc.* **21** (1969), 96–100.
51. S.I. Goldberg: 'Curvature and Homology', revised reprint of the 1970 edition, Dover Publications (1998).
52. A. Gray, Curvature identities for Hermitian and almost Hermitian manifolds, *Tôhoku Math. J.* **28** (1976), 601–612.
53. A. Gray and L. Hervella, The sixteen classes of almost Hermitian manifolds and their linear invariants, *Ann. Math. Pura Appl.* **123** (1980), 35–58.
54. P. Griffiths and J. Harris: 'Principles of Algebraic Geometry', Wiley-Interscience, (1978).
55. D. Guan: Examples of compact holomorphic symplectic manifolds which are not Kählerian II, *Invent. Math.* **121** (1995), 135–145.
56. J. Heber: Noncompact homogeneous Einstein spaces, *Invent. Math.* **133** (1998), 279–352.
57. F. Hirzebruch: 'Topological methods in algebraic geometry', Springer (1966).
58. N.J. Hitchin: Compact four-dimensional Einstein manifolds, *J. Differ. Geom.* **9** (1974), 435–441.
59. N.J. Hitchin: The moduli space of complex Lagrangian submanifolds, *Asian J. Math.* **3** (1999), 77–91.
60. N.J. Hitchin and J. Sawon: Curvature and characteristic numbers of hyperkähler manifolds, *Duke Math. J.* **106** (2001), 599–615.
61. G. Jensen: Homogeneous Einstein spaces of dimension 4, *J. Differ. Geom.* **3** (1969), 309–349.
62. D. Joyce: Compact hypercomplex and quaternionic manifolds, *J. Differ. Geom.* **35** (1992), 743–761.

63. D.D. Joyce: 'Compact manifolds with special holonomy', Oxford University Press (2000).
64. D. Kaledin: A canonical hyperkähler metric on the total space of a cotangent bundle, in 'Quaternionic structures in mathematics and physics' (eds S. Marchiafava, P. Piccinni, M. Pontecorvo), World Scientific (2001).
65. J. Kim, C. LeBrun and M. Pontecorvo: Scalar-flat Kähler surfaces of all genera, *J. Reine Angew. Math.* **486** (1997), 69–95.
66. F.C. Kirwan: 'Complex algebraic curves', LMS Student texts 23, Cambridge University Press (1992).
67. P. Kobak: Explicit doubly-Hermitian metrics, *Differ. Geom. Appl.* **10** (1999), 179–185.
68. S. Kobayashi and K. Nomizu: 'Foundations of Differential Geometry', Volume 2, Interscience (1969).
69. K. Kodaira: On the structure of compact complex analytic surfaces I, *Amer. J. Math.* **86** (1964), 751–798.
70. D. Kotschick: Orientations and geometrisations of compact complex surfaces, *Bull. London Math. Soc.* **29** (1997), 145–149.
71. C. LeBrun: Explicit self-dual metrics on $\mathbb{CP}^2\#\cdots\#\mathbb{CP}^2$, *J. Differ. Geom.* **34** (1991), 223–253.
72. C. LeBrun: Einstein metrics on complex surfaces, 'Geometry and Physics' (Aarhus 1995), Lect. Notes Pure Appl. Math. 184, Dekker (1997).
73. C. LeBrun: Four-dimensional Einstein manifolds, and beyond, 'Essays on Einstein Manifolds', Surveys in Differential Geometry VI, International Press (1999).
74. C. LeBrun: Ricci curvature, minimal volumes, and Seiberg–Witten theory, *Invent. Math.* **145** (2001), 279–316.
75. C. LeBrun and M. Singer: A Kummer-type construction of self-dual 4-manifolds, *Math. Ann.* **300** (1994), 165–180.
76. H.C. Lee: A kind of even dimensional differential geometry and its application to exterior calculus, *Amer. J. Math.* **65** (1943), 433–438.
77. Z. Lu: A note on special Kähler manifolds, *Math. Ann.* **313** (1999), 711–713.
78. D. McDuff and D. Salamon: 'Introduction to Symplectic Topology', 2nd edition, Oxford University Press (1998).
79. A.I. Malcev: On a class of homogeneous spaces, reprinted in Amer. Math. Soc. Translations, Series 1, Volume 9 (1962).
80. M. Migliorini: Special submanifolds of the twistor space, *Rend. Istit. Mat. Univ. Trieste* **26** (1994), 239–260.
81. J. Milnor: 'Topology from a Differentiable Viewpoint', Princeton University Press (1997).
82. J. Neissendorfer and L. Taylor: Dolbeault homotopy theory, *Trans. Amer. Math. Soc.* **245** (1978), 183–210.
83. A. Newlander and L. Nirenberg: Complex analytic coordinates in almost complex manifolds, *Annals Math.* **65** (1957), 391–404.
84. P. Nurowski and M. Przanowski: A four-dimensional example of Ricci-flat metric admitting almost-Kähler non-Kähler structure, *Class. Quantum Grav.* **16** (1999), L9–L13.
85. N.R. O'Brian and J.H. Rawnsley: Twistor spaces, *Ann. Glob. Anal. Geom.* **3** (1985), 29–58.
86. K. O'Grady: Desingularized moduli spaces of sheaves on a K3, *J. Reine Angew. Math.* **512** (1999), 49–117.

87. H. Pedersen and Y.S. Poon: Inhomogeneous hypercomplex structures on homogeneous manifolds, *J. Reine Angew. Math.* **516** (1999), 159–181.
88. H.V. Pittie: The nondegeneration of the Hodge–deRham spectral sequence, *Bull. Amer. Math. Soc.* **20** (1989), 19–22.
89. M. Pontecorvo: Complex structures on Riemannian 4-manifolds, *Math. Ann.* **309** (1997), 159–177.
90. Y-S. Poon: Compact self-dual manifolds with positive scalar curvature, *J. Differ. Geom.* **24** (1986), 97–132.
91. S. Salamon: Harmonic and holomorphic maps, 'Geometry Seminar Luigi Bianchi II', LNM 1164, Springer (1985).
92. S. Salamon: 'Riemannian Geometry and Holonomy Groups', Pitman Research Notes Math. 201, Longman (1989).
93. S. Salamon: Orthogonal complex structures, 'Differential Geometry and Applications', Brno University, 1996.
94. S. Salamon: On the cohomology of Kähler and hyper-Kähler manifolds, *Topology* **35** (1996), 137–155.
95. S. Salamon: Complex structures on nilpotent Lie algebras, *J. Pure Applied Algebra*, **157** (2001), 311–333.
96. S. Salamon and V. deSmedt: Self-dual metrics on 4-dimensional Lie groups, Proc. conf. Integrable Systems and Differential Geometry (Tokyo 2000), to appear.
97. H. Samelson: A class of complex-analytic manifolds, *Portugaliae Math.* **12** (1953), 129–132.
98. K. Sekigawa: On some 4-dimensional compact Einstein almost-Kähler manifolds, *Math. Ann.* **271** (1985), 333–337.
99. D. Snow, Invariant complex structures on reductive Lie groups, *J. Reine Angew. Math.* **371** (1986), 191–215.
100. C.H. Taubes: The existence of anti-self-dual conformal structures, *J. Differ. Geom.* **36** (1992), 163–253.
101. A. Tralle and J. Oprea: 'Symplectic manifolds with no Kähler structure', LNM 1661, Springer (1997).
102. F. Tricerri and L. Vanhecke: Curvature tensors on almost Hermitian manifolds, *Trans. Amer. Math. Soc.* **267** (1981), 365–398.
103. I. Vaisman: Symplectic curvature tensors, *Monat. Math.* **100** (1985), 299–327.
104. H.C. Wang: Complex parallizable manifolds, *Proc. Amer. Math. Soc.* **5** (1954), 771–776.
105. F.W. Warner: 'Foundations of differentiable manifolds and Lie groups', Scott Foresman (1971).
106. A. Weil: 'Introduction à l'étude des variétés kählériennes', Hermann (1958).
107. R.O. Wells: 'Differential analysis on complex manifolds', Springer (1979).
108. J. Winkelmann: 'The classification of three-dimensional homogeneous complex manifolds', LNM 1602, Springer (1995).
109. K. Yano, 'Differential geometry on complex and almost complex spaces', Pergamon Press (1965).
110. S.T. Yau: On the Ricci-curvature of a complex Kähler manifold and the complex Monge–Ampère equations, *Comment. Pure Appl. Math.* **31** (1978), 339–411.

8
Indices of vector fields and Chern classes for singular varieties

JOSÉ SEADE

Introduction

The theorem of Poincaré–Hopf about the total index of a vector field on a manifold is one of the fundamental results of mathematics in the twentieth century, giving rise to many different theories and developments in several branches of mathematics and science in general. In particular, it gave rise to obstruction theory and the theory of characteristic classes, such as the Chern classes of complex manifolds, which are very useful invariants. When we deal with spaces which are not manifolds but singular varieties, it is natural to ask whether one has similar theories to those for manifolds, in particular regarding vector fields and characteristic classes. For this, the first thing is to say what we mean by a vector field on a singular variety; then we need to define the concept of 'index' and the corresponding 'characteristic classes', and search for theorems in this context equivalent to those for manifolds.

To simplify the discussion, let us restrict to complex analytic varieties. Let V be such a variety and assume it is embedded in some complex manifold M; we can define a vector field on V to be the restriction to V of a vector field v on a neighbourhood U of V in M, such that whenever an integral curve of v intersects V, the whole curve is contained in V. In other words, V is an invariant set of v. The question is, given such a vector field v, how does one define its local index at a point P which is in the singular set of V? There are (at least) two different points of view. On one hand, we would like the index to give the 'correct number', i.e. the total sum should be the Euler–Poincaré characteristic $\chi(V)$, as in the case of manifolds. This point of view was introduced by M.H. Schwartz in [42, 43] for 'radial vector fields'; she used this to define certain characteristic classes, the *Schwartz classes*. These classes are defined via obstruction theory, in a similar way to the obstruction theory definition of the Chern classes. At the time, not much attention was paid to these classes, since it was not evident that they had any nice properties, such as naturality. Years later, Grothendieck and Deligne conjectured that compact (algebraic) varieties should have characteristic classes with functorial properties similar to those of the Chern classes. This was proved by MacPherson in [30], constructing the 'Chern classes for singular varieties'. These are classes in the homology of V, not in cohomology. Some years later, Brasselet and Schwartz [6] proved that the Alexander homomorphism for singular varieties carries the Schwartz classes into the MacPherson classes. It is easy to extend Schwartz's construction to vector fields and frames which are not radial. This was done recently in [8] and this allows us to give an easy description of the 'Chern–Schwartz–MacPherson' classes of singular varieties.

On the other hand, the index of Poincaré–Hopf has the nice property of being stable under perturbations. That is, if we perturb the space slightly, the singularities of the vector field may change, but their total index remains constant. We would also like the index to be 'stable' under perturbations when the space in question is not a manifold but a singular variety. In the simplest case, this can be explained as follows: if we have a family of manifolds V_t with vector fields v_t with isolated singularities, that degenerate to (V, P) and v, respectively, then we want the local index of v at P to be the sum of the Poincaré–Hopf indices of the singularities of v_t that degenerate to P. It turns out that this condition is not compatible with the previous condition, of getting the 'correct' number. The corresponding index was introduced in [45, 15, 47] when V is a local complete intersection; this is now called *the* GSV-*index*. The definition of the GSV-index requires (a priori) the singularities of V to be isolated, but this is not actually necessary, as shown in [8]. When the singularities of V are isolated, the difference between the GSV-index and the Schwartz index is the local Milnor number of V, independently of the vector field. Comparing the two indices globally [48, 8], we arrive to a generalization of the classical adjunction formula for complex curves in surfaces. If V is given by a regular section of a rank k holomorphic bundle E over a complex m-manifold M, $k < m$, then V has a canonical *virtual tangent bundle* $\tau(V) = (TM - E)|_V$. The Chern classes of this bundle are the Fulton–Johnson canonical classes of V, [13, 14], and it is natural to ask how are these classes related to the Chern–Schwartz–MacPherson classes of V. This problem has been studied in [50, 48, 35, 8, 2], and the answer in top degree is given by the above mentioned adjunction formula. It would be nice to have similar (geometric) formulae for the lower classes, cf. [8].

This work is an exposition of the above indices and characteristic classes. As mentioned before, there are several definitions of characteristic classes for singular varieties. Each one of them is defined in a relevant context and has its own interest and advantage. The Chern–Schwartz–MacPherson classes are related to the Schwartz index of a vector field in just the same way that the usual Chern classes are related to the Poincaré–Hopf local index. A similar relation holds for the GSV-index and the top dimensional Fulton–Johnson class, though it is not evident whether one has a similar relation for the lower classes. The starting point in all cases are the Chern classes (for manifolds and complex vector bundles in general). Hence we take these as the starting point for this work. We begin by reviewing, in Section 1, the Poincaré–Hopf theorem about the zeroes of a vector field on a smooth manifold; Sections 1 to 4 are essentially taken from [8]. In Section 2 we review the Chern classes of a complex manifold from the obstruction theory point of view, and we present them in a way which lends itself to generalization, making it easy to understand the analogy with the Chern classes for singular varieties, which take up the rest of this work.

In Sections 3 and 4 we describe the Schwartz classes, starting with the simplest case: the Schwartz index of a vector field. In Section 5 we consider the GSV-index; we describe the generalization in [8, 48] of the classical adjunction formula and we state some open problems concerning this index. In Section 6 we briefly describe MacPherson's construction of the Chern classes for singular varieties. One of the key ingredients for this is the local Euler obstruction. We describe this obstruction in some detail, and we give an outline of the proof in [7] of a theorem by Dubson [11], that relates the local Euler obstruction of an isolated complete intesection germ, with the Milnor number of a general hyperplane section; this uses the GSV-index introduced in Section 5. This section finishes with the theorem of [6], stating that the Schwartz classes coincide with MacPherson's classes. Thus one may call these the Chern–Schwartz–MacPherson classes. This is interesting because the Schwartz classes arise from the viewpoint of topology, they are a natural generalization of the Chern classes when these are defined via obstruction theory. On the other hand, MacPherson's classes are functorial, they are a natural generalization of the Chern classes when these are defined axiomatically. Finally, in Section 7 we briefly describe the Fulton–Johnson classes of singular varieties which are local complete intersections, and we discuss the relation of these classes with the Chern–Schwartz–MacPherson classes. This comparison gives rise to a generalization of the Milnor number for non-isolated singularities [8].

1 The Poincaré–Hopf index of a vector field

Let $v = (v_1, \dots, v_n)$ be a vector field in an open set $U \subset \mathbb{R}^n$. The vector field is said to be continuous, smooth, analytic, etc., according as its components $\{v_1, \dots, v_n\}$ are continuous, smooth, analytic, etc., respectively. A *singularity* P of v is a point where all of its components vanish, i.e. $v_i(P) = 0$ for all $i = 1, \dots, n$. The singularity is *isolated* if at every point x near P there is at least one component of v which is not zero.

The Poincaré–Hopf index of a vector field at an isolated singularity is its most basic invariant, and it has many interesting properties. Let us recall its definition and some properties.

Let v be a vector field as above and let P be an isolated singularity of v. Let S_ε be a small sphere around P. Then the Poincaré–Hopf index of v at P is the degree of the map $v/\|v\|$ from S_ε into the unit sphere in $\mathbb{R}^n$. If v and v' are two such vector fields, then their local indices at P coincide if and only if they are homotopic through a family of vector fields, defined on a neighbourhood of P and having each an isolated singularity at P.

If we now let M be a smooth manifold of dimension n, then a *vector field* on M is a section of its tangent bundle TM. A vector field on M is locally expressed as above, and the above definitions extend in the obvious way. A fundamental property of the index is the following theorem:

Theorem 1.1 [Poincaré–Hopf] *Let M be a closed, oriented n-manifold and let v be a continuous vector field on M with isolated singularities. Define the total index of v, denoted* $\mathrm{PH}(v, M)$, *to be the sum of all its local indices at the singular points. Then one has*

$$\mathrm{PH}(v, M) = \chi(V),$$

independently of v, where $\chi(V)$ is the Euler–Poincaré characteristic.

The proof of this theorem can be found in many textbooks, as for example [31], so we only give the idea of the proof here. It has two big steps: the first step is to show that the total index is independent of the vector field. This is a special case of a general theorem in algebraic topology, saying that the primary obstruction for constructing a cross-section of a vector bundle is independent of the choice of section, see [52]. The second step for proving this theorem is to construct an explicit vector field for which one can easily 'count' the local indices. This is done as follows: take a triangulation (K) of M, and construct a vector field v in the following way. At each vertex of (K), v has a zero. Now, each edge has two vertices, and we move simultaneously away from these vertices towards the centre of the edge, where v has one singularity. For each 2-face, we take the vector field on the boundary of the 2-face and we 'push' from all directions towards the centre, and so on. At the end we get a vector field on M with isolated singularities. These are: the vertices, the centre of each edge, the centre of each 2-face, the centre of each 3-face, and so on. Now it is just a matter of computing the local indices of v and adding them up. At each vertex the vector field is going out in all directions, so the local index is 1. At the centre of each edge, v is attracting in one direction and repulsive in all other directions, hence its local index is -1, for the 2-faces the index is 1, and so on. Hence the total index is the number of vertices, minus the edges, plus the faces and so on. Thus $\mathrm{PH}(v, M) = \chi(M)$.

If M is now an oriented manifold with boundary, one has a similar theorem:

Theorem 1.2 *Let M be a compact, oriented n-manifold with boundary ∂M, and let v be a non-singular vector field on a neighbourhood U of ∂M. Then:*

(i) v can be extended to the interior of M with isolated singularities.

(ii) The total index of v in M is independent of the way we extend it to the interior of M. In other words, the total index of v is fully determined by its behaviour near the boundary.

(iii) If v is everywhere transversal to the boundary and pointing outwards from M, then one has $\mathrm{PH}(v, M) = \chi(M)$. *If v is everywhere transversal to ∂M but it points inwards M, then* $\mathrm{PH}(v, M) = \chi(M) - \chi(\partial M)$.

The proof of the first statement is by ordinary obstruction theory. Following the 'stepwise' process of [52] one easily shows that the vector field can be extended without singularities to the $(n-1)$-skeleton of M. Then we extend it to the n-cells introducing (if necessary) a singular point for each n-cell. Statement (ii) is again a special case of a general result in algebraic topology, about the primary obstruction for constructing a cross-section [52], see also [34, 41]. Statement (iii) can be easily proved by applying to the 'double' of M the previous theorem for closed manifolds, using the fact that v can be moved by a homotopy to make it normal to ∂M.

It is worth saying that although $\mathrm{PH}(v, M)$ is determined by its behaviour near the boundary, it does depend on the topology of the interior of M, see exercise (iii) below (cf. [34, 41, 46]).

Let us consider a triangulation (K) of M compatible with the boundary, $(\hat{K})$ the barycentric subdivision of (K) and (D) the associated cellular dual decomposition of M. (We refer to [5] for background material concerning these concepts.) In the following, the triangulation (K) that we consider will be taken to be compatible with some subsets of M. For instance, let S be a compact connected (K)-subcomplex of the interior of M, and denote by $\tilde{\mathfrak{T}}$ the union of (closed) cells of (D) which are dual to simplices in S. The boundary $\partial\tilde{\mathfrak{T}}$ is the union of cells in $\tilde{\mathfrak{T}}$ which do not meet S.

Definition 1.3 *A cellular tube around S in M is a regular neighbourhood of S in M which is the union of dual cells (in M) relative to the triangulation (K).*

This notion generalizes the concept of tubular neighbourhood of a submanifold S. In that case, $\tilde{\mathfrak{T}}$ is a bundle on S, whose fibres are discs.

Remark 1.4 A cellular tube $\tilde{\mathfrak{T}}$ around S has the following properties:

(i) $\tilde{\mathfrak{T}}$ is a compact neighbourhood of S, containing S in its interior and such that $\partial\tilde{\mathfrak{T}}$ is a retract of $\tilde{\mathfrak{T}} - S$.
(ii) $\tilde{\mathfrak{T}}$ is a *regular* neighbourhood of S, i.e. $\tilde{\mathfrak{T}}$ retracts to S.

Let us denote by $\tilde{U}$ a neighbourhood of S in M. If the triangulation is sufficiently 'fine', then we can assume $\tilde{\mathfrak{T}} \subset \tilde{U}$. Let v be a continuous vector field on a neighbourhood $\tilde{U}$ of S in M, non-singular on $\tilde{U} - S$. Then the *Poincaré–Hopf index* of v at S, denoted $\mathrm{PH}(v, S)$, can be defined as $\mathrm{PH}(v, \tilde{\mathfrak{T}})$, where $\tilde{\mathfrak{T}}$ denotes any cellular tube in $\tilde{U}$ around S. This number $\mathrm{PH}(v, S)$ depends only on the behaviour of v near S and not on the choice of the neighbourhood $\tilde{U}$, nor on the tube $\tilde{\mathfrak{T}}$. Moreover, for this index it does not matter what actually happens on S, we only care what happens around S, but away from S.

Now let M be a compact oriented C^∞ manifold and v a continuous vector field on M, non-singular on the boundary. From the above considerations, we may assume that the set $\mathrm{Sing}(v)$ of singular points of v has only a finite number of components $(S_\alpha)_\alpha$. The theorem of Poincaré–Hopf obviously implies that, if

M has no boundary, then we have

$$\sum_{\alpha} \mathrm{PH}(v, S_\alpha) = \chi(M). \tag{1.5}$$

If M has a boundary ∂M, the sum $\sum_\alpha \mathrm{PH}(v, S_\alpha)$ depends only on the behaviour of v near ∂M. For example, if v is pointing outwards everywhere on ∂M, then we have the same formula (1.5), if v is pointing inwards everywhere on ∂M, the right hand side is $\chi(M) - \chi(\partial M)$. In particular, if the (real) dimension of M is even (as it will normally be the case in this work) and if v is everywhere transverse to ∂M, then we have again the same formula (1.5).

There is another way of defining the index, which brings us closer to the topic that we discuss in the next sections. Suppose first that M has no boundary and we try to construct a cross-section v of its tangent bundle TM. The stepwise process tells us we can always construct it up to the $(n-1)$-skeleton $M^{(n-1)}$ of M for some triangulation or cell decomposition. When we try to extend it to the n-skeleton, for each n-cell we have the vector field on its boundary, a $(n-1)$-sphere, defining an element in the homotopy group $\pi_{n-1}(S^{n-1}) \cong \mathbb{Z}$. The vector field can be extended to this cell iff the corresponding map is null-homotopic. Thus we have for each cell a local obstruction in $\mathbb{Z}$. This gives rise to a cochain of dimension n, which is in fact a cocycle, hence representing a cohomology class with integer (local) coefficients. The cohomology class that we obtain in this way is *the Euler class* of M, $e(M) \in H^n(M;\mathbb{Z})$, which evaluated on the orientation cycle $[M]$ gives a number: the Euler–Poincaré characteristic $\chi(M)$.

Suppose now that M has boundary ∂M and the vector field v is defined and non-singular around the boundary. We know that v can be extended to all of $\partial M \cup M^{(n-1)}$. For each n-cell of M we have a local obstruction in $\mathbb{Z}$. Thus we get a cocycle $e(M, v)$ which vanishes over the boundary, so it can be regarded as representing a class in the relative cohomology group $H^n(M, \partial M; \mathbb{Z})$. As a relative class $e(M, v)$ does depend on the choice of the vector field, generally speaking. This is the *Euler class of M relative to v*. Evaluating $e(M, v)$ on the orientation cycle of $(M, \partial M)$ we get a number, *the index of v* in M.

Exercises 1.6 (i) Write complete proofs of Theorems 1.1 and 1.2.

(ii) Prove the above formula (1.5).

(iii) Let M and M' be compact oriented n-manifolds, $n > 1$, with the same boundary $N = \partial M = \partial M'$, and let v be a non-singular vector field defined on a neighbourhood of N. Show that one has:

$$\mathrm{PH}(v, M) - \mathrm{PH}(v, M') = \chi(M) - \chi(M').$$

(iv) Let v and v' be continuous vector fields on a neighbourhood U of 0 in $\mathbb{C}^n$, with isolated singularity at 0. Assume that for each $x \in U$, the vectors $v(x)$ and $v'(x)$ are linearly dependent over $\mathbb{C}$. Show that if $n > 1$, then $\mathrm{PH}(v, 0) = \mathrm{PH}(v', 0)$. Conclude that in this case the index of v can be defined in terms of the map into $\mathbb{CP}^{n-1}$ defined by v. What can you say when $n = 1$?

2 The Chern classes

Let us recall the definition of the Chern classes via obstruction theory [52, 33, 22]. This can be done in full generality, however for simplicity we restrict to the case where the base space is an almost-complex $2m$-manifold M, so its tangent bundle TM is a complex vector bundle of rank m.

Definition 2.1 *An r-field is a set $F^{(r)} = \{v_1, \dots, v_r\}$ of r continuous vector fields defined on a subset in M. A singular point of $F^{(r)}$ is a point where the vectors (v_i) fail to be linearly independent. A non-singular r-field is also called an r-frame.*

Let $W_r(m)$ be the Stiefel manifold of complex r-frames in $\mathbb{C}^m$. Notice that we do not use r-frames which are necessarily orthonormal, just linearly independent, but this does not change the results, because every such framing is homotopic to an orthonormal one. We know (see [52]) that $W_r(m)$ is $(2m-2r)$-connected and its first non-zero homotopy group is $\pi_{2m-2r+1}(W_r(m)) \cong \mathbb{Z}$. The bundle of r-frames on M, denoted by $W_r(TM)$, is the bundle associated to the tangent bundle TM and whose fibre over $x \in M$ is the set of r-frames in $T_x(M)$ (diffeomorphic to $W_r(m)$). In the following, we fix the notation $q = m - r + 1$.

The Chern class $c^q(M) \in H^{2q}(M)$ is the first possibly non-zero obstruction to construct a section of $W_r(TM)$. Let us recall the standard obstruction theory process. Let σ be a k-cell of the given cell decomposition (D), contained in an open subset $\Omega \subset M$ on which the bundle $W_r(TM)$ is trivialized. If the section $F^{(r)}$ of $W_r(TM)$ is already defined over the boundary of σ, it defines a map:

$$\partial\sigma \cong S^{k-1} \xrightarrow{F^{(r)}} W_r(TM)|_\Omega \cong \Omega \times W_r(m) \xrightarrow{pr_2} W_r(m),$$

thus an element of $\pi_{k-1}(W_r(m))$. If $k \leqslant 2m - 2r + 1$, this homotopy group is zero, so the section $F^{(r)}$ can be extended to σ without singularity. If $k = 2m - 2r + 2 = 2q$, we meet an obstruction. So we can always construct a section $F^{(r)}$ of $W_r(TM)$ over the $(2q-1)$-skeleton of (D). When we try to extend $F^{(r)}$ to the $2q$-skeleton, we have an r-frame on the boundary of each cell σ, which defines an element $I(F^{(r)}, \sigma)$ in the homotopy group $\pi_{2q-1}(W_r(m)) \cong \mathbb{Z}$. The generators of $\pi_{k-1}(W_r(m))$ being consistent (see [52]), this defines a cochain

$$\gamma \in C^{2q}(M; \pi_{2q-1}(W_r(m))),$$

by $\gamma(\sigma) = I(F^{(r)}, \sigma)$, for each $2q$-cell σ, and then extend it by linearity. This cochain is actually a cocycle and the cohomology class that it represents is the qth Chern class $c^q(M)$ of M in $H^{2q}(M)$. It is independent of the various choices involved in its definition. Note that $c^m(M)$ coincides with the Euler class of the underlying real tangent bundle $T_{\mathbb{R}}M$.

There is another useful definition of the index $I(F^{(r)}, \sigma)$: let us write the frame $F^{(r)}$ as $(F^{(r-1)}, v_r)$, where the last vector is individualized, and suppose that $F^{(r)}$ is already defined on $\partial\sigma$. There is no obstruction to extend the

$(r-1)$-frame $F^{(r-1)}|_{\partial\sigma}$ to σ, because the dimension of the obstruction for such an extension is $2(m-(r-1)+1) = \dim\sigma + 2$. The $(r-1)$-frame $F^{(r-1)}|_{\sigma}$ generates a complex subbundle G^{r-1} of rank $(r-1)$ of $TM|_{\sigma}$, i.e.

$$TM|_{\sigma} \cong G^{r-1} \oplus H^q,$$

where H^q is an orthogonal complement of (complex) rank $q = m-(r-1)$. The obstruction to extend the last vector v_r inside σ as a non-zero section of H^q is given by an element of $\pi_{2q-1}((\mathbb{C}^q)^*) \cong \mathbb{Z}$, corresponding to the composition of the map $v_r : \partial\sigma \cong S^{2q-1} \to H^q|_{\Omega}$ with the projection on the fibre $(\mathbb{C}^q)^*$, where $(\mathbb{C}^q)^* = \mathbb{C}^q \setminus \{0\}$. Let us denote by $I_{H^q}(v_r, \sigma)$ the integer so obtained. The obstruction to extend the r-frame $F^{(r)}|_{\partial\sigma}$ inside σ as an r-frame tangent to M is the same as the obstruction to extend the last vector v_r inside σ as a non-zero section of H^q. In fact, there is a natural isomorphism $\pi_{2q-1}(W_r(m)) \cong \pi_{2q-1}((\mathbb{C}^q)^*)$ (for compatible orientations) and by this isomorphism, we have the equality of integers

$$I(F^{(r)}, \sigma) = I_{H^q}(v_r, \sigma).$$

Another choice of $F^{(r-1)}$ gives another choice of v_r and of H^q, but all H^q are homotopic and the index obtained is the same.

Remark 2.2 Given a $2q$-cell σ contained in an open set $\Omega \subset \mathbb{C}^m$, it is possible to proceed in a more precise way. Let us denote by s the $2(r-1)$-simplex of (K) whose dual is σ and by $\hat{s} = s \cap \sigma$ the barycentre of s. Consider a complex $(r-1)$-frame $F^{(r-1)}(\hat{s}) = (e_1, e_2, \dots, e_{r-1})$ at this point, which generates $T_{\hat{s}}(s)$, i.e. the (real) tangent plane in $\hat{s}$ to s (it has the right dimension). Considering, for $F^{(r-1)}|_{\sigma}$ an $(r-1)$-frame parallel to $F^{(r-1)}(\hat{s})$, (and considering if necessary a smoothing of the cell σ), the orthogonal complement H^q is then identified with the tangent bundle to σ and, in this case, $I_{H^q}(v_r, \sigma)$ is the usual $I(v_r, \sigma)$.

Suppose now that (L) is a subcomplex of (D), whose realization $|L|$ is also denoted by L. Assume that we are already given an r-frame $F^{(r)}$ on the $2q$-skeleton of L, denoted by $L^{(2q)}$. The same arguments as before say that we can always extend $F^{(r)}$ without singularity to $L^{(2q)} \cup D^{(2q-1)}$. If we want to extend this frame to the $2q$-skeleton of (D) we meet an obstruction for each corresponding cell which is not in (L). This gives rise to a cochain which vanishes on L and represents the relative Chern class

$$c^q(M, L; F^{(r)}) \in H^{2q}(M, L),$$

whose image, by the natural map into $H^{2q}(M)$, is the usual Chern class, but as a relative class it does depend on the choice of the frame $F^{(r)}$ on L.

If we have two frames $F_1^{(r)}$ and $F_2^{(r)}$ on $L^{(2q)}$, the difference between the corresponding classes is given by the difference cocycle of the frames on L: in the product $L \times I$, suppose $F_1^{(r)}$ is defined at the level $L \times \{0\}$ and $F_2^{(r)}$ is defined at the level $L \times \{1\}$, then the difference cocycle $d(F_1^{(r)}, F_2^{(r)})$ is well defined in

$$H^{2q}(L \times I, L \times \{0\} \cup L \times \{1\}) \cong H^{2q-1}(L),$$

as the obstruction to the extension of the given sections on the boundary of $L \times I$ ([52] §33.3). As shown in [52], we have the following formula:

$$c^q(M, L; F_2^{(r)}) = c^q(M, L; F_1^{(r)}) + \delta d(F_1^{(r)}, F_2^{(r)}),$$

where $\delta : H^{2q-1}(L) \to H^{2q}(M, L)$ is the connecting homomorphism. Also, for three frames $F_1^{(r)}$, $F_2^{(r)}$ and $F_3^{(r)}$ as above, we have

$$d(F_1^{(r)}, F_3^{(r)}) = d(F_1^{(r)}, F_2^{(r)}) + d(F_2^{(r)}, F_3^{(r)}). \tag{2.3}$$

Let S be a compact (K)-subcomplex of M, and $\tilde{U}$ a neighbourhood of S. Let $\tilde{\mathfrak{T}}$ be a cellular tube in $\tilde{U}$ around S. Take an r-field $F^{(r)}$ defined on $D^{(2q)}$, possibly with singularities. We suppose that the only singularities inside $\tilde{U}$ are located on S. This implies that $F^{(r)}$ has no singularity on $(\partial\tilde{\mathfrak{T}})^{(2q)}$ so there is a well-defined relative Chern class,

$$c^q(\tilde{\mathfrak{T}}, \partial\tilde{\mathfrak{T}}; F^{(r)}) \in H^{2q}(\tilde{\mathfrak{T}}, \partial\tilde{\mathfrak{T}}).$$

Taking the image of this class by the isomorphism $H^{2q}(\tilde{\mathfrak{T}}, \partial\tilde{\mathfrak{T}}) \cong H^{2q}(\tilde{\mathfrak{T}}, \tilde{\mathfrak{T}} - S)$ followed by the Alexander duality

$$\psi_M : H^{2q}(\tilde{\mathfrak{T}}, \tilde{\mathfrak{T}} - S) \xrightarrow{\sim} H_{2r-2}(S),$$

we get a class

$$\psi_M(c^q(\tilde{\mathfrak{T}}, \partial\tilde{\mathfrak{T}}; F^{(r)})) \in H_{2r-2}(S),$$

that we call the *Poincaré–Hopf class* $\mathrm{PH}(F^{(r)}, S)$ of $F^{(r)}$ at S.

Note that if $\dim S < 2r - 2$, then $\mathrm{PH}(F^{(r)}, S) = 0$.

The relation between the Poincaré–Hopf class of $F^{(r)}$ and the index we defined above is the following:

$$\mathrm{PH}(F^{(r)}, S) = \sum I(F^{(r)}, \sigma(s))s,$$

where the sum runs over the $2r-2$-simplices s of the triangulation of S and $\sigma(s)$ is the dual cell of s (of dimension $2q$). This relation is a consequence of the combinatorial definition of the Alexander duality [5]. In particular, when $r = 1$, then $F^{(1)} = \{v\}$ and $\mathrm{PH}(F^{(1)}, S) \in H_0(S)$ is identified with the integer $\mathrm{PH}(v, S)$ previously defined.

Let M be a compact almost-complex $2m$-manifold, possibly with boundary, and let $F^{(r)}$ be an r-field on the $(2q)$-skeleton of M, with singularities located on a compact subcomplex Σ in the interior of M. On the $(2q)$-skeleton of ∂M, we have a well-defined r-frame $F^{(r)}$. Let $(S_\alpha)_\alpha$ be the connected components of Σ. Then, from the definitions, we have

$$\sum_\alpha (i_\alpha)_* \mathrm{PH}(F^{(r)}, S_\alpha) = c_{r-1}(M; F^{(r)}) \quad \text{in } H_{2r-2}(M), \tag{2.4}$$

where $i_\alpha : S_\alpha \hookrightarrow M$ is the inclusion and we set

$$c_{r-1}(M; F^{(r)}) = \psi_M(c^q(M, \partial M; F^{(r)})) = c^q(M, \partial M; F^{(r)}) \frown [M, \partial M].$$

In particular, the sum of the Poincaré–Hopf classes is determined by the behaviour of $F^{(r)}$ near ∂M and does not depend on the extension to the interior of M. Note that we may assume that $F^{(r)}$ is non-singular on $D^{(2q-1)}$. If $r = 1$, $F^{(1)} = \{v\}$, and if v is everywhere transverse to ∂M, then (2.4) reduces to (1.5).

Exercises 2.5 (i) Let M^n be a compact, oriented manifold with boundary ∂M. Assume the tangent bundle TM is trivial over ∂M; let $F_1 = (v_1, \ldots, v_n)$ and $F_2 = (w_1, \ldots, w_n)$ be two n-frames on $TM|_{\partial M}$. Show that:

(i) All the v_i's have the same Poincaré–Hopf index in M and one has:

$$c_n(M, F_1) \frown [M, \partial M] = \mathrm{PH}(v_i, M).$$

(The analogous statement holds for F_2.)

(ii) For all $i, j \neq n$ one has

$$c_i(M, F_1) \cdot c_j(M, F_1) = c_i(M, F_2) \cdot c_j(M, F_2).$$

Hence, all the decomposable relative Chern numbers of M are indeed independent of the choice of framing, while the one involving the top degree class is given by the Poincaré–Hopf index of one of the vector fields in the framing.

3 The Schwartz index

Let us now define the (generalized) Schwartz index of a vector field on a singular variety. This was first done in [42, 43, 6] for 'radial' vector fields. The restriction of radiality was dropped, independently, in [23, 49] and in [48, 8]; see also [1, 3, 12].

We now consider a real analytic n-variety (i.e. a reduced real analytic space) V in a real analytic m-manifold M. The singular set of V is denoted by $\mathrm{Sing}(V)$ and the regular one $V_0 = V - \mathrm{Sing}(V)$. If V is reducible, we assume it is pure dimensional.

Let S denote either a compact connected subcomplex in the regular part of V, for some triangulation, or a compact connected component of $\mathrm{Sing}(V)$. By [29], we may always assume that $\mathrm{Sing}(V)$ is a subcomplex of M, and by [19] we can assume it has a regular neighbourhood $\mathcal{T}$ with smooth boundary. Hence there exists a vector field v_0, tangent to V_0 and normal to $\partial\mathcal{T}$, pointing outwards. (These are essentially the 'radial' vector fields of [42, 43].) Furthermore, if we let $\tilde{\mathcal{T}}$ be a regular neighbourhood of S in M with $\mathcal{T} = \tilde{\mathcal{T}} \cap V$, then this vector field extends naturally to a radial (i.e. everywhere transversal) vector field on the whole boundary $\partial\tilde{\mathcal{T}}$. The sum of the indices of the singularities of v_0 in $\tilde{\mathcal{T}}$ is

$$\mathrm{PH}(v_0, \tilde{\mathcal{T}}) = \chi(S),$$

and is independent of the ambient manifold M. This is a fundamental property of the Schwartz radial vector fields. Recall from Section 2 above, that for two continuous vector fields v_1 and v_2, non-singular on $\mathcal{T}-S$, we have the difference $d_S(v_1, v_2) \in H^{2n-1}(\partial\mathcal{T})$. To recall this, let N_S and N_S^* be compact regular neighbourhoods of S with smooth boundary, such that N_S^* is contained in the interior of N_S. Let τ be a vector field on N_S^* which is everywhere transverse to ∂N_S^* and points outwards from N_S^*. Let C_S be the cylinder bounded by ∂N_S and ∂N_S^*, and let ζ be the vector field on ∂C_S which is v on ∂N_S and τ on ∂N_S^*. The *difference* between v and τ at S, $d_S(v, \tau)$, is the total Poincaré–Hopf index of ζ in C_S, i.e. the number of zeros, counted with their local indices, of an extension of ζ to C_S. (We remark that C_S has $\partial N_S \cong \partial\mathcal{T}$ as a deformation retract.)

Definition 3.1 *Let $S \subset V$ be either a compact connected component of the singular set* Sing(V) *or a compact subcomplex of V_0. Let v be a continuous vector field , non-singular on $\mathcal{T} - S$, where $\mathcal{T}$ is a neighbourhood of S. The (generalized) Schwartz index of v at S, denoted* Sch(v, S), *is:*

$$\mathrm{Sch}(v, S) = \chi(S) + d_S(v_0, v),$$

where v_0 is a radial vector field pointing outwards from S.

Note that we have: (i) $\mathrm{Sch}(v_0, S) = \chi(S)$ for radial vector fields, (ii) the Schwartz index coincides with the Poincaré–Hopf index when $S \subset V_0$, and (iii) this index is independent of the ambient manifold M. From the definition and (2.3) we have, for two vector fields v_1 and v_2 as above,

$$\mathrm{Sch}(v_2, S) = \mathrm{Sch}(v_1, S) + d_S(v_1, v_2). \qquad (3.2)$$

For a continuous vector field v whose domain of definition is in the regular part $V_0 = V - \mathrm{Sing}(V)$ of V, we denote by $(S_\alpha)_\alpha$ the connected components of $\mathrm{Sing}(V) \cup \mathrm{Sing}(v)$, where $\mathrm{Sing}(v)$ is the set of points in V_0 where v vanishes. In what follows we assume that each component S_α is either in V_0 or in $\mathrm{Sing}(V)$ and that it admits a neighbourhood disjoint from each other. We have the following theorem [1, 8, 12, 42, 43, 48], which is a generalization of the theorem of Poincaré–Hopf.

Theorem 3.3 *Let V be a compact real analytic variety and let v be a continuous vector field on the regular part of V. Let S_α be defined as above, then we have*

$$\sum_\alpha \mathrm{Sch}(v, S_\alpha) = \chi(V),$$

independently of v.

Proof Let $(S_\alpha)_\alpha$ be as above. Assume first that near each S_i, v is transversal to the smooth boundary K_i of a regular neighbourhood N_i of S_i. Therefore

$$\mathrm{Sch}(v, S_i) = \chi(S_i),$$

by definition. Let $\overset{\circ}{N}_i$ be the interior of N_i and set $V^* = V - \{\overset{\circ}{N}_1 \cup \cdots \cup \overset{\circ}{N}_r\}$. Then V^* is a compact manifold with boundary $K = \{K_1 \cup \cdots \cup K_r\}$, the vector field v is transversal to K and points inwards. Therefore $\mathrm{PH}(v, V^*) = \chi(V^*) - \chi(K)$, by the theorem of Poincaré–Hopf for manifolds with boundary (1.5 above). On the other hand

$$\chi(V) = \chi(V^*) + \{\chi(S_1) + \cdots + \chi(S_r)\} - \chi(K).$$

Thus

$$\chi(V) = \mathrm{PH}(v, V^*) + \{\mathrm{Sch}(v, S_1) + \cdots + \mathrm{Sch}(v, S_r)\},$$

hence $\chi(V) = \mathrm{Sch}(v, V)$. Now suppose v is as above and let X be some other vector field on V, singular at $S_1, \ldots, S_r$ and possibly at some smooth points of V. Then

$$\mathrm{Sch}(X, S_i) = \chi(S_i) + d_i(X, v),$$

for each i, where $d_i(X, v)$ is the difference introduced in Section 2. Similarly, the Poincaré–Hopf index of X in V^* is

$$\mathrm{PH}(X, V^*) = \chi(V^*) - \chi(K) + \{d_1(v, X) + \cdots + d_r(v, X)\}.$$

Hence

$$\chi(V) = \mathrm{PH}(X, V^*) + \{\mathrm{Sch}(X, S_1) + \cdots + \mathrm{Sch}(X, S_r)\} = \mathrm{Sch}(X, V),$$

as claimed, because $d_i(X, v) + d_i(v, X) = 0$. □

It is worth saying that the original definition of the Schwartz index is not the one we presented here. This uses a Whitney stratification $\{V_i\}$ of $V \subset M$. If v is a vector field on a strata V_i and P is an isolated zero of v, then v can be extended *radially* to a vector field $\tilde{v}$ on a neighbourhood of P in M. The Schwartz index of v at P is, by definition, the Poincaré–Hopf index of $\tilde{v}$ at P. If S is a connected component of $\mathrm{Sing}(V)$ and v is a *stratified vector field* on V (see 4.3 below) with isolated singularities on S, then the sum of the Schwartz indices of v at these points equals the index $\mathrm{Sch}(v, S)$ in 3.1. above.

Exercises 3.4 Let V be the singular hypersurface in $\mathbb{C}^4$ defined by

$$\{x^2 + y^2 + z^2 + w^r = 0\}.$$

Let F be the vector field $F = (2y, -2x, rw^{r-1}, -2z)$.

(i) Compute the Poincaré–Hopf index of F at $0 \in \mathbb{C}^4$.

(ii) Show that F is everywhere tangent to V.

(iii) Compute the Schwartz index of F at $0 \in V$.

(iv) Find similar formulae for the singularity $\{x^p + y^q + z^r + w^s = 0\}$, and the hamiltonian vector fields:

$$F_1 = (qy^{q-1}, -px^{p-1}, sw^{s-1}, -rz^{r-1}), \quad F_2 = (-rz^{r-1}, sw^{s-1}, px^{p-1}, -qy^{q-1}).$$

4 The Schwartz classes

Let us now define the Schwartz classes of a singular variety, the top one corresponding to the total Schwartz index of a vector field. We follow [8]. We consider a compact, complex analytic n-variety (i.e. a reduced complex analytic space) V embedded in a complex m-manifold M. As before, the singular set of V is denoted $\mathrm{Sing}(V)$ and the regular one $V_0 = V - \mathrm{Sing}(V)$. We assume V is reduced and it is pure dimensional. We denote by $[V]$ the sum of the fundamental classes of the irreducible components of V. Denote by $\{V_i\}$ a Whitney stratification of M compatible with V and $\mathrm{Sing}(V)$.

Let S denote a compact connected subset of V either contained in the regular part V_0, or a component of $\mathrm{Sing}(V)$.

We suppose now that the triangulation (K) of M is compatible with the stratification of M and with S, and we denote by (D) a dual cell decomposition of M obtained from a barycentric subdivision of (K). We assume (K) makes M a PL manifold and we endow (K) with an orientation compatible with the orientation of the stratification $\{V_i\}$. This determines an orientation for (D) (see [5]).

Let $\tilde{U}$ be an open neighbourhood of S in M, denote $U = \tilde{U} \cap V$, and assume that $U - S$ is in V_0. The union of (D)-cells which are dual of (K)-simplices contained in S is a regular tube $\tilde{\mathcal{T}}$ around S in M, contained in $\tilde{U}$ (taking a subtriangulation if necessary). On the other hand, $\partial\tilde{\mathcal{T}}$ is transverse to V_0. The intersection $\mathcal{T} = \tilde{\mathcal{T}} \cap V$ is no longer a cellular tube in V, but it still satisfies the properties (i) and (ii) of Remark 1.4. We write $\partial\mathcal{T} = V \cap \partial\tilde{\mathcal{T}}$, this is a hypersurface in V_0.

Definition 4.1 *A neighbourhood $\mathcal{T}$ of S in V, contained in U and satisfying the properties (i) and (ii) of Remark 1.4 will be called a tube in U around S.*

We already used the Alexander isomorphism, for $0 \leqslant q \leqslant m$,

$$\psi_M : H^{2q}(\tilde{\mathcal{T}}, \tilde{\mathcal{T}} - S) \cong H^{2q}(\tilde{\mathcal{T}}, \partial\tilde{\mathcal{T}}) \xrightarrow{\sim} H_{2m-2q}(S),$$

when S is a compact subcomplex of the complex m-manifold M. In the situation $S \subset V \subset M$ considered now, there is an Alexander homomorphism (in general not an isomorphism), for any $0 \leqslant p \leqslant n = \dim_{\mathbb{C}} V$,

$$\psi_V : H^{2p}(\mathcal{T}, \mathcal{T} - S) \cong H^{2p}(\mathcal{T}, \partial\mathcal{T}) \to H_{2n-2p}(S),$$

defined in the following way [5]: let us denote by $[\mathcal{T}, \partial\mathcal{T}]$ the fundamental class of $\mathcal{T}$ in $H_{2n}(\mathcal{T}, \partial\mathcal{T})$, then

$$\psi_V(c) = r_*(c \frown [\mathcal{T}, \partial\mathcal{T}]),$$

where the cap-product is the internal one

$$H^{2p}(\mathcal{T}, \partial\mathcal{T}) \times H_{2n}(\mathcal{T}, \partial\mathcal{T}) \to H_{2n-2p}(\mathcal{T}),$$

and $r_* : H_{2n-2p}(\mathcal{T}) \to H_{2n-2p}(S)$ is induced by the retraction $r : \mathcal{T} \to S$.

The Alexander homomorphism ψ_V corresponds, at the chain level, to the composition

$$C^{2p}_{(\hat{K})}(\mathcal{T}, \partial\mathcal{T}) \xrightarrow{\ \tau\ } C^{2q}_{(D)}(\tilde{\mathcal{T}}, \partial\tilde{\mathcal{T}}) \xrightarrow{\ \psi_M\ } C^{(K)}_{2m-2q}(S),$$

with $2m - 2q = 2n - 2p$, where the map τ is defined by $\langle \tau(c), \sigma \rangle = \langle c, \sigma \frown V \rangle$ for a $2q$-cell σ in (D) and the subscripts and superscripts indicate the complex considered (see [5]). The map τ defines a map

$$\tau : H^{2p}(\mathcal{T}, \partial\mathcal{T}) \to H^{2q}(\tilde{\mathcal{T}}, \partial\tilde{\mathcal{T}}),$$

which, via the isomorphisms $H^{2p}(\mathcal{T}, \partial\mathcal{T}) \cong H^{2p}(\mathcal{T}, \mathcal{T} - S) \cong H^{2p}(U, U - S)$ in V and the corresponding ones in M, provides a homomorphism, still denoted

$$\tau : H^{2p}(U, U - S) \to H^{2q}(\tilde{U}, \tilde{U} - S) \tag{4.2}$$

and called Thom–Gysin homomorphism. In general, it is neither injective, nor surjective. There is a commutative diagram:

$$\begin{array}{ccccc}
H^{2q}(\tilde{U}, \tilde{U} - S) & \xrightarrow[\cong]{} & H^{2q}(\tilde{\mathcal{T}}, \partial\tilde{\mathcal{T}}) & \xrightarrow[\cong]{\psi_M} & H_{2m-2q}(S) \\
\uparrow{\scriptstyle \tau} & & \uparrow{\scriptstyle \tau} & & \uparrow{\scriptstyle =} \\
H^{2p}(U, U - S) & \xrightarrow[\cong]{} & H^{2p}(\mathcal{T}, \partial\mathcal{T}) & \xrightarrow{\psi_V} & H_{2n-2p}(S).
\end{array}$$

M. H. Schwartz proved that, given the Whitney stratification $\{V_i\}$ of M, one can construct a continuous vector field v_0 on M, which is tangent to the strata, with isolated singularities and *radial* in a sense which she made precise. One of the fundamental properties of a radial vector field v_0 is that, if p is an isolated singularity of v_0, then the indices at p of v_0 with respect to M and to the stratum containing p are the same. We refer to [42, 43, 6] for details. Applying this construction to the stratification $\{V_i\}$, we obtain a radial vector field tangent to V, i.e. leaving V invariant.

The above considerations can be generalized to the 'higher' Schwartz–MacPherson classes as follows. With the previous stratification and triangulation of M, let us remark that the cells of (D) are transverse to each stratum, so that, for every stratum V_i of complex dimension d and every cell of (real) dimension $2q = 2(m - r + 1)$, $V_i \cap \sigma$ is a cell of dimension $2k = 2(d - r + 1)$. In particular $V_i \cap \sigma$ is empty when $d < r - 1$.

For each point $x \in M$, let $T_x M$ be the tangent space of M at x. Let V_i be the stratum that contains x and let $T_x V_i$ be the corresponding tangent space, so that $T_x V_i$ is a subspace of $T_x M$. Let us denote $\tilde{T}(M) = \bigcup T_x V_i \subset TM$. Thus one has a projection $\pi : \tilde{T}(M) \to M$, which is the restriction to $\tilde{T}(M)$ of the usual projection $TM \to M$.

Definition 4.3 *Let A be a subspace of M. A stratified vector field on A is a section of $\tilde{T}(M)$ over A, i.e. a continuous section v of TM on A such that at each point $x \in V_i \cap A$, $v(x)$ is tangent to the stratum V_i. A stratified r-field on A is an r-field $F^{(r)} = \{v_1, \ldots, v_r\}$ consisting of stratified vector fields $v_1, \ldots, v_r$. A stratified r-frame is a non-singular stratified r-field.*

Let us recall that the obstruction dimension to construct an r-frame tangent to M is $2q = 2(m - r + 1)$, and the index $I(F^{(r)}, \sigma) \in \pi_{2q-1}(W_r(m))$ is well defined for every $2q$-cell σ. The obstruction dimension to construct an r-frame tangent to the stratum V_i of (complex) dimension d is $2k = 2(d - r + 1)$, so, if $F^{(r)}$ is a stratified r-field, the index $I(F^{(r)}, V_i \cap \sigma) \in \pi_{2k-1}(W_r(d))$ is well defined, for the same cell σ, by the above considerations of dimensions. In general these two indices are different.

The Schwartz classes are the obstruction to constructing such stratified frames on V. We will denote the r-fields by $F^{(r)} = (F^{(r-1)}, v_r)$, individualizing the last vector field in $F^{(r)}$. Then one knows [42, 6]:

Lemma 4.4 *One can construct on the $(2q-1)$-skeleton $(D)^{(2q-1)}$ of (D) a stratified r-frame, called a radial r-frame, $F_0^{(r)} = (F_0^{(r-1)}, v_r)$ which satisfies:*

(i) $F_0^{(r-1)}$ is the restriction to $(D)^{(2q-1)}$ of an $(r-1)$-frame, still denoted by $F_0^{(r-1)}$, defined on the $2q$-skeleton $(D)^{(2q)}$.

(ii) $F_0^{(r)}$ extends to $(D)^{(2q)}$ as an r-field with isolated singularities, which are the singularities of v_r.

(iii) If $F_0^{(r)}$ has singularities in a $2q$-cell σ which intersects several strata of M, then all these singularities are in the stratum V_i of the lowest dimension.

(iv) The index of $F_0^{(r)}$ in σ : $I(F_0^{(r)}, \sigma)$ is equal to the index of its restriction along V_i : $I(F_0^{(r)}, V_i \cap \sigma)$. If, furthermore, the complex dimension of V_i is $r-1$ (the minimal possible value), then $I(F_0^{(r)}, a) = 1$.

Remark 4.5 In fact, we can take all singularities of $F_0^{(r)}$ as the centres of the cells σ, so that there is only one singular point in each cell.

The radial r-frames $F_0^{(r)}$ are pointing outwards from certain regular neighbourhoods of the strata V_i and from the cellular tube $\tilde{\mathfrak{T}}_V$ around V, this motivates the terminology.

The radial r-frame $F_0^{(r)}$ determines a $2q$-cochain $\tilde{c}_q \in C^{2q}(\tilde{\mathfrak{T}}_V, \partial\tilde{\mathfrak{T}}_V)$ on M as follows: If σ is a $2q$-cell that does not intersect V, then $\tilde{c}^q(\sigma) = 0$; If σ intersects V, then

$$\tilde{c}^q(\sigma) = I(F_0^{(r)}, \sigma).$$

Then one extends $\tilde{c}^q$ as a cochain by linearity.

It is proved in [42, 6] that this cochain is actually a cocycle, representing a cohomology class $\tilde{c}^q(V) \in H^{2q}(\widetilde{\mathbb{C}T}_V, \partial\tilde{\mathfrak{T}}_V) \cong H^{2q}(M, M - V)$. This class does not depend on the choices of the Whitney stratification of M, the triangulations, nor the r-frame $F^{(r)}$, so long as it is radial (see [42] and [44]). This result is also an easy consequence of Theorem 6.7 below.

Definition 4.6 *The class $\tilde{c}^q(V) \in H^{2q}(M, M - V)$ is the qth Schwartz class of V.*

Now denote by S a compact connected (K)-subcomplex of V such that $S \cap D^{(2q)}$ is either a subset of V_0 or a component of $\mathrm{Sing}(V)$. Let us denote by U a neighbourhood of S in V such that $U - S$ still intersects $(D)^{(2q)}$ in V_0. We write $p = n - r + 1$, thus we have $q - p = m - n$.

It follows from 4.4 that there exist stratified r-fields on $(D)^{(2q)} \cap U$ whose singularities are all located on S. Let $F_1^{(r)}$ and $F_2^{(r)}$ be two such r-fields, and let us consider a tube $\mathcal{T}$ in U around S. There is a well-defined secondary characteristic class $d(F_1^{(r)}, F_2^{(r)}) \in H^{2p-1}(\partial\mathcal{T})$, called the *difference* and defined as in Section 2. Let $\delta : H^{2p-1}(\partial\mathcal{T}) \to H^{2p}(\mathcal{T}, \partial\mathcal{T})$ be the connecting homomorphism and let $\psi_V : H^{2p}(\mathcal{T}, \partial\mathcal{T}) \to H_{r-1}(S)$ be the Alexander homomorphism. We set

$$d_S(F_1^{(r)}, F_2^{(r)}) = \psi_V \delta\, d(F_1^{(r)}, F_2^{(r)}).$$

Definition 4.7 *For an r-frame $F^{(r)}$ on $(D)^{(2q)} \cap (U - S)$, we define the Schwartz class $\mathrm{Sch}(F^{(r)}, S)$ of $F^{(r)}$ at S to be the class in $H_{2r-2}(S)$ given by:*

$$\mathrm{Sch}(F^{(r)}, S) = \begin{cases} \mathrm{PH}(F^{(r)}, S) & \text{if } S \cap (D)^{(2q)} \subset V_0, \\ c_{r-1}(S) + d_S(F_0^{(r)}, F^{(r)}) & \text{if } S \subset \mathrm{Sing}(V), \end{cases}$$

where PH *is computed in V_0 and $F_0^{(r)}$ is a radial frame.*

In particular, for a radial frame $F_0^{(r)}$, $\mathrm{Sch}(F_0^{(r)}, S)$ is the Schwartz class $c_{r-1}(S) \in H_{2r-2}(S)$.

From the definition and (2.3), we get, for two r-frames $F_1^{(r)}$ and $F_2^{(r)}$ on $(D)^{(2q)} \cap (U - S)$,

$$\mathrm{Sch}(F_2^{(r)}, S) = \mathrm{Sch}(F_1^{(r)}, S) + d_S(F_1^{(r)}, F_2^{(r)}).$$

Let us consider now a neighbourhood U of $\mathrm{Sing}(V)$ in V. We know already that there exist stratified r-fields on $(D)^{(2q)} \cap U$ whose singularities are all in $\mathrm{Sing}(V)$. Elementary obstruction theory [52] then tells us that every such r-field can be extended to all of $(D)^{(2q)} \cap V_0$ with a singular set which is a subcomplex of V_0. More generally, let Σ be a compact (K)-subcomplex in V_0 disjoint from a neighbourhood U_1 of $\mathrm{Sing}(V)$ in V. We denote by (S_α) the connected components of $\mathrm{Sing}(V) \cup \Sigma$ and set $V^* = V - U_1$. Let i_α and ι be the inclusions $S_\alpha \hookrightarrow V$ and $V^* \hookrightarrow V$, respectively. The second one induces a homomorphism ι_* in homology with compact supports. The following theorem follows from (2.4), the Schwartz construction and arguments similar to those for Theorem 3.3.

Theorem 4.8 *Let V be a compact complex analytic n-variety embedded in a complex m-manifold M and let Σ be a subcomplex in V_0 as above. For any stratified r-frame $F^{(r)}$ on $(D)^{(2q)} \cap (V_0 - \Sigma)$, $q = m - r + 1$, we have*

$$\sum_\alpha (i_\alpha)_* \mathrm{Sch}(F^{(r)}, S_\alpha) = c_{r-1}(V).$$

Thus, decomposing the previous summation according to the fact that S_α is in Sing(V) *or in Σ, we get*

$$\sum_{S_\alpha \subset \mathrm{Sing}(V)} (i_\alpha)_* \mathrm{Sch}(F^{(r)}, S_\alpha) + \iota_* c_{r-1}(V^*; F^{(r)}) = c_{r-1}(V),$$

where the sum is taken over the connected components of Sing(V)*. In particular, for a radial r-frame $F_0^{(r)}$, we have:*

$$c_{r-1}(V) = \sum_{S_\alpha \subset \mathrm{Sing}(V)} (i_\alpha)_* c_{r-1}(S_\alpha) + \iota_* c_{r-1}(V^*; F_0^{(r)}).$$

In other words, to define the Schwartz class $\tilde{c}^q(V) \in H^{2q}(M, M - V; \mathbb{Z})$ we consider some (any) stratified r-frame $F^{(r)}$, $q = m - r + 1$, on $(D)^{(2q)} \cap \tilde{U}$, then $F^{(r)}$ is the obstruction for extending it to a stratified r-frame on $V \cap (D)^{(2q)}$.

Exercise 4.9 Write a complete proof of 4.8.

5 The GSV-index

We now define the GSV-index of a vector field on a singular variety, introduced in [45]. The basic references for this section are [15, 48]. Other references are [1, 3, 7, 8, 9, 10, 21, 26, 47]. For simplicity, we consider first the case of a complex analytic germ which is an isolated complete intersection singularity, though everything generalizes (with the appropriate modifications) to the non-isolated singularity case [8]. Later in this section we consider isolated singularities which are not complete intersections, following [7]. The same ideas work for isolated real analytic germs of odd real dimension, and to some extent in the even dimensional case too, though there are additional difficulties in this case, see [1, 17]. We refer to [28, 32] for background material on singularities.

Let us denote by (V, P) the germ of an n-dimensional, isolated complete intersection singularity, defined by a function

$$f = (f_1, \ldots, f_k) : \mathbb{C}^{n+k} \to \mathbb{C}^k,$$

and let v be a continuous vector field on V, singular only at P. If $n = 1$, we further assume V is irreducible (see the exercises below). Since P is an isolated singularity of V, it follows that the gradient vector fields $\{\nabla f_1, \dots, \nabla f_k\}$ are linearly independent everywhere away from P. Moreover, they are never tangent to V. Hence the set $\{v, \nabla f_1, \dots, \nabla f_k\}$ is a $(k+1)$-frame on $V - \{P\}$. Let $K = V \cap S_\varepsilon$, be the link of P in V, see [32, 18, 28]. It is an oriented, real manifold of dimension $(2n-1)$, and the above frame defines a map

$$\phi_v = (v, \nabla f_1, \dots, \nabla f_k) : K \to W_{k+1}(n+k),$$

into the Stiefel manifold of complex $(k+1)$-frames in $\mathbb{C}^{n+k}$. Every map from a closed oriented $(2n-1)$-manifold into $W_{k+1}(n+k)$ factors through the sphere S^{2n-1}. (*Exercise:* prove this statement.) Hence ϕ_v represents an element in the homotopy group $\pi_{2n-1} W_{k+1}(n+k) \cong \mathbb{Z}$, so ϕ_v is classified by its degree.

Definition 5.1 *The* GSV*-index of v at $P \in V$,* GSV(v, P)*, is the degree of the above map ϕ_v.*

This index depends not only on the topology of V near P, but also on the way V is embedded in the ambient space. For instance [15], the singularities in $\mathbb{C}^3$ defined by

$$\{x^2 + y^7 + z^{14} = 0\} \quad \text{and} \quad \{x^3 + y^4 + z^{12} = 0\},$$

are orientation preserving homeomorphic, but the GSV-index of the radial vector field is 79 in the first case and 67 in the latter.

We recall that one has a Milnor fibration associated to the function f, see [32, 18, 28], and the Milnor fibre F can be regarded as a compact $2n$-manifold with boundary $\partial F = K$. Moreover (see [15]), there is an ambient isotopy of the sphere S_ε taking K into ∂F, which can be extended to a collar of K, which goes into a collar of ∂F in F. Hence v can be regarded as a non-singular vector field on ∂F.

Theorem 5.2 *(i) The* GSV*-index of v at P equals the Poincaré–Hopf index of v in the Milnor fibre:*

$$\mathrm{GSV}(v, P) = \mathrm{PH}(v, F).$$

(ii) If v is everywhere transversal to K, then

$$\mathrm{GSV}(v, P) = 1 + (-1)^n \mu.$$

(iii) One has:

$$\mathrm{GSV}(v, P) = \mathrm{Sch}(v, P) + (-1)^n \mu,$$

where μ is the Milnor number of P. In particular, if P is a regular point of V, then the GSV*-index coincides with the usual index of Poincaré–Hopf.*

The proof of the first statement is given with details in [15] for hypersurface singularities, but the same proof works in general with a minor modification. The idea is that the gradient vector fields $\{\nabla f_1, \dots, \nabla f_k\}$ are all linearly independent everywhere on F, so the degree of ϕ_v can be regarded as the obstruction for extending v to a tangent vector field on F. Statement (ii) follows from statement (i) together with Theorem 1.2 above and the fact that, by [32, 18, 28], the Euler–Poincaré characteristic of F is $1+(-1)^n\mu$. Statement (iii) is left as an exercise. The idea is to prove it first for a radial vector field using statement (ii), and then use exercise (iii) in Section 1 to prove the general case.

We also have a Poincaré–Hopf type theorem for this index:

Theorem 5.3 *Let V be now a compact, complex analytic variety with isolated singularities $P_1,\ldots,P_r$, which are all isolated (geometric) complete intersection germs. Let v be a continuous vector field on V, singular at the P_i's and probably at some other smooth points $Q_1,\ldots,Q_s$ of V. Let* GSV(v,V) *denote the total* GSV *index of v, i.e. the sum of the local* GSV*-indices at the P_i's and the usual Poincaré–Hopf indices at the Q_i's. Then one has:*

$$\mathrm{GSV}(v,V)=\chi(V)+(-1)^n\sum_{i=1}^{r}\mu(P_i),$$

where $\mu(P_i)$ is the Milnor number of P_i.

This theorem follows easily from (3.3) and (5.2) above.

We notice that this definition of the GSV-index works equally well for singularities which are only geometric complete intersections [28], not necessarily algebraic complete intersections. We also note that our definition above does require P to be an isolated singularity. When the singularities are not isolated, there is a similar definition of the GSV-index given in [8], but one needs elaborated techniques in differential geometry to show that this is well defined.

Remark 5.4 When the vector field in question is holomorphic one also has a definition of this index in terms of homological algebra [16]. Also in this case, it is known [3] that there is a lower bound for the GSV-index, i.e. the GSV-index of a holomorphic vector field on V cannot be arbitrary, it has to be more than 'something'. The question is: what is this 'something', cf. exercise (ii) below.

More generally, let 0 be an isolated singularity in a complex analytic space Y. We say that this singularity is *smoothable* if there exists a complex analytic space X and a flat analytic map,

$$F:X\to\mathbb{C},$$

such that $F^{-1}(0)$ is Y and $F^{-1}(t)$ is non-singular for all t near 0. Assume for simplicity that X is embedded in an open subset U of $\mathbb{C}^N$. We know by [24, Th. 1.1] that for every $\varepsilon\gg\eta>0$ sufficiently small, the restriction

$$F:F^{-1}(\mathbb{S}_\eta)\cap\mathbb{B}_\varepsilon\to\mathbb{S}_\eta,$$

is a fibre bundle over $\mathbb{S}_\eta$. Therefore $\chi(\mathbb{F}_t)$, *the Euler–Poincaré characteristic* of each fibre

$$\mathbb{F}_t = F^{-1}(t) \cap \mathbb{B}_\varepsilon, \quad t \in \mathbb{S}_\eta,$$

is independent of t.

Now denote by v a continuous vector field on Y with an isolated singularity at 0. Since, by the Ehresmann fibration lemma, the intersection of Y with the boundary sphere $\mathbb{S}_\varepsilon$ of $\mathbb{B}_\varepsilon$ is isotopic to the intersection of $F^{-1}(t)$ with this sphere, it follows that we can think of v as being a vector field around the boundary $\partial\mathbb{F}_t$ of $\mathbb{F}_t$. By Theorem 1.2 above, we know that we can extend v to the interior of $\mathbb{F}_t$ with a finite number of singularities, each of which has its local Poincaré–Hopf index; the sum of all these local indices is the total Poincaré–Hopf index of v in $\mathbb{F}_t$, that we denote by

$$\mathrm{PH}(v, \mathbb{F}_t).$$

One has:

Lemma 5.5 *The number* $\mathrm{PH}(v, \mathbb{F}_t)$ *is independent of the choice of t and of the extension of v to the interior of this fibre. In particular, if v is everywhere transversal to the local link of* 0 *in Y, then one has:*

$$\mathrm{PH}(v, \mathbb{F}_t) = \chi(\mathbb{F}_t).$$

Proofs If v is everywhere transversal to the local link of 0 in Y, which is isotopic to the boundary of the fibres $\mathbb{F}_t$, then (4.2) is an immediate consequence of the theorem of Poincaré–Hopf for manifolds with boundary, together with the fact [24] that the $\mathbb{F}_t$'s are the fibres of a fibre bundle. The proof in general follows from this theorem together with the well-known fact (see [52], also [46, lemma 2.3] or exercise (iii) in section 1 above) that given compact (oriented, connected) manifolds M and M' with the same boundary $\partial M = \partial M'$, and a non-singular vector field v on a neighbourhood of the boundary, one has

$$\mathrm{PH}(v, M) - \chi(M) = PH(v, M') - \chi(M').$$

□

We define *the* GSV*-index of v* in Y relative to the smoothing given by F by:

$$\mathrm{GSV}(v; Y; F) = \mathrm{PH}(v, \mathbb{F}_t),$$

and this is well defined by (5.5). For $v = r$, a radial vector field on Y, one has

$$\mathrm{GSV}(r; Y; F) = \chi(\mathbb{F}_t).$$

It is worth noting that this index does depend on the choice of the analytic map F chosen as a smoothing of Y, and not only on Y and X. However, it is shown in [7] that if the smoothing is given by a general linear form, then this index determines the local Euler obstruction of X at 0 (see Section 6 below). Hence the GSV-index is independent of the smoothing when this is given by a linear form. We also notice that if $(Y, 0)$ is a complete intersection germ, then the smoothing is essentially unique, because the base space of the universal deformation is connected, and one has:

$$\mathrm{GSV}(r; Y; F) = 1 + (-1)^n \mu,$$

where n is the dimension of Y and μ is its Milnor number at 0, because the fibre $\mathbb{F}_t$ has the homotopy type of a bouquet of μ spheres of real dimension n, by [32, 18].

Exercises 5.6 (i) Prove 5.3. (Hint: If you remove from V neighbourhoods of the $P_i's$ and you attach copies of the corresponding Milnor fibres, you get a closed manifold V', whose Euler–Poincaré characteristic is $\chi(V) + (-1)^n \times \sum_{i=1}^{r} \mu(P_i)$.)

(ii) Use the Puisseaux parametrization for complex analytic plane curves to show that the index of a holomorphic vector field on an irreducible plane curve is at least $1 - \mu$, where μ is the Milnor number. Can you say something similar when V is a curve in higher codimension? (this is not known yet). In fact, if a holomorphic vector field in an open set in $\mathbb{C}^N$ has an isolated singularity at a point P, then it is well known that its local Poincaré–Hopf index is necessarily positive, because it can be defined as the dimension of a certain vector space. If v is a holomorphic vector field on an isolated, complex analytic local complete intersection germ (V, P) and v is only singular at P, then its GSV-index is necessarily bounded below, by [3]. The problem is to determine what this bound is. I believe this is $1 + (-1)^n \mu$, where n is the dimension of V and μ is its local Milnor number at P; this is true when $n = 1$ and V has codimension 1.

(iii) For vector fields on a curve C which is not irreducible, one has two types of 'GSV' indices: on one hand, for each irreducible component of C one has a GSV-index defined as above, as the degree of the corresponding map; now we can add all these integers and get a number, say $\mathrm{GSV}^+(v, P)$. On the other hand, one can define another index by thinking of v as a vector field on the boundary of the Milnor fibre F and defining $\mathrm{GSV}^-(v, P)$ to be the Poincaré–Hopf index of v in F. It is shown (independently) in [10, 21] that if C is a plane curve, then these indices are related by the formula:

$$\mathrm{GSV}^+(v, P) = \mathrm{GSV}^-(v, P) - 2(I),$$

where I is the intersection number of the irreducible components. This is actually an easy consequence of (and equivalent to) a well-known formula in [32] relating the local Milnor number of C with the sum of the Milnor numbers of the various branches of C, see [1]. *Question*: Is there a similar formula when the codimension of C is more than 1? (All I know is that the difference of these two indices is an even number, by [1].)

(iv) Compute the GSV-index for the examples in Section 3.

(v) (cf. [1]) Let (V, P) be an isolated complete intersection singularity germ and let v and v' be vector fields on V, singular only at P. Show that v and

v' have the same GSV-index iff they have the same Schwartz index. Use this to show that either of these indices classifies the homotopy classes of vector fields on V with an isolated singularity at P (two such vector fields are said to be homotopic if they are homotopic through vector fields on V, each having an isolated singularity at P).

6 The MacPherson classes of a singular variety

The MacPherson classes for singular varieties have functorial properties similar to those of the Chern classes of manifolds. They are in fact characterized by certain naturality properties described in Section 1 of [30]. These classes actually live in (the usual integral) homology, not in cohomology; when the variety in question is a manifold, Poincaré duality takes these into the usual Chern classes. More precisely, let X be a reduced complex algebraic variety over $\mathbb{C}$. A *constructible set* in X is a subset obtained from the subvarieties of X through the usual set-theoretic operations. A function $\alpha : X \to \mathbb{Z}$ is *constructible* if there exists a partition of X by constructible sets X_i so that α is constant on each X_i. The set $C(X)$ of constructible functions on X is obviously an additive abelian group. Given a closed subset W in X, we denote by 1_W the characteristic function of W, i.e., it is 1 on W and 0 on $X - W$. MacPherson showed that there is a unique covariant functor $\mathbf{F}$ from the category of compact complex algebraic varieties to abelian groups, such that $\mathbf{F}(X) = C(X)$ and whose value f_* on a proper map $f : X \to Y$, satisfies:

$$f_*(1_W)(p) = \chi(f^{-1}(p) \cap W),$$

for every subvariety W of X and $p \in Y$, where χ is the (topological) Euler–Poincaré characteristic. He proved:

Theorem 6.1 [Deligne–Grothendieck conjecture] *There exists a (unique) natural transformation c_* from the functor $\mathbf{F}$ to homology, such that if X is smooth and if we let $c(X)$ be the total Chern class of X, then*

$$c_*(1_X) = c(X) \cap [X],$$

where $[X]$ is the orientation cycle.

The (total) Chern–MacPherson class of X is defined to be $c_*(1_X)$. MacPherson's theorem is indeed more precise than this, for he actually tells us how to construct the transformation c_*. To explain this we need to introduce the *Chern–Mather classes* and the *local Euler obstruction.* Both of these concepts arise from the *Nash transform* of X, which we now introduce.

First we let $(V, 0)$ be an equidimensional complex analytic singularity germ of dimension d in an open set $U \subset \mathbb{C}^N$. Let $G(d, N)$ denote the Grassmanian of complex d-planes in $\mathbb{C}^N$. On the regular part of V, $V_{\mathrm{reg}} := V - V_{\mathrm{sing}}$, where V_{sing} is the singular set, there is a map $\sigma : V_{\mathrm{reg}} \to U \times G(d,N)$, defined by $\sigma(x) = (x, T_x(X_{\mathrm{reg}}))$. The Nash transformation $\tilde{V}$ of V is the closure of $\mathrm{Im}(\sigma)$ in $U \times G(d,N)$. It is a (usually singular) complex analytic space, endowed with an analytic projection map $\nu : \tilde{V} \to V$ which is a biholomorphism away from $\nu^{-1}(V_{\mathrm{sing}})$. Let us denote by $U(d,N)$ the tautological bundle over $G(d,N)$, and denote by $\mathbb{U}$ the corresponding trivial extension bundle over $U \times G(d,N)$. We denote by π the projection map of this bundle. Let $\tilde{T}$ be the restriction of $\mathbb{U}$ to $\tilde{V}$, with projection map π. The bundle $\tilde{T}$ on $\tilde{V}$ is called the Nash bundle of V.

Consider a complex analytic stratification $(V_i)_{i \in A}$ of $U \subset \mathbb{C}^N$ satisfying the Whitney conditions adapted to V. A *stratified vector field* v on V is a continuous section of the restriction $TU|V$ of TU to V, such that $v(x) \in T_x(V_i)$ for every $x \in V_i \cap V$. A *radial vector field* in a neighbourhood of $\{0\}$ in V is a stratified vector field so that there is ε_0 such that for every $0 < \varepsilon \leqslant \varepsilon_0$, $v(x)$ is pointing outwards from the ball $\mathbb{B}_\varepsilon$ over the boundary $\mathbb{S}_\varepsilon := \partial \mathbb{B}_\varepsilon$. Recall that, by the Bertini–Sard theorem, for ε small enough, the spheres $\mathbb{S}_\varepsilon$ are transverse to the strata $(V_i)_{i \in A}$. One has the following lemma ([6], Proposition 9.1), that we state without proof:

Lemma 6.2 *Let v be a stratified, non-zero vector field on $L \subset V$. Then v can be lifted as a section $\tilde{v}$ of $\tilde{T}$ over $\nu^{-1}(L)$.*

Proposition-Definition 6.3 [Proposition 10.1 of [6]; see [30] for the original definition] *Let v be a radial vector field on $V \cap \mathbb{S}_\varepsilon$ and $\tilde{v}$ the lifting of v on $\nu^{-1}(V \cap \mathbb{S}_\varepsilon)$. The local Euler obstruction (or simply the Euler obstruction) $\mathrm{Eu}_0(V)$ is defined to be the obstruction to extending $\tilde{v}$ as a nowhere zero section of $\tilde{T}$ over $\nu^{-1}(V \cap \mathbb{B}_\varepsilon)$.*

We remark that if $\tilde{v}$ is a nowhere zero section of $\tilde{T}$ defined over $\nu^{-1}(V \cap \mathbb{S}_\varepsilon)$ and transversal to $\nu^{-1}(V \cap \mathbb{S}_\varepsilon)$, for some $\varepsilon > 0$ sufficiently small, then $\tilde{v}$ is homotopic to a section of $\tilde{T}$ over $\nu^{-1}(V \cap \mathbb{S}_\varepsilon)$ obtained by lifting a radial vector field of V at 0. Therefore the Euler obstruction $\mathrm{Eu}_0(V)$ equals the obstruction to extend $\tilde{v}$ as a nowhere zero section of $\tilde{T}$ over $\nu^{-1}(V \cap \mathbb{B}_\varepsilon)$. Hence, to calculate $\mathrm{Eu}_0(V)$ it is only necessary to construct a nowhere zero section of $\tilde{T}$, defined on $\nu^{-1}(V \cap \mathbb{S}_\varepsilon)$ and transversal to $\nu^{-1}(V \cap \mathbb{S}_\varepsilon)$, for which one can calculate the obstruction to extend it to $\nu^{-1}(V \cap \mathbb{B}_\varepsilon)$.

In the case of a complete intersection with isolated singularity, we have the following theorem of [11], giving a nice interpretation of the Euler obstruction (see also [20, 25, 40]). This result is in fact a special case of a result in [7], in the spirit of the Lefschetz theorem for hyperplane sections. In fact it is shown in [7] that the Euler obstruction at a (possibly non-isolated) singularity germ (V, P) is determined by invariants of a general hyperplane section of V passing near P.

Theorem 6.4 *Let V be a complex analytic complete intersection in $\mathbb{C}^N$ with an isolated singularity at 0. The local Euler obstruction of V at 0, $\mathrm{Eu}_0(V)$,* *equals* $1+(-1)^{d-1}\mu^{(d-1)}$, *where* $\mu^{(d-1)}$ *is the Milnor number at* 0 *of a general hyperplane section of* V.

We now give a sketch of the proof in [7] for this case. The first step is to observe that if H is a general hyperplane in $\mathbb{C}^N$ passing through 0, then the term $1+(-1)^{d-1}\mu^{(d-1)}$, equals the GSV-index of a radial vector field r on the complete intersection germ $(V\cap H, 0)$. This is obvious from the discussion in Section 5 above. In other words, if we let H_t be a translate of H, passing near 0 but not through 0, and if we let r be a vector field defined on a neighbourhood of $V\cap H_t\cap\mathbb{S}_\varepsilon$ in $V\cap H_t\cap\mathbb{B}_\varepsilon$ and transversal to $V\cap H_t\cap\mathbb{S}_\varepsilon$, then (6.4) is equivalent to proving that $\mathrm{Eu}_0(V)$ equals the Poincaré–Hopf total index of r in $V\cap H_t\cap\mathbb{B}_\varepsilon$, which is a manifold. We know, from Section 1, that such a vector field r can always be extended to all of $V\cap H_t\cap\mathbb{B}_\varepsilon$ with finitely many singularities $Q_1,\ldots,Q_r$. Since the Nash transform is a biholomorphism away from 0, r can be lifted to a section $\tilde{r}$ of the Nash bundle $\tilde{T}$ over $\nu^{-1}(V\cap H_t\cap\mathbb{B}_\varepsilon)$ with zeroes at the points $\tilde{Q}_i:=\nu^{-1}(Q_i)$, $i=1,\ldots,r$. Then the key step for proving (6.4) consists in showing that $\tilde{r}$ can be further extended to the whole Nash bundle $\tilde{T}$ with no more zeroes but the $\tilde{Q}_i's$, in such a way that it is always transversal to the boundary $\nu^{-1}(V\cap\mathbb{S}_\varepsilon)$ and in a neighbourhood of each $\tilde{Q}_i$, the section $\tilde{r}$ is the sum of two sections: one of them obtained by parallel translating the given section $\tilde{r}$ on $\nu^{-1}(V\cap H_t\cap\mathbb{B}_\varepsilon)$, the other being always normal to $\nu^{-1}(V\cap H_t\cap\mathbb{B}_\varepsilon)$ in $\nu^{-1}(V\cap\mathbb{B}_\varepsilon)$. This implies (6.4).

Let us now define the Chern–Mather classes of X, a compact, complex analytic irreducible variety of dimension d, embedded in a complex manifold M of dimension N. The above construction of the Nash transform and the Nash bundle can be carried on to this situation in the obvious way. We obtain the *Nash transformation* $\nu:\tilde{X}\to X$, which is a biholomorphism over the regular set X_{reg}, and the *Nash bundle* $\pi:\tilde{T}\to\tilde{X}$. Let $c(\tilde{T})\in H^*(\tilde{X})$ be the total Chern class of this bundle, i.e. the sum of all its Chern classes, where $c_0(\tilde{T})$ is 1, by definition, and the other classes can be defined as we did in Section 2 for the tangent bundle. Let $\beta: H^*(\tilde{X})\to H_*(\tilde{X})$ be the Poincaré duality homomorphism, given by the cap-product with the orientation class $[\tilde{X}]$, see [5]. Then *the Chern–Mather classes* of X are defined by:

$$c_M(X)=\nu_*(\beta(c(\tilde{T}))).$$

These are classes in the homology of X, $H_*(X)$, which is isomorphic to the relative cohomology $H^*(M, M-X)$ via Alexander duality.

We have the following lemma of [30]:

Lemma 6.5 *Let* τ *be the map from the group of algebraic cycles in* X *to* $C(X)$, *the constructible functions, defined by*

$$\tau\Big(\sum n_iV_i\Big)(p)=\sum n_i\mathrm{Eu}_p(V_i),$$

where $\mathrm{Eu}_p(V_i)$ *is the local Euler obstruction of* V_i *at the point* p. *Then* τ *is a well-defined isomorphism.*

Then, the main result in [30] is the following,

Theorem 6.6 *$c_M T^{-1}$ satisfies the requirements for c_* in Theorem 6.1. That is, $c_* = c_M T^{-1}$.*

This defines the Chern–MacPherson classes in terms of the Chern–Mather classes and the local Euler obstruction. We conclude this section by stating the beautiful theorem of [6], which unifies all these classes.

Theorem 6.7 *Let X be a d-dimensional, compact, complex algebraic variety embedded in a complex manifold M of dimension N. For each $r = 1, \dots, d$ and $q = N - r + 1$, let $c_{r-1}(V) \in H_{2r-2}(V)$ be the Chern–MacPherson class of V. Let $\psi_M : H^{2q}(M, M - V) \tilde{\to} H_{2r-2}(V)$ be the homomorphism given by Alexander duality (which may be neither injective nor surjective). Then*

$$\psi_M(\tilde{c}^q(V)) = c_{r-1}(V),$$

where $\tilde{c}^q(V)$ is the qth Schwartz class.

Thus, following [8], we call these classes the *Chern–Schwartz–MacPherson classes for singular varieties.*

7 The Chern classes of the virtual tangent bundle

To finish this article we shall briefly discuss a different type of characteristic class for singular varieties, which also arise naturally when the varieties satisfy certain conditions. The basic references for this section are [8, 26, 50, 48]. Let us now denote by V a compact subvariety of dimension n in a complex manifold M of dimension $m = n + k$. From now on we assume that there exists a holomorphic vector bundle $E \to M$ of rank k over M and a holomorphic section s of E, generically transverse to the zero section, so that $V = s^{-1}(0)$. Thus V is a local geometric complete intersection and the restriction $E|_{V_0}$ coincides with the (holomorphic) normal bundle N_{V_0} of the regular part V_0 of V. Let us set $N = E|_V$. We call $\tau(V) = TM|_V - N$ *the virtual tangent bundle* of V. We note that if V has no singularities, then $\tau(V)$ is a virtual bundle which is equivalent (in K-theory) to the tangent bundle of V, hence the name for $\tau(V)$.

For example, the above conditions are satisfied in the following cases, with a naturally given bundle E:

(i) V a hypersurface in M $(k = 1)$. In this case, we may take as E the line bundle determined by the divisor V.

(ii) V a geometric complete intersection in M (i.e. defined by k equations $f_\lambda = 0$, $(\lambda = 1, \dots, k)$, where the f_λ's denote holomorphic functions globally defined on M). In this case, we may take as E the trivial bundle.

(iii) V a geometric (projective algebraic) complete intersection in the projective space $\mathbb{CP}^m$, i.e. V is the zero set of k homogeneous polynomials F_λ,

($\lambda = 1, \ldots, k$). It is only locally a geometric complete intersection in the previous sense, while it is globally the intersection of the k algebraic hypersurfaces $F_\lambda = 0$. In this case, we may take as E the bundle $L^{d_1} \oplus \cdots \oplus L^{d_k}$, where L denotes the hyperplane bundle and the d_λ's the degrees of the homogeneous polynomials F_λ defining V.

The virtual bundle $\tau(V)$ has Chern classes, given by the formula:

$$c(\tau(V)) = c(TM|_V) \cdot c(N)^{-1},$$

where $c(TM|_V)$ and $c(N)$ are the total Chern classes of these bundles. Hence the p-th Chern class $c^p(\tau(V))$ is the coefficient of t^p in the expansion of

$$\left(1 + \sum_{i=1}^{m} t^i c^i(TM|_V)\right)\left(1 + \sum_{j=1}^{k} t^j c^j(N)\right)^{-1}.$$

The Chern classes of the virtual tangent bundle $\tau(V)$ are the *virtual Chern classes* of V, see [8]. When V is an algebraic complete intersection, these classes coincide with the *Fulton–Johnson canonical classes* [13, 14]. An interesting problem is to study the relationship between these classes and the Chern–Schwartz–MacPherson classes of V, cf. [2, 8, 35–39, 50, 53].

We note that the above discussion about the virtual tangent bundle holds even if V has non-isolated singularities. The following theorem holds in this generality, with the appropriate modifications. This result was first proved in [26] for holomorphic vector fields with isolated singularities; essentially the same proof holds for continuous vector fields [48, 8].

Theorem 7.1 *Let $V \subset M$ be as above and assume (for simplicity) that V has only isolated singularities. Let v be a continuous vector field on the regular part of V with isolated singularities. Then one has:*

$$\mathrm{GSV}(v, V) = c^n(TM|_V - N) \frown [V].$$

The following result is immediate from (5.3) and (7.1). When $n = k = 1$ this is a reformulation of the classical adjunction formula for singular curves in surfaces. This generalizes also to the non-isolated singularities case. When $k = 1$, so that V is a hypersurface in M, this formula is similar to those in [36–39].

Theorem 7.2 *Let V be a subvariety of a complex manifold M as above. Then:*

$$\chi(V) = c^n(TM|_V - N) \frown [V] + (-1)^{n+1} \sum_{i=1}^{r} \mu(P_i),$$

where the sum is taken over the local Milnor numbers at the singular points of V.

Proof For each component S of Sing(V), take a radial vector field pointing outwards from S, and extend these to have a vector field v on V whose singular set Sing(v) consists of Sing(V) and a finite number of compact connected sets in V_0. For a component S of Sing(v) in V_0, we have $\mathrm{Sch}(v, S) = \mathrm{PH}(v, S) = \mathrm{Vir}(v, S)$. Hence the theorem follows from Theorems 4.3, 5.3 and 7.1.

In other words, the total GSV-index is the dual of the top degree Chern class of the virtual tangent bundle of V. That is, a vector field on V, with isolated singularities, gives a localization of the virtual class $c^n(TM|_V - N)$ at the singularities of the vector field, with a certain contribution to the total class given by each singular point of V. Such a vector field also gives a localization of the 0-degree Chern–Schwartz–MacPherson class of V, which corresponds to the Euler–Poincaré characteristic of V, and the difference of these localizations is given by the local Milnor numbers of V, independently of the vector field. This is explained carefully in [8], where it is shown that similar considerations apply to the higher order classes. For this, one considers r-frames defined on an appropriate skeleton of V for a certain cell decomposition. Such a frame determines localizations of both the corresponding Schwartz class and virtual class; there is a contribution for each of these classes coming from each connected component of V_{sing}, giving rise to the 'generalized Milnor classes' of [8], see also [2, 35–59, 50, 53].

References

1. M. Aguilar, J. Seade and A. Verjovsky: Indices of vector fields and topological invariants of real analytic singularities, *J. Reine Angew. Math.* **504** (1998), 1–28.
2. P. Aluffi: Chern classes for singular hypersurfaces, *Trans. Amer. Math. Soc.* **351** (1999), 3989–4026.
3. Ch. Bonatti and X. Gómez-Mont: The index of a holomorphic vector field on a singular variety, *Asterisque* **222** (1994), 9–35.
4. J.P. Brasselet: From Chern classes to Milnor classes – a history of characteristic classes for singular varieties, *Adv. Stud. Pure Math.* **29** (2000), 31–52.
5. J.P. Brasselet: Définition combinatoire des homomorphismes de Poincaré, Alexander et Thom pour une pseudo-variété, in 'Caractéristique d'Euler–Poincaré', *Astérisque* **82–83** (1981), 71–91.
6. J.P. Brasselet et M-H. Schwartz: Sur les classes de Chern d'un ensemble analytique complexe, in 'Caractéristique d'Euler-Poincaré' *Astérisque* **82–83** (1981), 93–147.
7. J.P. Brasselet, D.T. Lê and J. Seade: Indices of vector fields and the Euler obstruction, *Topology*, **39** (2000), 1193–1208.
8. J.P. Brasselet, D. Lehmann, J. Seade and T. Suwa: Milnor classes of local complete intersections, *Trans. Amer. Math. Soc.* **354** (2002) 1351–1371.
9. M. Brunella: Feuilletages holomorphes sur les surfaces complexes compactes, *Ann. Sci. École Norm. Sup.* **30** (1997), 569–594.
10. M. Brunella: Some remarks on indices of holomorphic vector fields, *Publicacions Matematiques* **41**, no. 2 (1997), 527–544.
11. A. Dubson: Classes caractéristiques des variétés singulières, *C.R. Acad. Sc. Paris* **287** (1978), 237–240.

12. W. Ebeling and S.M. Gusein-Zade: On the index of a vector field at an isolated singularity, in 'The Arnoldfest' (Toronto, On., 1997), 141–152, Fields Inst. Commun. 24, Amer. Math. Soc. (Providence, 1999).
13. W. Fulton: 'Intersection Theory', Springer (Berlin, 1984).
14. W. Fulton and K. Johnson: Canonical classes on singular varieties, *Manuscripta Math.* **32** (1980), 381–389.
15. X. Gómez-Mont, J. Seade and A. Verjovsky: The index of a holomorphic flow with an isolated singularity, *Math. Ann.* **291** (1991), 737–751.
16. X. Gómez-Mont: An algebraic formula for the index of a vector field on a hypersurface with an isolated singularity, *J. Alg. Geom.* **7** (1998), 731–752.
17. X. Gómez-Mont and P. Mardešić: The signature of the relative Jacobian determinant and the index of a vector field tangent to a hypersurface, *Ann. Inst. Fourier* **47** (1997), 1523–1539.
18. H. Hamm: Lokale topologische Eigenschaften komplexer Räume, *Math. Ann.* **191** (1971), 235–252.
19. M. Hirsch: Smooth regular neighborhoods, *Annals of Math.* **76** (1962), 524–530.
20. M. Kato: Singularities and some global topological properties, Proc. R.I.M.S. Singularities Symposium, Kyoto University (1978).
21. B. Khanedani and T. Suwa: First variation of holomorphic forms and some applications, *Hokkaido Math. J.* **26** (1997), 323–335.
22. M. Kervaire: Relative characteristic classes, *Amer. J. Math.* **79** (1957), 517–558.
23. H. King and D. Trotman: Poincaré–Hopf theorems on stratified sets, preprint.
24. D.T. Lê: Some remarks on relative monodromy, in 'Real and Complex Singularities' (P. Holm, ed), Proc. Nordic Summer School, Oslo 1976, Sijthoff and Noordhoff (1977), 397–403.
25. D.T. Lê et B. Teissier: Variétés polaires locales et classes de Chern des variétés singulières, *Annals of Math.* **114** (1981), 457–491.
26. D. Lehmann, M. Soares and T. Suwa: On the index of a holomorphic vector field tangent to a singular variety, *Bol. Soc. Bras. Mat.* **26** (1995), 183–199.
27. D. Lehmann and T. Suwa: Residues of holomorphic vector fields relative to singular invariant subvarieties, *J. Differ. Geom.* **42** (1995), 165–192.
28. E. Looijenga: 'Isolated Singular Points on Complete Intersections', LMS Lect. Note Series 77, Cambridge University Press (Cambridge, 1984).
29. S. Lojasiewicz: Triangulation of semi-analytic sets, *Annali Sc. Norm. Sup. Pisa* **18** (1964), 449–474.
30. R.D. MacPherson: Chern classes for singular algebraic varieties, *Annals of Math.* **100** (1974), 423–432.
31. J. Milnor: 'Topology from the Differentiable Viewpoint', University Press of Virginia (Charlottesville, 1965).
32. J. Milnor: 'Singular Points of Complex Hypersurfaces', Annals Math. Studies 61, Princeton University Press (Princeton, 1968).
33. J. Milnor and J. Stasheff: 'Characteristic Classes', Annals Math. Studies 76, Princeton University Press (Princeton, 1974).
34. M. Morse: Singular points of vector fields under general boundary conditions, *Amer. J. Math.* **51** (1929), 165–178.
35. T. Ohmoto, T. Suwa and S. Yokura: A remark on the Chern classes of local complete intersections, *Proc. Japan Acad* **73** (1997), 93–95.
36. A. Parusiński: A generalization of the Milnor number, *Math. Ann.* **281** (1988), 247–254.

37. A. Parusiński and P. Pragacz: A formula for the Euler characteristic of singular hypersurfaces, *J. Algebraic Geom.* **4** (1995), 337–351.
38. A. Parusiński and P. Pragacz: Characteristic numbers of degeneracy loci, *Contemp. Math.* **123** (1991), 189–198.
39. A. Parusiński and P. Pragacz: Characteristic classes of hypersurfaces and characteristic cycles, *J. Algebraic Geom.* **10** (2001), 63–79.
40. R. Piene: Cycles polaires et classes de Chern pour les variétés projectives singulières, in 'Introduction à la théorie des singularités' (Lê D.T., ed), Tome II, Travaux en cours 37, Hermann (Paris, 1988), 7–34.
41. Ch. Pugh: Generalized Poincaré index formulas, *Topology* **7** (1968), 217–226.
42. M.H. Schwartz: Classes caractéristiques définies par une stratification d'une variété analytique complexe, *C.R. Acad. Sci. Paris* **260** (1965), 3262–3264, 3535–3537.
43. M.H. Schwartz: 'Champs radiaux sur une stratification analytique complexe', Travaux en cours 39, Hermann (Paris, 1991).
44. M.H. Schwartz: Classes obstructrices des ensembles analytiques, Travaux en cours, Hermann, to appear.
45. J. Seade: The index of a vector field on a complex surface with singularities, in 'The Lefschetz Centennial Conference' (A. Verjovsky, ed), Contemp. Math. 58, Part III, Amer. Math. Soc. (Providence, 1987), 225–232.
46. J. Seade: The index of a vector field under blow-ups, *Bol. Soc. Mat. Mexicana* **37** (1992), 449–462.
47. J. Seade and T. Suwa: A residue formula for the index of a holomorphic flow, *Math. Ann.* **304** (1996), 621–634.
48. J. Seade and T. Suwa: An adjunction formula for local complete intersections, *International J. Math.* 9 (1998), 759–768.
49. S. Simon: Champs totalment radiaux sur une structure de Thom–Mather, *Ann. Inst. Fourier* (1995), 1423–1447.
50. T. Suwa: Classes de Chern des intersections complètes locales, *C.R. Acad. Sci. Paris* **324** (1996), 67–70.
51. T. Suwa: 'Indices of vector fields and residues of singular holomorphic foliations'. Actualités Mathématiques, Hermann, Paris, 1998.
52. N. Steenrod: 'The Topology of Fibre Bundles', Princeton University Press (Princeton, 1951).
53. S. Yokura: On Characteristic classes of complete intersections, in 'Algebraic geometry: Hirzebruch 70'. Contemp. Math. 241, Amer. Math. Soc. (Providence, 1999), 349–369.

Appendix: Research publications of B. F. Steer

Brian Steer has contributed to the mainstream of topology and geometry for four decades, from his use of algebraic topology in the 1960s to explore the geometry of manifolds, to the applications of gauge theory on which he is working today. He has always had the courage to follow what he saw as the most exciting mathematics of the time, and he has encouraged his students to do the same.

The evolution of his interests can be discerned in his list of publications. Along with his many individual contributions to mathematics, this list records the many productive collaborations in which he has been involved. His first collaboration was with André Haefliger during his time as a postdoc in Geneva. In the 1960s Brian also worked with Ed Brown on exotic smooth structures and Stiefel manifolds, with Michael Boardman on Hopf invariants, and with Peter Hilton on foundational issues in cohomology.

Since the 1970s most of Brian Steer's joint work has been with his former students. With Michael Crabb he investigated the geometry of vector bundles. With José Seade he worked on analytic questions concerning the framing of quotients of 3-dimensional Lie groups. And he is writing a series of papers with Howard Fegan (one in collaboration with Lorne Whiteway) studying first-order operators on special manifolds. Brian worked with Ben Nasatyr and Andrew Wren on aspects of the geometry of parabolic bundles and complex orbifolds. And in recent work with Christian Gantz and Olivier Collin, he has produced striking applications of gauge theory to the study of low-dimensional manifolds. He has also worked on Yang–Mills theory with Mikio Furuta.

Although this book is not a survey of the mathematics in which Brian Steer has been involved, the diversity of the ideas described here does reflect the breadth of his contributions to geometry and topology. It also reflects the generosity with which he has shared his many insights.

Transgression in sphere-bundles, *Topology* **2** (1963), 1–10.

Extensions of mappings into H-spaces, *Proc. London Math. Soc. (3)* **13** (1963), 219–272.

Generalized Whitehead products, *Quart. J. Math. Oxford (2)* **14** (1963), 29–40.

Symmetry of linking coefficients (with A. Haefliger), *Comment. Math. Helv.* **39** (1965), 259–270.

A note on Stiefel manifolds (with E.H. Brown), *Amer. J. Math.* **87** (1965), 215–217.

Axioms for Hopf invariants (with J.M. Boardman), *Bull. Amer. Math. Soc.* **72** (1966), 992–994.

On Hopf invariants (with J.M. Boardman), *Comment. Math. Helv.* **42** (1967), 180–221.

Une interprétation géométrique des nombres de Radon-Hurwitz, *Ann. Inst. Fourier* **17** (1967), 209–218.

Note on the Hurewicz homomorphism in the space of homotopy equivalences of S^n preserving a base-point, *Quart. J. Math. Oxford (2)* **19** (1968), 245–248.

On fibred categories and cohomology (with P.J. Hilton), *Topology* **8** (1969), 243–251.

On the embedding of projective spaces in Euclidean space, *Proc. London Math. Soc. (3)* **21** (1970), 489–501.

On immersing complex projective $(4k+3)$-space in Euclidean space, *Quart J. Math. Oxford (2)* **22** (1971), 339–345.

Vector-bundle monomorphisms with finite singularities (with M.C. Crabb), *Proc. London Math. Soc. (3)* **30** (1975), 1–39.

Orbits and the homotopy class of a compactification of a classical map, *Topology* **15** (1976), 383–393.

The elements of π_3^s represented by invariant framings of quotients of $\widetilde{\mathrm{SL}}_2(\mathbb{R})$ by certain discrete subgroups (with J. Seade), *Adv. in Math.* **46** (1982), 221–229.

A note on the eta function for quotients of $\mathrm{PSL}_2(\mathbb{R})$ by co-compact Fuchsian groups (with J. Seade), *Topology* **26** (1987), 79–91.

On the "strange formula" of Freudenthal and de Vries (with H.D. Fegan), *Proc. Camb. Phil. Soc.* **105** (1989), 249–252.

Complex singularities and the framed cobordism class of compact quotients of 3-dimensional Lie groups by discrete subgroups, *Comment. Math. Helv.* **65** (1990), 349–374.

Spectral symmetry of the Dirac operator for compact and noncompact symmetric pairs (with H.D. Fegan and L. Whiteway), *Pacific J. Math.* **145** (1990), 211–221.

Seifert fibred homology 3-spheres and the Yang–Mills equations on Riemann surfaces with marked points (with M. Furuta), *Adv. Math.* **96** (1992), 38–102.

Spectral symmetry of the Dirac operator in the presence of a group action (with H.D. Fegan), *Trans. Amer. Math. Soc.* **335** (1993), 631–647.

Narasimhan–Seshadri theorem for parabolic bundles: an orbifold approach (with E.B. Nasatyr), *Phil. Trans. Roy. Soc. London Ser. A* **353** (1995), 137–171.

Orbifold Riemann surfaces and the Yang–Mills–Higgs equations (with E.B. Nasatyr), *Ann. Scuola Norm. Sup. Pisa Cl. Sci. (4)* **22** (1995), 595–643.

First order operators on manifolds with a group action (with H.D. Fegan), *Canad. J. Math.* **48** (1996), 758–776.

Grothendieck topology and the Picard group of a complex orbifold (with A. Wren), *Contemp. Math.* **239** (1999), 1–262.

Instanton Floer homology for knots via 3-orbifolds (with O. Collin), *J. Diff. Geom.* **51** (1999), 149–202.

Gauge fixing for logarithmic connections over curves and the Riemann–Hilbert problem (with C. Gantz), *J. London Math. Soc. (2)* **59** (1999), 479–490.

First order differential operators on a locally symmetric space (with H.D. Fegan), *Pacific J. Math.* **194** (2000), 83–96.

Stable parabolic bundles over elliptic surfaces and over Riemann surfaces (with C. Gantz), *Canad. Math. Bull.* **43** (2000), 174–182.

The Donaldson–Hitchin–Kobayashi correspondence for parabolic bundles over orbifold surfaces (with A. Wren), *Canad. J. Math.* **53** (2001), 1309–1339.

Index